Statistical Papers in Honor of
GEORGE W. SNEDECOR

George W. Snedecor

Statistical Papers
in Honor of
GEORGE W. SNEDECOR

Edited by T. A. Bancroft

Assisted by Susan Alice Brown

THE IOWA STATE UNIVERSITY PRESS

AMES, IOWA

Table of Contents

Preface

THIS VOLUME is presented as a tribute to George Waddel Snedecor, founder and first director of the Statistical Laboratory at Iowa State University. Its contributors include friends, former students, and former colleagues of Professor Snedecor—able statisticians who were asked to select their own topics and provide their own scientific refereeing. Unfortunately, not all invited contributors found it possible to meet a limited deadline for publication. Furthermore, the projected size of the volume restricted the number of invitations which could be extended.

In adopting a policy of freedom of choice of topic, it was realized that there would be considerable variability of statistical methodology and application presented. However, we believe that this collection will be of interest to statisticians and scientists who know Professor Snedecor personally and to those who have become acquainted with him through various editions of his outstanding book, *Statistical Methods*. Together, these groups should include nearly every statistician and many scientists in the world, for the impact of George Snedecor's statistical work on many scientific investigations in various fields is indeed worldwide.

The brief presentation of biographical information about Professor Snedecor which introduces this volume includes honors and accomplishments which are truly notable for one individual. Since plans for this volume were formulated, Iowa State University has added yet another honor to this list. On May 18, 1970, the building which houses Iowa State's statistical center was formally named Snedecor Hall. A copy of the address delivered by ISU President W. Robert Parks on this occasion is included.

Thanks are due Merritt Bailey, Director of Book Publishing, Iowa State University Press, and his co-workers, for their technical expertise and cooperation in expediting the publication of this volume. All of us who have had a part in its preparation have done so to express our admiration and affection for George Waddel Snedecor.

Iowa State University   T. A. BANCROFT
Ames, Iowa

vii

George W. Snedecor: A Chronology

October 20, 1881—born in Memphis, Tennessee

1899-1901—attended Alabama Polytechnic Institute

1901-1905—attended University of Alabama

1905—received B.S., University of Alabama (mathematics and physics)

1905-1907—instructor, Selma Military Academy

1907-1910—professor of mathematics, Austin College, Sherman, Texas

1908—married Gertrude Crosier

1910-1913—graduate assistant in physics, University of Michigan

1913—received A.M., University of Michigan (physics)

1913-1914—assistant professor of mathematics, Iowa State University

1914-1930—associate professor of mathematics, Iowa State University

1915—first courses in statistics offered at Iowa State University

1924—attended a ten-week course conducted by Henry A. Wallace on rapid machine calculation of correlation coefficients, partial correlation, and the calculation of regression lines

1925—coauthor, with Wallace, of "Correlation and Machine Calculation," a bulletin which attained worldwide distribution

1927—in charge, with A. E. Brandt, of a newly created Mathematics Statistical Service at Iowa State University

1931-1947—professor of mathematics, Iowa State University

1931—first degree in statistics at Iowa State University, an M.S., was awarded through the Department of Mathematics to his student, Gertrude M. Cox

1931—brought R. A. Fisher to Iowa State University for the first time, as a visiting professor of statistics during the summer

1933—elected Fellow, American Association for the Advancement of Science

1933—organized the Statistical Laboratory as an institute under the president's office

1933-1947—director of Statistical Laboratory, Iowa State University

1933—appointed first Station Statistician of the Iowa Agricultural Experiment Station

1934—*Analysis of Variance* published by the Iowa State University Press

1935—named head of the newly created statistical section of the Agricultural Experiment Station

1935—chairman of the Osborne Club (local staff research club)

1936—brought R. A. Fisher to Iowa State University for the second time, as a visiting professor of statistics during the summer

1936—established the custom of tea drinking at a weekly statistics seminar which provided an opportunity for dialogue between staff and students on research problems presented by persons in substantive areas

1936—president of Sigma Xi (Iowa State University Chapter)

1937—first edition of *Statistical Methods* published by Iowa State University Press

1938—second edition of *Statistical Methods* published by Iowa State University Press

1938—established a cooperative agreement between the Statistical Laboratory and the United States Department of Agriculture which provided funds for the expansion of laboratory personnel and projects

1938—brought W. G. Cochran to Iowa State University as a visiting professor—he joined the staff in 1939; Charles P. Winsor joined the staff

1939—Statistical Laboratory was moved into newly built Service Building (now known as Snedecor Hall)

1939—elected Fellow, American Statistical Association

1940—third edition of *Statistical Methods* published by Iowa State University Press

1941—guest lecturer, North Carolina Institute of Statistics, summer

1942—elected Fellow, Institute of Mathematical Statistics

1943—established a cooperative agreement between the Statistical Laboratory and the Bureau of the Census; Master Sample Project began

1944—instituted a contractual project with the United States Weather Bureau

1945–1958—editor of the Queries Section of *Biometrics*

1946—guest lecturer, North Carolina Institute of Statistics, summer

1946—fourth edition of *Statistical Methods* published by Iowa State University Press

1947—retired as director of the Statistical Laboratory, Iowa State University

1947—Department of Statistics organized at Iowa State University

1947-1958—professor of statistics, Iowa State University

1947—vice-president of American Statistical Association

1947—guest lecturer at statistical summer session, Virginia Polytechnic Institute

1948—visiting research professor of statistics, Alabama Polytechnic Institute, spring

1948—president of American Statistical Association

1948—introduced a pioneering basic introductory course sequence in statistics for undergraduates at Iowa State University, especially designed for students majoring in natural and social sciences

1949—guest lecturer, North Carolina Institute of Statistics, summer

1950—limited first edition of *Everyday Statistics* published by William C. Brown, Dubuque

1951—limited second edition of *Everyday Statistics* published by William C. Brown, Dubuque

1951—guest lecturer, North Carolina Institute of Statistics, summer

1953—consultant in experimental statistics, Alabama Polytechnic Institute and Florida University, January–June

1954—professor of experimental statistics, Alabama Polytechnic Institute, January–March

1954—elected honorary Fellow, British Royal Statistical Society

1955—consultant in experimental statistics, Woman's College of North Carolina, January–April

1955—awarded faculty citation, Iowa State University

1956—consultant in agricultural statistics, Brazil, January–June

1956—fifth edition of *Statistical Methods* published by Iowa State University Press

1956—received honorary D.Sc., North Carolina State University

1957-58—visiting professor, Department of Experimental Statistics, North Carolina State University

1958—received honorary D.Sc., Iowa State University

1958-1963—consultant in experimental statistics (part time), U.S. Navy Electronics Laboratory, San Diego

1959-present—professor emeritus, Iowa State University

1967—sixth edition of *Statistical Methods* published by Iowa State University Press

1969—Service Building renamed Snedecor Hall; formal ceremonies held May 18, 1970

1970—awarded Samuel S. Wilks memorial medal

Foreword

Remarks by W. ROBERT PARKS at the Dedication of Snedecor Hall
Monday, May 18, 1970

———————

IT IS EMINENTLY APPROPRIATE that we dedicate this building to George W. Snedecor. Indeed, for the building that houses the Department of Statistics and the Statistical Laboratory, there could have been no other name than Snedecor Hall. The name Snedecor is synonymous with statistics at Iowa State University.

It is almost a disservice to a man who is a giant in his field, a man of great intellect and personal warmth, to recite his professional accomplishments. One might say that to do so borders on the statistical. Yet, I feel that it is important on this occasion that we review this man's long and distinguished career and his extensive contribution to scientific knowledge.

It was Iowa State's good fortune that George Snedecor came to this campus in 1913 as an assistant professor of mathematics to begin an association that has continued for nearly 60 years. By the time he arrived here, he had earned two degrees and had eight years of teaching experience in Alabama, Texas, and Michigan. Two years later he began teaching the first formal statistics courses at Iowa State.

A landmark in the use of statistics in research came in 1924 when Professor Snedecor instituted Saturday afternoon seminars for research workers in agriculture. What meetings those must have been! Henry A. Wallace—the Iowa State alumnus who was then an editor with *Wallaces Farmer* and later was to become Secretary of Agriculture, Secretary of Commerce, and Vice President of the United States—asked Professor Snedecor to establish the seminars. Wallace had a keen interest in corn breeding experiments and in the use of business machines for research computations, and he led those Saturday afternoon discussions.

W. ROBERT PARKS is President of Iowa State University, Ames.

Computational machinery was brought to the campus for use in the seminars—one of the first uses of business machines designed for commercial computations in the analysis of research data. This early undertaking predated by many years the use of electronic computers for research computations and was eventually to result in the organization of a separate Computation Center at Iowa State.

The Wallace-Snedecor association also led to the publication of a bulletin, "Correlations and Machine Calculation," which promptly attained worldwide distribution, remained in use for many years, and is a prized item for those who possess a copy.

This developing interest in statistical methods was served in 1927 by the formation of a Mathematics Statistical Service, the first formal step toward organization of the Statistical Laboratory. Professor Snedecor and A. E. Brandt were in charge. Calculating punched-card tabulating equipment was installed, thus initiating computational service to research workers.

In Britain at the time was a man who was to become known as the greatest figure in the history of statistics and one of the greatest figures in the history of scientific method. Because Professor Snedecor had an appreciation for the importance of the work of Sir Ronald Fisher, Iowa State became the first American university to recognize the British statistician. Professor Snedecor brought Fisher to the Iowa State campus as visiting lecturer in the summers of 1931 and 1936. Iowa State was the first U.S. institution to grant him an honorary degree.

In 1933 the Statistical Laboratory was organized, the first statistical center of its kind in the United States. Professor Snedecor was its first director and continued in that position for 14 years. Its organization provided the impetus for other universities to establish similar research and service institutes. Over the years more than twenty institutions, in the United States and in other countries, have sought advice here on the organization of statistical centers.

In 1937 there came another landmark, the publication of a book by Professor Snedecor called *Statistical Methods*. That book is now in its sixth edition; has sold more than 100,000 copies; and has been translated into Spanish, Hindi, Japanese, and Rumanian, with a French translation now in progress and an Indian reprint having been published in English. This is an amazing performance for a technical book. Professor Snedecor's book has introduced several generations of statisticians and research workers throughout the world to the topic and stands as a classic in its field.

Professor Snedecor pioneered in the development and utilization of sound statistical applications in experimentation, especially in biology and agriculture. He was appointed statistician with the Agriculture Experiment

Station, and he instituted the first cooperative agreement between the Statistical Laboratory and the U.S. Department of Agriculture for work on statistical problems. Similar agreements were made later with the Bureau of the Census and the U.S. Weather Bureau. These agreements have supported research in survey sampling, particularly in the improvement of livestock and crop estimating procedures, the U.S. census, and climatology.

Under one of those agreements Professor Snedecor directed the development of the National Master Sample of Agriculture for use by the Census Bureau and the Bureau of Agricultural Economics. This was a million dollar project at a time when a million dollars was a more staggering figure than it is today. The Survey Section of the Statistical Laboratory continues that work and has accumulated a sizable store of basic data and materials on the rural portions of every county in the United States.

When Professor Snedecor reached the age of 65 in 1947, he gave up his administrative duties as director of the Statistical Laboratory and head of the statistics department* of the Agriculture Experiment Station. Before doing so, however, he was active in the planning of a new teaching department—the Department of Statistics in what is now the College of Sciences and Humanities.

He was not retired in any sense of the word, however. He continued on the faculty for 11 more years; and in the months when he was not at work on this campus, he was teaching and consulting at other universities around the country and in South America.

Even when Professor Snedecor left this campus in 1958—and he was seventy-seven years old then—he had not retired because for the next five years he was a consultant in experimental statistics to the U.S. Navy Electronics Laboratory in San Diego, California.

His former students, his colleagues, and his book carried the Snedecor name far and wide. The recognition that came to him brought Iowa State an international reputation for leadership in the field of statistics. Among his honors are these:

● He was the first person in the field of agricultural research ever to head the American Statistical Association when he became its president in 1948.

● He was elected a Fellow of the American Statistical Association, the Institute of Mathematical Statistics, the American Association for the Advancement of Science, and an honorary Fellow of the British Royal Statistical Society.

---

*At the time, this was known as the Statistical Section.

● He was awarded honorary doctor of science degrees by North Carolina State University and by Iowa State University.

All of this, then, is the chronological record of this great man's years at Iowa State. But such an accounting speaks only of his professional accomplishments; it says nothing of the personal qualities that earned the respect and admiration of those who studied under him or worked with him.

There is on file at this university a remarkable collection of letters from former students and colleagues who were personally acquainted with Professor Snedecor and from others whose only acquaintance is through the pages of his book, *Statistical Methods*. These letters come from persons at major universities, large corporations, and government agencies all across the United States, and from Europe and the Far East. The letters testify to the mark Professor Snedecor has made on his field and on the people who have known him. I would like to mention some of the things that are contained in the letters.

An alumnus reported that in 1913 he attended his first college mathematics class, a course in algebra. It happened to be the first course taught here by Professor Snedecor. In those days students were allowed at the end of a course to name the professor from whom they wished to take the next course in the sequence. This student said that he selected Professor Snedecor for the next course and all subsequent mathematics courses and later he returned to Iowa State to receive master's and doctor's degrees in mathematics.

Another said that when he was a beginning graduate student, Professor Snedecor quizzed him in order to understand the student as an individual and to determine his aims and goals. The student stated his professional aim and then, in his own words, "Professor Snedecor set my sights above that modest goal."

Here is the recollection of one who said coming to Iowa State as a student was his first venture away from home: "Professor Snedecor showed his interest in me, gave his hospitality and warmth to me, and inspired me."

A British statistician wrote, "His teaching, and the enthusiasm that he put into it, was a delight to hear. Yet I think his greatest memorial is the very fine team of people whom he collected at Ames and whose tradition continues in one of the greatest of the world's statistics departments."

Little wonder then that the foresight of such a man should have placed Iowa State University in the forefront of statistical developments—a position that has been maintained and strengthened in the years since by Dr. T. A. Bancroft and his staff.

I want today to express this university's appreciation to Dr. Bancroft and his associates for upholding the high standard of excellence so long ago established by Professor Snedecor. I also want to express our pleasure at having with us:

● Professor Gertrude Cox, the first student to receive a master's degree in statistics at Iowa State, who did her thesis under the direction of Professor Snedecor and who later served on our faculty.

● Professor William Cochran, a former faculty member and now a co-author of Professor Snedecor's book, who will present a special seminar here tonight.

● Professor James Snedecor, son of the man we honor today. We hope that you will convey the fondest wishes of all of us to your father.

Iowa State University has honored George W. Snedecor before, but no greater honor can it bestow than the recognition that is represented in the action we take here today. It gives me great pleasure to have a part in this ceremony which formally recognizes that this building, the home of statistics at Iowa State University, shall henceforth be known as Snedecor Hall.

Statistical Papers in Honor of
GEORGE W. SNEDECOR

1

Selection of Predictor Variables in Linear Multiple Regression

R. L. ANDERSON, D. M. ALLEN, and F. B. CADY

1. Introduction

THE CLASSICAL multiple linear regression model is

$$\underset{(n \times 1)}{\mathbf{Y}} = \underset{(n \times r)}{\mathbf{X}} \underset{(r \times 1)}{\boldsymbol{\beta}} + \underset{(n \times 1)}{\boldsymbol{\varepsilon}}, \tag{1.1}$$

where $\mathbf{Y}$ is a vector of responses; $\mathbf{X}$ is a known, full-rank matrix of prediction variables; $\boldsymbol{\beta}$ is the unknown weight vector corresponding to $\mathbf{X}$; and $\boldsymbol{\varepsilon}$ is a vector of random variables having expected value 0 and variance-covariance matrix $\mathbf{I}\sigma^2$. Once $\mathbf{Y}$ has been observed, we denote the observed values by $\mathbf{y}$. The usual statistical problem is to estimate $\boldsymbol{\beta}$ using $\mathbf{y}$. This estimate is the value of $\mathbf{b}$ that minimizes $(\mathbf{y} - \mathbf{Xb})'(\mathbf{y} - \mathbf{Xb})$ and is $\mathbf{b} = (\mathbf{X'X})^{-1}\mathbf{X'y}$. This minimum value is $SSE = \mathbf{y'}[I - \mathbf{X(X'X)}^{-1}\mathbf{X'}]\mathbf{y}$. As a notational convenience we will hereafter regard $\mathbf{b}$ and SSE as functions of $\mathbf{Y}$ rather than $\mathbf{y}$. In this formulation we consider only models which are linear in the parameters $\boldsymbol{\beta}$. In this model the X's may represent different functional forms of the same basic variables, e.g. $X_1 = w$, $X_2 = z$, $X_3 = w^2$, $X_4 = z^2$, $X_5 = wz$.

Regression analysis has two major uses:

1. *Control a system or environment.* If b_i is the value of the ith value in $\mathbf{b}$, b_i is the estimator of the effect on Y of changing X_i by one unit while holding the other X's fixed. Hence, one can estimate values of the X's needed to attain a given value of Y.

RICHARD L. ANDERSON is Professor and Chairman, Department of Statistics, University of Kentucky, Lexington.

DAVID M. ALLEN is Assistant Professor Department of Statistics, University of Kentucky, Lexington.

FOSTER B. CADY is Professor of Biological Statistics, Cornell University, Ithaca.

3

2. *Predict a future Y for a given new set of X's.* The future value of Y is denoted by Y_0. The usual predictor of Y_0 is $\hat{Y}_0 = \mathbf{X}_0' \mathbf{b}$, where $\mathbf{X}_0'$ is the $(1 \times r)$ vector of new X's and $\mathbf{b}$ is the estimator of β obtained from the original set of X's.

In this paper we will be concerned only with the problem of prediction. One aspect of this prediction problem is the selection of the best set of predictor variables (including various functional forms) to be used in a multiple linear regression model to determine $\hat{Y}_0$. The predictive ability of alternative prediction models must be evaluated; i.e., we need a procedure that will furnish evidence of the usefulness of the prediction function in new situations. This procedure then would form the basis for the selection of the best predictor variables.

In the prediction problem the experimenter's primary concern is not the nature of the relationship between a dependent variable and the predictor variables. His main concern is the determination of the performance of this prediction function to predict an observation that is yet to be measured. The usual criterion for determining the best prediction function is the residual sum of squares—the sum of the squared differences between the observations and the predicted values of future observations having the same expected value. To obtain a small residual sum of squares, a number of potentially important variables are identified and a full model, which includes all relevant functions of these variables, is estimated. The full model is considered by many experimenters to be the best for prediction.

However, the following two facts indicate that the residual sum of squares is not an entirely satisfactory criterion if we are interested in minimizing prediction error:

1. Most multiple regression analyses in which the number of predictor variables r is large reveal that the estimated standard errors of some of the estimated regression coefficients exceed the magnitude of the coefficients themselves. This indicates that there probably would be little bias in the predicted $\hat{Y}_0$ if some of these regressors were omitted from the model.
2. Walls and Weeks (1969) have shown that the standard error of $\hat{Y}_0$ is always increased when an extra predictor variable is added to the prediction function.

2. Comparisons of Selection Procedures

Coupling the above facts with economic considerations of collecting data on additional X's and the increased computing required, we will usually be led to the use of a prediction function based on less than the total number of potential predictor variables, where each used function of a predictor variable in the model is considered as a separate predictor variable. Currently, many experimenters select the variables to be used in the prediction equation from a relatively large number of potential variables by means of the forward selection procedure (FS) or the Efroymson (1960), stepwise, technique (SW). The FS procedure first selects the predictor variable with the highest correlation with y, i.e., that variable resulting in the smallest residual sum of squares. At subsequent stages the predictor variable having the highest partial correlation with y, given those predictor variables previously included, is entered. The SW procedure differs in that at each stage each previously included predictor variable is again evaluated as if the variable had been the last one to be entered. Another sequential selection procedure is to start with the full model and then eliminate predictor variables one at a time until a satisfactory prediction equation has been obtained (backward elimination, BE). These various procedures are described in detail by Draper and Smith (1966). None of these sequential selection procedures ensure that the subset of predictor variables with the minimum residual sum of squares will be found. However, when the number of predictor variables to be included is specified, an algorithm by LaMotte and Hocking (1970) will give that subset with the minimum residual sum of squares.

Two examples are presented to show that full models, and to some extent models selected by sequential selection procedures, may lead to worse predictions than is currently believed (Table 1.1). Weiner and Dunn (1966) present the analyses of five studies of hospital cases for which linear regression was used to predict in which of two populations a given patient belonged. The prediction equations were derived from index samples of 50 from each of the two populations; the remaining patients were used to indicate the predictive ability in follow-up samples. In all studies 25 predictor variables were available. Four procedures were used to select the 2, 5, and 10 best predictor variables for the index sample; then the resulting prediction functions were tested on the follow-up samples. The FS procedure was superior to the other three methods. The percentages of misclassifications using FS are presented in Table 1.1. The misclassification rate is much larger for the follow-up sample unless two predictor variables are used; also the performance for this follow-up sample is usually best when fewer than all 25 possible predictor variables are used. If as many as 5 predictor variables are used, the experimenter would form an overoptimistic judgment of the future usefulness of the linear discriminant, based on the results for this sample alone.

Laird and Cady (1969) combined yield data on corn (tons per hectare)

Table 1.1
Discrimination Analysis of Hospital Cases*

Study	Follow-up Samples		No. of Variates	
	Population 1	Population 2	Continuous	Discrete
I	250	250	1	24
II	20	30	2	23
III	35	55	0	25
IV	123	250	1	24
V	23	73	25	0
Index sample, each study	50	50		

Percentages Misclassified†

Number of Predictor Variables‡	Sample	Study					Ave.
		I	II	III	IV	V	
25	index	(16)	12	(31)	(10)	20	18
	follow-up	37	27	52	21	(32)	34
10	index	18	(8)	32	(10)	(16)	17
	follow-up	35	(19)	(44)	21	33	30
5	index	26	16	35	17	22	23
	follow-up	(32)	26	46	22	36	32
2	index	33	27	39	19	28	29
	follow-up	37	30	45	(19)	35	33

*Taken from Weiner and Dunn (1966).
†Minimal value for given study and sample are within parentheses.
‡Selected by the FS procedure for less than 25 variables.

from a large number of applied-nitrogen fertilizer studies over a four-year period. They used 36 predictor variables in a full model (linear and quadratic functions of applied nitrogen and soil nitrogen; previous crop performance; amount of excess moisture; drought, hail, blight, and weed indexes; soil characteristics; interactions). Reduced models were selected by SW and BE procedures; variables were added or dropped based on 5% significance levels. For SW no additional variable would produce a significant decrease in SSE; for BE all included variables had significant regression coefficients.

A total of 76 experiments had been conducted, each with four levels of applied nitrogen, giving a total of 304 means. A preliminary regression analysis based on all 304 observations was carried out for each of the three selection procedures, including the full model (FM); 18 of 36 predictor variables were retained by the SW procedure and 25 by the BE procedure. The residual mean squares were as follows:

FM	SW	BE
0.382	0.390	0.373

These experiments were next divided into two groups by taking every other experiment (starting with the first one) in group 1 and the other experiments in group 2, giving 38 experiments (152 observations) for each group. Each of the three models (FM, SW with the above 18 predictor variables, and BE with the above 25 predictor variables) was then used to determine a linear prediction function for each group. These prediction functions were then used to predict the yields (Y) for each group (the group used to obtain the prediction function is the index group and the other is the follow-up group).

This procedure was repeated three more times, as follows:

Run 2: random assignment to groups, with each year equally represented in each group and one predictor variable (amount of hail) guaranteed to have a reasonable range in each group.

Run 3: completely random assignment.

Run 4: group 1 had a wide range of certain site variables, whereas group 2 had a narrow range.

For each run the residual sums of squares for both the index and follow-up groups are presented in Table 1.2, which reveals a tremendous increase in the residual mean square for the follow-up group when the FM was used. The increase for the SW procedure is much less, and for the BE procedure it is midway between the FM and SW. The performance of the SW and BE procedures might have been slightly poorer if the selection of predictor variables had been based on the index sample for each run instead of the combined sample; however, it is almost certain that the FM is far from the best prediction model.

3. Mean Square Error Criterion

The above examples indicate the need for developing a different criterion and a subsequent procedure for the selection of potential predictor variables. Anderson (1957) used a mean square error criterion to show that the true model is not necessarily the best model for prediction.

Table 1.2

**Index and Follow-up Residual Sums of
Squares for Three Different Linear Prediction Models**

		Index and Follow-up Groups*					
Run	Model	(1, 1)	(2, 1)	(2, 2)	(1, 2)	(1, 1) + (2, 2)†	(2, 1) + (1, 2)
1	FM	39	209	22	617	61	826
	SW	59	104	35	133	94	237
	BE	47	167	25	317	72	484
2	FM	33	278	27	349	60	627
	SW	55	92	38	109	93	201
	BE	41	171	35	114	76	285
3	FM	22	183	30	228	52	411
	SW	34	109	53	138	87	247
	BE	29	189	47	167	76	356
4	FM	49	203	19	374	68	577
	SW	67	147	30	63	97	210
	BE	61	121	24	88	85	209
Ave	FM	36	218	24	392	60	610
SS	SW	54	113	39	111	93	224
	BE	44	162	33	172	77	334
Ave	FM					.26	2.01
MS‡	SW					.35	0.74
	BE					.31	1.10

*First number is index group and second number is follow-up group.
†Residual sum of squares for (1,1) and (2,2) is the usual SSE.
‡Mean square is SS divided by total degrees of freedom: 230, 266 and 252 respectively
for (1,1) + (2,2) and 304 for (2,1) + (1,2).

Assume the true model is

$$Y_i = \beta_0 + \beta_1 x_{1i} + \beta_2 x_{2i} + \varepsilon_i; \qquad i = 1, 2, \ldots, n, \qquad (3.1)$$

where $\sum x_{1i} = 0$ and $\sum x_{2i} = 0$, $\text{Var}(\varepsilon_i) = \sigma^2$ and $\text{Cov}(\varepsilon_i, \varepsilon_j) = 0$, $i \neq j$.
Now let us transform the x's to a set of orthogonal variables, x_1 and z_2;
therefore, $z_{2i} = x_{2i} - \delta x_{1i}$, where $\delta = \sum x_{1i} x_{2i} / \sum x_{1i}^2$. Hence the model
now becomes

$$Y_i = \gamma_0 + \gamma_1 x_{1i} + \beta_2 z_{2i} + \varepsilon_i, \qquad (3.2)$$

where $\gamma_0 = \beta_0$ (in this example) and $\gamma_1 = \beta_1 + \delta \beta_2$. Now suppose we
use a prediction function based only on x_1; i.e., $\hat{Y}_{01} = c_0 + c_1 x_{10}$, where
$c_0 = \bar{y}$ and $c_1 = \sum x_{1i} Y_i / \sum x_{1i}^2$. Because of the orthogonal nature of the

variables, c_0 and c_1 are unbiased estimators of $\gamma_0 = \beta_0$ and γ_1; however, $\hat{Y}_{01}$ is biased by $\beta_2 z_{20}$, where $z_{20} = x_{20} - \delta x_{10}$. An unbiased predictor based on the true model is $\hat{Y}_{02} = c_0 + c_1 x_{10} + b_2 z_{20}$, where

$$b_2 = \sum z_{2i} Y_i / \sum z_{2i}^2$$

is an unbiased estimator of β_2. The addition of $b_2 z_{20}$ eliminates the bias but increases the variance of prediction by $z_{20}^2 \, \text{Var}(b_2) = z_{20}^2 \sigma^2 / \sum z_{2i}^2$. Hence if we use a quadratic loss function, a comparison of the two estimators involves the squared bias of $\hat{Y}_{01}$ versus the added variance of $\hat{Y}_{02}$, i.e., $\beta_2^2 z_{20}^2$ vs $\sigma^2 z_{20}^2 / \sum z_{2i}^2$.

In other words, if mean square error is the criterion of efficiency of prediction, the truncated prediction model is superior to the true model if

$$\beta_2^2 < \sigma^2 / \sum z_{2i}^2 \quad \text{or} \quad |\beta_2| < \sigma / \sqrt{\sum z_{2i}^2}. \tag{3.3}$$

One should not use predictor variables for which the magnitude of the regression coefficient is expected to be less than the standard error of its estimator. Anderson demonstrated the above by comparing first-degree and second-degree prediction models (1957) and various orders of factorial models (1960). These results are generalized below to the multiple linear regression model (1.1).

We will compare the least squares predictor using the full model (1.1), consisting of r predictor variables with a least squares predictor based on $r - k$ predictor variables. The criterion for comparison will be the mean square error of prediction. Let

$$\mathbf{X}_{(n \times r)} = \left[\mathbf{X_a} \mid \mathbf{X_k} \right]_{(n \times a)\,(n \times k)}; \qquad \mathbf{X}_0'_{(1 \times r)} = \left[\mathbf{X_{0a}'} \mid \mathbf{X_{0k}'} \right]_{(1 \times a)\,(1 \times k)};$$

$$\boldsymbol{\beta}_{(r \times 1)} \quad \left[\begin{array}{ll} \boldsymbol{\beta_a} & (a \times 1) \\ \boldsymbol{\beta_k} & (k \times 1) \end{array} \right],$$

where $\mathbf{X_a}$ represents the matrix of predictor variables to be used and $\mathbf{X_k}$ the matrix of predictor variables to be deleted; $\mathbf{X_{0a}'}$ and $\mathbf{X_{0k}'}$ are the corresponding vectors for a new observation. The vector $\boldsymbol{\beta}$ is similarly partitioned. There is no loss of generality in placing the predictor variables to be deleted last in the model.

The predictor using the FM is

$$\hat{Y}_{0r} = \mathbf{X}_0' (X'X)^{-1} X'\mathbf{Y}. \tag{3.4}$$

However, for comparison purposes we need an expression for $\hat{Y}_{0r}$ in terms of $\mathbf{X_a}$. We reparametrize the model as

$$E(\mathbf{Y}) = (\mathbf{X_a} | \mathbf{Z}) \left(\frac{\gamma}{\beta_k} \right), \tag{3.5}$$

where $\mathbf{Z} = [\mathbf{I} - \mathbf{X_a}(\mathbf{X_a'X_a})^{-1}\mathbf{X_a'}]\mathbf{X_k}$ and $\gamma = \boldsymbol{\beta_a} + (\mathbf{X_a'X_a})^{-1}\mathbf{X_a'X_k}\boldsymbol{\beta_k}$. The columns of Z are orthogonal to the columns of X_a. Correspondingly, we reparametrize the model of Y_{0r} as

$$E(Y_{0r}) = (\mathbf{X_{0a}'}|\mathbf{Z_0'}) \left(\frac{\gamma}{\boldsymbol{\beta_k}}\right), \tag{3.6}$$

where

$$\mathbf{Z_0'} = \mathbf{X_{0k}'} - \mathbf{X_{0a}'}(\mathbf{X_a'X_a})^{-1}\mathbf{X_a'X_k}.$$

An alternate expression for the predictor using the FM is

$$\hat{Y}_{0r} = \mathbf{X_{0a}'}(\mathbf{X_a'X_a})^{-1}\mathbf{X_a'Y} + \mathbf{Z_0'}(\mathbf{Z'Z})^{-1}\mathbf{Z'Y}. \tag{3.7}$$

The predictor mean square error for $\hat{Y}_{0r}$ is

$$\begin{aligned} MSE_r = E(Y_0 - \hat{Y}_{0r})^2 &= \sigma^2 + \mathbf{X_{0a}'}(\mathbf{X_a'X_a})^{-1}\mathbf{X_{0a}}\sigma^2 \\ &+ \mathbf{Z_0'}(\mathbf{Z'Z})^{-1}\mathbf{Z_0}\sigma^2. \end{aligned} \tag{3.8}$$

The prediction using the submodel is

$$\hat{Y}_{0a} = \mathbf{X_{0a}'}(\mathbf{X_a'X_a})^{-1}\mathbf{X_a'Y}. \tag{3.9}$$

The prediction mean square error for $\hat{Y}_{0a}$ is

$$MSE_a = E(\mathbf{Y_0} - \hat{Y}_{0a})^2 = \sigma^2 + \mathbf{X_{0a}'}(\mathbf{X_a'X_a})^{-1}\mathbf{X_{0a}}\sigma^2 + (\mathbf{Z_0'}\boldsymbol{\beta_k})^2. \tag{3.10}$$

We prefer the FM for prediction if

$$MSE_r < MSE_a. \tag{3.11}$$

Relations equivalent to (3.11) are

$$(\mathbf{Z_0'}\boldsymbol{\beta_k})^2 - \mathbf{Z_0'}(\mathbf{Z'Z})^{-1}\mathbf{Z_0}\sigma^2 > 0 \tag{3.12}$$

and

$$\lambda = \frac{(\mathbf{Z_0'}\boldsymbol{\beta_k})^2}{2\mathbf{Z_0'}(\mathbf{Z'Z})^{-1}\mathbf{Z_0}\sigma^2} > \frac{1}{2}. \tag{3.13}$$

If $k = 1$, then $\mathbf{Z_0'}$ is a scalar and cancels from λ. In this case 2λ is simply the ratio of $\boldsymbol{\beta_k^2}$ to the variance of its least squares estimator [see (3.3)]. Since λ is a function of unknown parameters, its value is not known. As an approximation we might base our decision on λ evaluated at the estimates of the parameters.

In addition to our earlier assumptions we now assume that Y has a normal distribution. The usual test statistic for the null hypothesis H_0: $\mathbf{Z_0'}\boldsymbol{\beta_k} = 0$ is

$$W = \frac{(\mathbf{Z_0'}\mathbf{b_k})^2(n - r)}{\mathbf{Z_0'}(\mathbf{Z'Z})^{-1}\mathbf{Z_0}SSE}, \tag{3.14}$$

where $\mathbf{b_k}$ is a vector containing the last k elements of $\mathbf{b}$. Let $F(f_1, f_2, \eta)$ denote the noncentral F distribution with f_1 and f_2 degrees of freedom and noncentrality parameter η. We recognize that W is distributed as $F(1, n - r, \lambda)$. Toro-Vizcarrondo and Wallace (1968) show that a uniformly most powerful test of the hypothesis $H_0: \lambda \leq 1/2$ against the alternative hypothesis $H_1: \lambda > 1/2$ is also based on W. The test is given by

$$\text{Accept } H_0: \text{ if } W < W_\alpha$$
$$\text{Reject } H_0: \text{ if } W \geq W_\alpha$$

where W_α is the $(1 - \alpha)$th quantile of $F(1, n - r, 1/2)$. Wallace and Toro-Vizcarrondo (1969) give the .50, .75, .90, and .95 quantiles of $F(f_1, f_2, 1/2)$ for reasonably extensive values of f_1 and f_2. The significance level α of this test can be interpreted as the probability of using the full model for prediction when in fact the two predictors are equally good. Thus $\alpha = .50$ seems to be an appropriate significance level for practical use. We may compare two or more submodels to the FM. The distribution of W does not depend on a and k. Thus we can validly compare submodels with different numbers of variables to the FM. Unfortunately, the submodel that compares most favorably with the FM is not necessarily the best.

The comparison of two or more submodels directly is a more difficult problem than comparing a submodel to the FM. For this problem we work with (3.12) evaluated at sample quantities. Specifically, we choose the submodel having the smallest value of

$$(\mathbf{Z_0'b_k})^2 - \mathbf{Z_0'(Z'Z)^{-1}Z_0} \; SSE/(n - r) \tag{3.15}$$

The expression (3.15) is actually $MSE_a - MSE_r$ evaluated at sample values. Once $\mathbf{Z_0}$ is determined, both terms of (3.15) can be found easily by any regression program able to test hypotheses about arbitrary linear combinations of the parameters. Here we must retain $\mathbf{Z_0}$ even when $k = 1$. At this time we do not know of a simple test for comparing two submodels.

For either a FM-submodel or a submodel-submodel comparison, the vector(s) $\mathbf{Z_0}$ must be computed. If one has access to a multivariate linear model program that calculates residuals and allows differential weighting of the observations, then $\mathbf{Z_0}$ is obtained very simply. Enter

$$\begin{pmatrix} \mathbf{X_{0a}'} \\ \mathbf{X_a} \end{pmatrix}$$

as the predictor variables and

$$\begin{pmatrix} \mathbf{X_{0k}'} \\ \mathbf{X_k} \end{pmatrix}$$

as the dependent variables, but give the first row a zero weight. The residual vector corresponding to the first row is Z'_0. If the program is sufficiently flexible, the Z_0 vectors for multiple future values and multiple submodels can be obtained in one run.

There are two features of our method of comparing submodels that distinguish it from the various sequential selection procedures. Both the values of the predictor variables associated with the future observation and the magnitude of the estimated variance are directly utilized in the criterion. In the Efroymson (1960) SW procedure, for example, the values of the estimated partial correlations determine the variable to enter the prediction equation at a given stage.

If many submodels need to be compared, it would be desirable to compare less than all $2^r - 1$ possible submodels. To reduce the amount of computation, we propose the following sequential procedure. Start with the one variable submodel having the smallest value of the criterion (3.15). At each subsequent step include the predictor variable that results in the smallest value of (3.15) in conjunction with those previously entered. Whenever there are three or more predictor variables in the prediction equation, check to see if the criterion can be reduced by deleting a predictor variable previously included, before considering variables for inclusion. If so, delete the predictor variable giving the largest reduction in (3.15). The process terminates when the criterion cannot be made smaller by either adding or deleting a predictor variable. Unlike other sequential algorithms based on changes in the residual sum of squares where the stopping rule is arbitrary, the proposed mean square error procedure has a well-defined stopping point. However, like other sequential procedures, no guarantee exists that an absolute minimum is obtained.

The procedure outlined has been described for one future predicted value. If there are multiple future observations to be predicted, then repeat the process for each future observation. Recall that the predictor variables may be different for each predicted observation. This procedure would be followed whenever computationally feasible. An experimenter might desire a single prediction equation for multiple future observations. In this case the submodel with the smallest average mean square error over all the future observations would be selected. Such an averaging process would be advantageous when data exist for a one-time consideration of a large number of potential predictor variables but it is desired to measure a limited number of predictor variables for use in the prediction equation. For example, in the Laird and Cady (1969) fertilizer example a number of potential predictor variables were measured during the conduct of the series of experiments over locations and years. Based on these data, a prediction equation is desired so that farmers' soil samples submitted in the future would only have to be accompanied by a minimum amount of information concerning environmental and soils information other than soil nutrient status which would be determined by the soil-testing laboratory. Consequently, a prediction equation would have to be available before the actual values of the predictor variables were available for prediction. When the future X values are not available, several values can be selected which span a reasonable range of the potential predictor variables. These values may be the points of the factor space generated by a fractional factorial (coordinates on a grid system) of the basic variables.

4. Example

To illustrate the use of criterion (3.15), 11 of the 76 experiments from the Laird and Cady (1969) analysis were selected. The following five predictor variables were used to represent the full model:

X_1 = applied nitrogen, units of 40 kg/ha
X_2 = applied nitrogen squared, X_1^2
X_3 = total soil nitrogen, percentage by weight × 100
X_4 = applied nitrogen by total soil nitrogen interaction, $X_1 X_3$
X_5 = hail damage, subjectively determined by field observation using a 0–6 scale; 0 = no damage

The full data including y, the yield of corn grain in tons per hectare, are given in Table 1.3. The simple correlation matrix of all the variables is given in Table 1.4. The values of the five predictor variables to be used in nine predictions are given in Table 1.5.

Table 1.3

Partial Listing of Data from Laird and Cady (1969)

Experiment	X_1	X_2	X_3	X_4	X_5	y
204	0	0	120	0	1	1.82
	1	1	120	120	1	3.71
	2	4	120	240	1	4.80
	3	9	120	360	1	5.48
213	0	0	54	0	2	1.74
	1	1	54	54	2	3.15
	2	4	54	108	2	4.01
	3	9	54	162	2	4.17

Table 1.3 (*continued*)

Experiment	X_1	X_2	X_3	X_4	X_5	y
217	0	0	95	0	0	3.14
	1	1	95	95	0	4.04
	2	4	95	190	0	3.71
	3	9	95	285	0	4.38
302	0	0	98	0	0	1.99
	1	1	98	98	0	2.45
	2	4	98	196	0	2.15
	3	9	98	294	0	2.18
315	0	0	100	0	3	0.36
	1	1	100	100	3	1.46
	2	4	100	200	3	2.43
	3	9	100	300	3	3.19
316	0	0	73	0	0	0.52
	1	1	73	73	0	1.66
	2	4	73	146	0	2.76
	3	9	73	219	0	3.06
408	0	0	64	0	4	1.00
	1	1	64	64	4	1.60
	2	4	64	128	4	1.78
	3	9	64	192	4	1.87
412	0	0	46	0	0	0.99
	1	1	46	46	0	2.53
	2	4	46	92	0	3.56
	3	9	46	138	0	3.61
419	0	0	129	0	0	1.37
	1	1	129	129	0	2.86
	2	4	129	258	0	4.56
	3	9	129	387	0	5.88
508	0	0	94	0	0	1.73
	1	1	94	94	0	2.90
	2	4	94	188	0	3.73
	3	9	94	282	0	4.37
511	0	0	54	0	0	0.92
	1	1	54	54	0	1.79
	2	4	54	108	0	3.28
	3	9	54	162	0	4.10

Table 1.4

Simple Correlation Matrix

	X_1	X_2	X_3	X_4	X_5	y
X_1	1.00	0.96	0.00	0.88	0.00	0.69
X_2		1.00	0.00	0.85	0.00	0.63
X_3			1.00	0.38	−0.15	0.27
X_4				1.00	−0.06	0.74
X_5					1.00	−0.28
y						1.00

The FS and BE procedures described by Draper and Smith (1966) were carried out to select the predictor variables to be used in a prediction equation. Both procedures gave the same results with a significance level of 5% used as a stopping rule; X_4 and X_5 were selected as the predictor variables.

For the 32 possible models, prediction criterion (3.15) was calculated for each future observation listed in Table 1.5. Values in Table 1.6 are the rankings among selected models for each column or future observation. The last column shows the rankings of prediction criterion (3.15) summed over the nine future observations. Both models 100011 (includes X_0, X_4, and X_5) and 110101 have the lowest value of (3.15) for two future observations, but no one model appears to do well consistently for all future observations. The model selected by the FS and BE procedures, 100011, predicts well for certain future observations but not for others.

Using the total value of criterion (3.15) as a predictor variable selection procedure would lead to a prediction model including variables X_1, X_2, X_4, and X_5. A prediction model including predictor variables X_1, X_4, and

Table 1.5

Values of Future Predictor Variables

	X_1	X_2	X_3	X_4	X_5
1	0	0	50	0	0
2	0	0	50	0	3
3	0	0	125	0	0
4	0	0	125	0	3
5	3	9	50	150	0
6	3	9	50	150	3
7	3	9	125	375	0
8	3	9	125	375	3
9	1.5	2.25	85	127.5	.9

X_5 gave a total value for (3.15) of nearly the same value as the model with X_1, X_2, X_4, and X_5. These differences between the SW and BE selection procedures and the selection procedure using prediction criterion (3.15) are interesting, but it is noted that the results are based on a limited number of data. Table 1.6 indicates that better results would be obtained by using the best predictor model for each future observation rather than the best predictor model using the total or average value of criterion (3.15).

Table 1.6

Ranks of Selected Models for Each Future Observation on the Basis of Prediction Criterion (3.15)

Model						Future Observation									
X_0	X_1	X_2	X_3	X_4	X_5	1	2	3	4	5	6	7	8	9	Total
1	1	1	0	1	1	11	3	9	7	9	8	9	8	1	1
1	1	0	0	1	1	18	9	5	3	7	6	17	11	19	2
1	1	1	1	0	1	14	5	17	11	8	7	7	7	2	3
1	1	1	1	1	1	15	7	19	14	11	9	10	9	3	4
1	1	0	1	0	1	8	1	24	19	19	16	1	4	12	5
1	0	0	0	1	1	22	13	2	1	1	3	23	16	9	6
1	1	0	1	1	1	16	8	21	15	12	10	12	10	10	7
1	1	1	0	1	0	1	19	25	5	21	17	8	21	6	11
1	1	0	0	0	1	20	6	4	4	29	25	24	19	21	14
1	1	1	0	0	1	12	2	8	10	28	22	26	24	7	15
1	1	0	0	1	0	4	22	12	12	10	21	6	25	20	16
1	0	0	0	1	0	6	24	7	17	16	2	11	26	14	18
1	1	1	0	0	0	2	20	26	6	18	30	28	3	8	24
1	1	0	0	0	0	5	23	13	13	25	31	27	1	23	25
1	0	0	0	0	1	32	30	30	28	15	19	31	32	22	30
1	0	0	0	0	0	21	32	29	30	26	1	32	30	24	32

References

Anderson, R. L. 1957. Some statistical problems in the analysis of fertilizer response data. In *Economic and technical analysis of fertilizer innovations and resource use*, ed. E. L. Baum et al., ch. 17. Ames: Iowa State College Press.

———. 1960. Some remarks on the design and analysis of factorial experiments. In *Contributions of probability and statistics*, ed. I. Olkin et al., ch. 5. Stanford: Stanford Univ. Press.

Draper, N. R., and H. Smith. 1966. *Applied regression analysis*. New York: Wiley.

Efroymson, M. A. 1960. Multiple regression analysis. In *Mathematical methods for digital computers*, vol. 1, ch. 17. New York: Wiley.

Kennedy, W. J., Jr. 1969. Model building for prediction in regression analysis based on repeated significance tests. Unpubl. Ph.D. thesis, Iowa State Univ., Ames.

Laird, R. J., and F. B. Cady. 1969. Combined analysis of yield data from fertilizer experiments. *Agron. J.* 61: 829–34.

Lamotte, L. R., and R. R. Hocking. 1970. Computational efficiency in the selection of regression variables. *Technometrics* 12: 83–93.

Toro-Vizcarrondo, C. E., and T. D. Wallace. 1968. A test of the mean square error criterion for restrictions in linear regression. *J. Am. Statist. Assoc.* 63: 558–72.

Wallace, T. D., and C. E. Toro-Vizcarrondo. 1969. Tables for the mean square error test for exact linear restrictions in regression. *J. Am. Statist. Assoc.* 64: 1649–63.

Walls, R. C., and D. L. Weeks. 1969. A note on the variance of a predicted response in regression. *Am. Statistician* 23(3): 24–26.

Weiner, J. M., and O. J. Dunn. 1966. Elimination of variates in linear discrimination problems. *Biometrics* 22: 268–75.

2

Some Recent Advances in Inference Procedures Using Preliminary Tests of Significance

T. A. BANCROFT

1. Introduction

IN THE APPLICATION of statistical methodology the experimenter is often uncertain of some one or more of the assumptions required to validate a desired inference. The desired inference may be either a test of a hypothesis, the estimation of a parameter, or a prediction. As an example of the estimation of a parameter, suppose that it is desired to estimate the population regression coefficient β_1 in the model $y = \beta_1 x_1 + \beta_2 x_2 + e$, when it is suspected but not known with certainty that $\beta_2 = 0$. It is assumed that e is NID$(0, \sigma^2)$, that x_1, x_2, and y are measured from their respective means, and that the values of x_1 and x_2 are fixed from sample to sample. Given a sample of size n, the experimenter may decide to fit the complete model and, based on the outcome of the preliminary test of significance of the additional reduction in sum of squares due to fitting x_2 after fitting x_1, then decide to estimate β_1 from either the fitted regression $\hat{y} = b_1 x_1 + b_2 x_2$ or $\hat{y} = b_1' x_1$. The inference of major interest, in this case estimation, made subsequent to the preliminary test of significance is of course conditional on the outcome of the preliminary test of significance.

Again, a second rather extensive use of preliminary tests involves their

T. A. BANCROFT is Director and Head, Statistical Laboratory and Department of Statistics, Iowa State University, Ames.

Journal paper J-6804 of the Iowa Agriculture and Home Economics Experiment Station, Ames, Project 169.

This work was partially supported by National Science Foundation Grant GP-9046.

19

use in deciding whether or not to pool a possible but uncertain estimate of a population parameter, obtained from the same or a second sample, with a known proper estimate of the particular population parameter of interest. After using a preliminary test to decide whether to pool or not, the experimenter then will ordinarily wish to obtain an interval estimate for the parameter or to test some hypothesis of primary interest concerning the parameter.

Although inference procedures incorporating preliminary tests of significance have been used intuitively for some time by research workers and applied statisticians, only in recent years and for a limited number of uses have attempts been made to evaluate their properties. Bibliographies of papers in this area up to five years ago are given by Kitagawa (1963) and Bancroft (1964, 1965). Subsequent relevant papers include Kale and Bancroft (1967), Srivastava and Bancroft (1967), and Han and Bancroft (1968). Some uses of preliminary tests of significance based on results obtained in earlier papers are given in the text by Bancroft (1968). The use of preliminary tests of significance is discussed briefly by Kruskal (1968).

2. Inference for Incompletely Specified Models Using Preliminary Tests of Significance

In making inferences about a model from sample data, there may exist one of the following types of uncertainty: (1) given a single sample of size n, there may be uncertainty regarding one or more of the model assumptions and (2) given two samples of sizes n_1 and n_2 respectively, there may be uncertainty as to whether both may be from the same model. (The latter type may be extended to more than two samples.)

Illustration (2.1). An illustration of (1) would involve a procedure for estimating θ_1 or testing some hypothesis about θ_1 when it is known that the single sample of size n came from either $f_{12}(X; \theta_1, \theta_2)$ or $f_1(X; \theta_1)$, where the latter distribution is the same as the former with $\theta_2 = 0$. An inference procedure for θ_1 is a member of a class of inference procedures designated as *inferences for incompletely specified models*.

Illustration (2.2). Additional illustrations of (1) involve inference procedures for incompletely specified linear models, including fixed, random, and mixed experimental design models; ordinary multiple regression models; and polynomial regression. In investigations involving designed experiments we may assume, for example, that the observations Y are $NID[E(Y), \sigma^2]$, where $E(Y) = \mu + \alpha_i + \beta_{ij} + \ldots$. There may be one or more "doubtful" terms in $E(Y)$; e.g., it may be uncertain as to whether the set of parameters defined by β_{ij} should occur in $E(Y)$. An inference procedure for the other parameters involved, incorporating a preliminary test of the hypothesis that $\beta_{ij} = 0$, is also a member of the class designated above as inferences for incompletely specified models.

Illustration (2.3). In ordinary multiple regression we may assume that the observations Y are $\mathrm{NID}[E(Y), \sigma^2]$, where $E(Y) = \beta_0 + \beta_1 X_1 + \beta_2 X_2 + \dots$. There may be one or more doubtful terms on the right of the equality sign, and preliminary test(s) may be used as an aid in determining the number of independent variates to be used in the model for predicting Y. Polynomial regression is of course a special case of ordinary multiple regression where the independent variates have a natural rank. Prediction incorporating preliminary test(s) in either of these two types of regression belongs to the class of inference procedures for incompletely specified models.

Illustration (2.4). An illustration of (2) might involve a procedure for estimating μ_1 or testing some hypothesis concerning μ_1, given a sample of size n_1 observations that are $\mathrm{NID}(\mu_1, \sigma^2)$ and a sample of size n_2 observations that are $\mathrm{NID}(\mu_2, \sigma^2)$, where it is suspected but not known for certain that $\mu_1 = \mu_2$. An inference procedure for μ_1 incorporating a preliminary test of the hypothesis $\mu_1 = \mu_2$ to decide whether or not to pool the samples is a member of the class of inference procedures for incompletely specified models.

In using inference procedures incorporating preliminary test(s) of significance the experimenter behaves as if the null hypothesis of the preliminary test is true if it is not rejected. This is contrary to the inference philosophy advocated in many texts in applied statistics to the effect that failure to reject a null hypothesis does not imply that it is true. Such texts imply that failure to reject the null hypotheses can be attributed to a small sample size and/or lack of power. Berkson (1942) disagrees with this philosophy and recommends acceptance as well as rejection of the null hypothesis in accordance with the outcome of the significance test.

3. Estimation after a Preliminary Test of Significance

Consider illustration (2.1). Assume that

1. $\hat{\theta}_1$ is the estimator for θ_1, possessing desirable properties assumed available when the model is $f_1(X; \theta_1)$.
2. $\hat{\theta}_{12}$ is the estimator for θ_1, possessing desirable properties assumed available when the model is $f_{12}(X; \theta_1, \theta_2)$.
3. θ_1^* is the estimator for θ_1 for the incompletely specified model, i.e., where a preliminary test of significance is used to ascertain whether

$f_1(X; \theta_1)$ or $f_{12}(X; \theta_1, \theta_2)$ should be used in the subsequent estimation of θ_1.

4. T_p is the preliminary test of significance criterion for testing $\theta_2 = 0$ at probability level α_1, possessing desirable properties assumed available when the model is $f_{12}(X; \theta_1, \theta_2)$.

Then $\theta_1^* = I\hat{\theta}_{12} + (1 - I)\hat{\theta}_1$, where $I = 1$ if $T_p \geq T_p(\alpha_1)$ and $I = 0$ if $T_p < T_p(\alpha_1)$.

For estimation of θ_1 it will be assumed that the characteristics of primary interest are the bias and mean square error. To obtain the bias we first find

$$E(\theta_1^*) = E\lfloor\hat{\theta}_{12}|T_p \geq T_p(\alpha_1)\rfloor P_r[T_p \geq T_p(\alpha_1)]$$
$$+ E[\hat{\theta}_1|T_p < T_p(\alpha_1)] \cdot P_r[T_p < T_p(\alpha_1)].$$

Then $\text{Bias}(\theta_1^*) = E(\theta_1^*) - \theta_1$. Similarly,

$$MSE(\theta_1^*) = E(\theta_1^{*2}) - [E(\theta^*)]^2 + (\text{Bias } \theta_1^*)^2,$$

where

$$E(\theta_1^{*2}) = E[\hat{\theta}_{12}^2|T_p \geq T_p(\alpha_1)]P_r[T_p \geq T_p(\alpha_1)]$$
$$+ E[\hat{\theta}_1^2|T_p < T_p(\alpha_1)] \cdot P_r[T_p < T_p(\alpha_1)].$$

The bias and mean square error will in general be functions of certain population parameters (here θ_1 and θ_2), the size of the sample, and the probability level α_1 of the preliminary test of significance. By constructing tables and graphs of the relative efficiency of $\hat{\theta}^*$ to $\hat{\theta}_{12}$ and $\hat{\theta}_1$ as functions of the unknown parameters, it should be possible to make recommendations for a choice of α_1 for which $\hat{\theta}^*$ will be preferred over $\hat{\theta}_{12}$ and $\hat{\theta}_1$, at least for certain ranges of values of the parameters involved.

4. Test after a Preliminary Test of Significance

Consider illustration (2.1) again. Assume that

1. T_1 is the test criterion for a null hypothesis regarding θ_1 at probability level α_2, possessing desirable properties assumed available when the model is $f_1(X; \theta_1)$.
2. T_{12} is the test criterion for a null hypothesis regarding θ_1 at probability level α_3, possessing desirable properties assumed available when the model is $f_{12}(X; \theta_1, \theta_2)$.
3. T_1^* is the test criterion for a null hypothesis regarding θ_1 for the incompletely specified model; i.e., when a preliminary test of significance is used to ascertain whether $f_1(X; \theta_1)$ or $f_{12}(X; \theta_1, \theta_2)$ should be used in the subsequent test regarding θ_1.

4. T_p is the preliminary test of significance criterion for testing that $\theta_2 = 0$ at probability level α_1, possessing desirable properties assumed available when the model is $f_{12}(X; \theta_1, \theta_2)$.

Then $T_1^* = IT_{12} + (1 - I)T_1$, where $I = 1$ if $T_p \geq T_p(\alpha_1)$ and $I = 0$ if $T_p < T_p(\alpha_1)$.

We now wish to obtain the power of the test T_1^* of the null hypothesis $H_0: \theta_1 = \theta_0$ for the incompletely specified model. (Note that T_1, T_{12}, and T_1^* may be considered either one-sided or two-sided tests.) For simplicity let us consider an upper one-sided test of H_0. In such case we reject H_0 if either

$$T_p \geq T_p(\alpha_1) \quad \text{and} \quad T_{12} \geq T_{12}(\alpha_3)$$

or

$$T_p < T_p(\alpha_1) \quad \text{and} \quad T_1 > T_1(\alpha_2).$$

The probability P of rejecting H_0, which in general is the power of the test procedure, is a function of the size of the sample, the parameters θ_1 and θ_2, and the levels of significance employed, α_1, α_2, and α_3. In the special case when $\theta_1 = \theta_0$, this power is equal to the size of the test, i.e., the probability of Type I error.

The power P may be obtained as the sum of two components corresponding to the mutually exclusive alternatives headed by *either* and *or* in the above two inequalities, i.e., as

$$P_1 = P_r[T_p \geq T_p(\alpha_1) \quad \text{and} \quad T_{12} \geq T_{12}(\alpha_3)]$$

and

$$P_2 = P_r[T_p < T_p(\alpha_1) \quad \text{and} \quad T_1 \geq T_1(\alpha_2)],$$

then $P = P_1 + P_2$.

The power in general will be a function of the size of the sample n; θ_1 and θ_2; and the levels of significance α_1, α_2, and α_3. For a particular problem, n will be given, while θ_1 and θ_2 are generally unknown. It seems appropriate to choose $\alpha_2 = \alpha_3$, say, equal to 0.05. By constructing tables and graphs of the power of T_1^* as functions of the unknown parameters involved, it will be possible to make comparisons with the power of alternative test procedures such as that of T_{12} or T_1 used separately and alone. Based on such studies, recommendations may be made regarding a proper choice of α_1 for the preliminary test procedure. Since preliminary tests used and final inferences made may differ in their properties from problem to problem, it will be necessary to make a detailed study for each type of such inference procedure to ensure proper recommendations regarding a choice of α_1.

5. Alternative Inference Procedures Useful for Situations Involving Uncertainty in Model Specifications

Alternative inference procedures involving estimates or tests may be those based on the usual single specification model alone or those obtained by deriving an inference methodology analogous to those in the consecutive decision theory of the multiple decision theory discussed by Lehmann (1957). This formulation of a consecutive decision theory for several applied problems has some features in common with the inference theory for incompletely specified models described in sections 3 and 4. In both, a single sample is available for investigating two (possibly more) questions of interest, the second dependent upon the outcome of the first. Both also make use of Neyman-Pearson inference theory to obtain usable results in applied situations. However, Lehmann's consecutive decision theory, as illustrated by the few examples considered, is based on conditioning on certain functions of the observations and obtaining a final inference possessing certain predetermined optimum properties.

For the estimation case involving a doubtful estimate and a proper estimate, Huntsberger (1955) discusses the use of a differential weighting procedure which always makes use of the doubtful estimate as well as the proper estimate of the population parameter. Veale and Huntsberger (1969) present a modification of Huntsberger's weighting procedure, in which the weighted estimator is used in preference to the proper estimator only if a preliminary test criterion falls within a specified region. This latter estimation procedure is recommended for estimation of a mean when one observation of a sample may be spurious. Other inference procedures, involving both a final estimate or a final test, for such situations might be attacked by using a Bayesian approach. Finally, in the face of uncertainty in model specifications some experimenters may consider using a nonparametric or distribution-free inference procedure.

6. Recent Advances in General Theory

Cohen (1965) has discussed the problem of the admissibility of an inference procedure involving a preliminary test followed by an estimate of primary interest. He gives a proof for a theorem stating that such an inference procedure is not admissible. On the other hand, Cohen (1968) discusses the problem of admissibility of an inference procedure involving a preliminary test followed by a test of primary interest. For this latter case he gives a proof of a theorem stating that such an inference procedure is admissible.

While the property of admissibility in general is desirable, in applied statistics it is not always a critical requirement. In this respect it is somewhat analogous to the property of unbiasedness. It is well known that inference procedures involving the use of preliminary tests are usually biased. However, it is also well known that biased inferences may be preferred on occasion to unbiased ones. In particular, in applied situations the experimenter in most cases would not consider using a preliminary test of some hypothesis, say $H_0: \theta = \theta_0$, unless he was reasonably sure that θ was at least within a small neighborhood of θ_0. A priori information obtained from previous experiences with similar data and/or theoretical knowledge of the substantive field involved should provide him with this assurance in many situations. For situations in which such assurance may not be particularly strong, it should be possible to devise a protection procedure. For example, the experimenter may decide to use a preliminary-test–based estimator whose relative efficiency to the usual single specification estimator is high, say higher than a certain preassigned minimum value. Using this value, the experimenter, for the case of pooling means when variance is unknown, may use the method provided by Han and Bancroft (1968) and select a proper significance level to ensure the preassigned relative efficiency, with the expectation of realizing a higher relative efficiency than the minimum.

The Cohen (1965) result is negative in nature; i.e., no attempt was made for this situation to provide an alternative inference procedure which would be admissible. If the experimenter uses only completely specified models, his estimates may not always be efficient. For certain values of the parameters involved, studies for certain cases have shown that the mean square error of estimates based on the incompletely specified model is less than that based on the completely specified model. Analogous statements for increase in power can be made for certain cases involving a test of primary inference after a preliminary test. In performing the preliminary test $H_0: \theta = \theta_0$, the experimenter expects θ_0 to be at least in a small neighborhood of θ. Hence, in general the experimenter would want an inference procedure which would perform well in such a small region with protection against unexpected deviations.

Arnold (1970) has shown by an alternative method that the inference procedure given by Srivastava and Bancroft (1967) for possibly pooling two correlations is one in which a preliminary-test type of estimate is uniformly better than the usual estimate.

7. Recent Advances in Methodology

Ruhl (1967) discussed several methods of pooling correlated data when observations are available on the random variables (x, y) and it is desired to estimate $E(y) = \mu_y$. It is assumed that (x, y) follows a bivariate normal distribution with parameters $(\mu_x, \mu_y, \rho, \sigma_x^2, \sigma_y^2)$. The sampling method allows for $n \geq 0$ independent observations on (x, y), $n_x \geq 0$ additional independent observations on x alone, and $n_y \geq 0$ additional independent observations on y alone. Assuming known ρ, σ_x^2, σ_y^2, the bias and mean square error of two estimators based on a preliminary test H_0 are derived. If $H_0: \mu_y = \mu_x$ is accepted by the preliminary test, then both PT and PTS (two types of preliminary estimators) are weighted averages of the respective means of the observations on x and y. If $H_A: \mu_y \neq \mu_x$ is accepted by the preliminary test, then the estimator PT is simply the mean of the observations on y; whereas PTR is a classical regression estimator of $\bar{y}_n$ on $\bar{x}_n$ with μ_x estimated from the sample. PTR is seen to have a smaller mean square error in most cases since it makes more efficient use of the available data. The two estimators PT and PTR are also compared to the estimation scheme which uses a regression estimator all the time and does not include a preliminary test.

Johnson (1967) has discussed the use of a preliminary test in deciding whether or not to pool a second doubtful sample regression with a sample regression known to be a proper estimate of the target population regression model. It is assumed that the resulting fitted regression obtained subsequent to the preliminary test would be used for prediction purposes. The case for two multiple regressions is considered as well as that for two simple regressions. Recommendations are attempted in each case for the level of the preliminary test α_1.

Carrillo (1969) considers the problem of estimating a weighted linear function of population variances $\sum_{i=1}^{n} w_i \sigma_i^2$ with known weights. It is assumed that there are k samples of n_i independent observations each from $N(\mu_i, \sigma_i^2)$. He proposes an estimation procedure based on a preliminary test of significance carried out to classify the variances into homogeneous groups. The bias and the efficiency of the proposed estimation procedure, as compared with the estimation procedure currently in use, were obtained. The theory is illustrated by applications to survey sampling.

Kennedy (1969) has discussed the use of preliminary tests in model building in multiple regression for prediction with unknown error variance. In using this procedure, it is assumed that the independent variates are ranked in "order of importance" either by the experimenter or possibly by use of an independent sample. He considers two procedures involving sequential preliminary tests in deciding whether or not to delete each independent variate in a doubtful subset of the full model containing all the independent variates. On the basis of the results of this study it is recommended that the testing begin with the independent variate of lowest rank in the doubtful set. Again, it is recommended that in general the significance level of the preliminary test, assumed to be the same for all such tests, be set at $\alpha = 0.25$ to control the magnitude of the mean square error of the predictand ($\hat{Y}$) relative to the magnitude of the mean square error of the predictant $\hat{Y}$ using the full model. Kennedy's results are applicable to polynomial regression for which there is a natural ordering of the independent variates, i.e., of powers of the independent variate X.

Bancroft (1953), Lemus (1955), and Mead (1968) obtained results for inference for the incompletely specified fixed linear model used in the analysis of variance. This study is similar to that made by Bozivich et al. (1956) for the random and mixed models. For the former study, investigations of size and power revealed that the preliminary test should be made at the probability level of $\alpha_1 = 0.25$ or in some cases even at $\alpha_1 = 0.50$.

Gupta and Srivastava (1969) have obtained the bias and the mean square error of an estimation procedure after two preliminary tests of significance in analysis of variance for the fixed model.

Mehta and Gurland (1969a) have discussed the use of a preliminary test of zero population correlation in a sample from a bivariate normal population with equal population variances to decide whether to use the ordinary t test for equal population means from independent samples or a paired comparison test applicable in the presence of correlation. Again, Mehta and Gurland (1969b) have discussed the use of a preliminary test of the hypothesis of equal population variances to decide whether or not to pool a doubtful estimate of a target population variance with a proper estimate of that population variance. In the same paper an alternative approach similar to that discussed by Huntsberger (1955) is used, which makes use of a differential weighting method always making use of both the doubtful and the proper estimate to estimate the target population variance.

Gun (1967) investigated the consequences of using a preliminary test to decide whether or not to use an estimator for the ijth population cell mean based on a two-way classification model containing an interaction term, or an estimator based on the same model without the interaction term. He derived the distribution of the preliminary test estimator, its bias and mean square error. He also studied how the bias and mean square error are affected by the number of observations in each cell, assumed to be equal, and α, the level of significance of the preliminary test. The paper suggests several methods for a choice of α, including a Bayes approach, a minimax approach, and a minimax regret approach. Again, Gun (1969) makes use of a Bayes approach in considering the problem of whether or not to pool two sample means, depending upon the outcome of a preliminary test. He makes use of a gamma-type prior distribution for the noncentrality parameter appearing in the mean square error of the preliminary-test estimator. In this latter paper Gun also makes use of this same approach for the preliminary-test estimators for cell means considered in his 1967 publication.

Cochran (1970) has recently considered the performance of a preliminary test of significance sometimes used in observational studies. He summarizes the results as follows:

The means $\bar{y}_1$, $\bar{y}_2$ of a response variable are being compared in two independent samples of size n. It is feared that the comparison may be biased because a confounding variable x may have different levels (means) in the two populations. A t-test of $(\bar{x}_1 - \bar{x}_2)$ is made. If t is not significant, $(\bar{y}_1 - \bar{y}_2)$ is used as an estimate of the population difference δ_y; if t is significant, $(\bar{y}_1 - \bar{y}_2)$ is adjusted by an analysis of covariance. This procedure is found to reduce the remaining bias in the estimate of δ_y to a value that does not seriously disturb tests of significance or confidence probabilities based on this estimate. If the estimation of small values of δ_y is important, however, the remaining bias may still be a sizeable percentage of δ_y, except with large samples.

8. Example of Inference for the Incompletely Specified Fixed Linear Model

Brownlee (1955), using data first reported by Sternhell (1958), gives an analysis of variance table for a two-way classification model with two observations per cell. The observations were measures of total acidities of three types of brown coals by absorption, using three methods of analysis. The fixed model assumed was

$$y_{ijk} = \mu + \alpha_i + \beta_j + (\alpha\beta)_{ij} + e_{ijk},$$
$$i = 1, 2, 3; \quad j = 1, 2, 3; \quad k = 1, 2,$$

where $\sum_i \alpha_i = \sum_j \beta_j = \sum_i (\alpha\beta)_{ij} = \sum_j (\alpha\beta)_{ij} = 0$, $e_{ijk} \sim \text{NID}(0, \sigma^2)$.

An analysis of variance table was obtained as shown in Table 2.1. Using the analysis in the table, Sternhell and Brownlee conclude that there

Table 2.1
Analysis of Variance without Pooling

Source of Variation	Degrees of Freedom	Sum of Squares	Mean Squares	Expected Mean Squares	Calculated F
Type	2	1.0024	0.5012	$\sigma_e^2 + 6K_T^2$	29.50**
Method	2	0.1244	0.0622	$\sigma_e^2 + 6K_M^2$	3.66
Interaction	4	0.0145	0.0036	$\sigma_e^2 + 2K_{TM}^2$	0.21
Within (error)	9	0.1530	0.0170	σ_e^2	...

Table 2.2

Analysis of Variance with Pooling

Source of Variation	Degrees of Freedom	Sum of Squares	Mean Squares	Expected Mean Squares	Calculated F
Type	2	1.0024	0.5012	$\sigma_e^2 + 6K_T^2$	38.85**
Method	2	0.1244	0.0622	$\sigma_e^2 + 6K_M^2$	4.82*
Error	13	0.1675	0.0129	σ_e^2	...

are highly significant differences among the three types but no significant differences are demonstrated among the three methods.

Suppose that the experimenter, from previous experience with similar data, suspects but is not sure that the expected mean square for Interaction is the same as that for Within (Error). In such case he may decide to calculate $F = 0.0036/0.0170 = 0.21$ as a preliminary test of the null hypothesis $H_0: K_{TM}^2 = 0$. Since this F value is not significant at the 0.25 or even at the 0.75 level, he accepts the null hypothesis and pools the Interaction and Within (Error) sum of squares to obtain the revised analysis of variance shown in Table 2.2. We note from this table that there are now significant differences among the three methods at the 0.05 probability level, hence additional information has been provided by the use of this preliminary test procedure. After encountering an $F = 0.21 < 1$ for the test of Interaction in Table 2.1, the experimenter may wish to verify the correctness of the other original model assumptions.

References

Arnold, Barry C. 1970. An alternative derivation of a result due to Srivastava and Bancroft. *J. Roy. Statist. Soc.*, Ser. B, 32: 265–67.

Bancroft, T. A. 1953. Certain approximate formulas for the power and size of a general linear hypothesis incorporating a preliminary test of significance. Unpubl. preliminary report, Statistical Laboratory, Iowa State Univ. Ames.

———. 1964. Analysis and inference for incompletely specified test(s) of significance. *Biometrics* 20: 427–42.

———. 1965. Inference for incompletely specified models in the physical sciences. Bull. Intern. Stat. Inst., Proc. 35th Session, vol. 41, book 1, pp. 497–515.

———. 1968. *Topics in intermediate statistical methods*, vol. 1. Ames: Iowa State Univ. Press.

Berkson, J. 1942. Tests of significance considered as evidence. *J. Am. Statist. Assoc.* 37: 325–35.

Bozivich, H., T. A. Bancroft, and H. T. Hartley. 1956. Power of analysis of variance test procedures for incompletely specified models. *Ann. Math. Stat.* 27: 1017–43.

Brownlee, K. A. 1955. *Statistical theory and methodology in science and engineering*, 2nd ed. New York: Wiley.

Carrillo, A. 1969. Estimation of variance after preliminary tests of significance. Unpubl. Ph.D. thesis, Iowa State Univ., Ames.

Cochran, W. G. 1970. Performance of a preliminary test of comparability in observational studies. Tech. Rept. 29, Office of Naval Research, Statistics and Probability Program.

Cohen, Arthur. 1965. Estimates of linear combinations of the parameters in the mean vector of a multivariate distribution. *Ann. Math. Stat.* 36: 78–87.

———. 1968. A note on the admissibility of pooling in the analysis of variance. *Ann. Math. Stat.* 39: 1744–46.

Gun, A. M. 1967. Use of a preliminary test in the estimation of factorial means. *Calcutta Stat. Assoc. Bull.* 16: 49–72.

———. 1969. On the significance level of preliminary tests in some "TE" procedures. *Ann. Inst. Statist. Math.* 21: 373–76.

Gupta, V. P., and S. R. Srivastava. 1969. Bias and mean square of an estimation procedure after two preliminary tests of significance in ANOVA model I. *Sankhya*, Ser. A, vol. 31, pt. 3.

Han, Chien-Pai, and T. A. Bancroft. 1968. On pooling means when variance is unknown. *J. Am. Statist. Assoc.* 63: 1333–42.

Huntsberger, D. V. 1955. A generalization of a preliminary testing procedure for pooling data. *Ann. Math. Stat.* 26: 734–43.

Johnson, John. 1967. Pooling regressions and a statistical outlier methodology for lines. Unpubl. Ph.D. thesis, Iowa State Univ., Ames.

Kale, B. K., and T. A. Bancroft. 1967. Inferences for some incompletely specified models involving normal approximations to discrete data. *Biometrics* 23: 335–48.

Kennedy, William. 1969. Model building for prediction in regression analysis based on repeated significance tests. Unpubl. Ph.D. thesis, Iowa State Univ., Ames.

Kitagawa, Tosio. 1963. Estimation after preliminary tests of significance. *Univ. Calif. Pub. Stat.* 3: 147–86.

Kruskal, William. 1968. Significance, tests of. In *International Encyclopedia of the Social Sciences*, vol. 14, pp. 238–50, New York: Crowell.

Lehmann, E. L. 1957. A theory of some multiple decision problems, II. *Ann. Math. Stat.* 28: 547–72.

Lemus, F. 1955. Approximations to distributions in certain analysis of variance tests. Unpubl. M.S. thesis, Iowa State Univ., Ames.

Mead, R. J. 1968. Size and power of analysis of variance test procedures for incompletely specified fixed models. Unpubl. M.S. thesis, Iowa State Univ., Ames.

Mehta, J. S., and J. Gurland. 1969a. Testing equality of means in the presence of correlation. *Biometrika* 56: 119–26.

———. 1969b. On utilizing information from a second sample in estimating variance. *Biometrika* 56: 527–32.

Ruhl, Donna. 1967. Preliminary test procedures and Bayesian procedures for pooling correlated data. Unpubl. Ph.D. thesis, Iowa State Univ., Ames.

Srivastava, S. R., and T. A. Bancroft. 1967. Inferences concerning a population correlation coefficient from one or possibly two samples subsequent to a preliminary test of significance. *J. Roy. Statist. Soc.* Ser. B, 29: 282–91.

Sternhell, S. 1958. Chemistry of brown coals, IV. Further aspects of the chemistry of hydroxyl groups in Victorian brown coals. *Australian J. Applied Sci.* 9: 376–78.

Veale, James R., and D. V. Huntsberger. 1969. Estimation of a mean when one observation may be spurious. *Technometrics* 11 (2): 331–39.

3

A Choice Test by
Two Panels of Young Trees

C. I. BLISS

1. Introduction

IN A VOLUME honoring the major role that George Snedecor has played in introducing statistical methods into biological research, I would like to discuss an example from soil science of these techniques. This is a study of the response of tree seedlings to the soil of two plantations in north-western Connecticut, separated by 20 miles and differing in elevation by about 450 feet. This study was both the Ph.D. dissertation of its author, George I. Garin at Yale University, and a project of the Connecticut Agricultural Experiment Station, published in 1942 as Bulletin 454. The unique feature of this research provoking further analysis is the relation of root abundance to a number of soil characteristics. Although each was analyzed separately, they were not independent of one another, as the author noted.

The two plantations were on formerly cultivated but abandoned land covered with heavy sod, one on Merrimac loamy sand and the other on Charlton fine sandy loam. One- to two-year-old tree seedlings of 17 species were intermingled when planted in April 1933 and were spaced at 6 × 6 foot intervals, each in a cleared 3-foot circle. In July 1940, 8 vigorous seedlings of each of three coniferous species (white pine, red pine, and Norway spruce) and of two deciduous species (white ash and red oak) were selected for study on each plantation. These constituted two panels of young trees, ranging in mean height from 2.1 to 6.9 feet per species.

C. I. BLISS is with the Connecticut Agricultural Experiment Station and Yale University, New Haven.

To determine the distribution of their roots, three vertical transects were made in a square about each tree at distances of 3, 2, and 1 foot from its stem. At each soil horizon (A, B_1, and B_2) two soil samples were collected from one or more transects—one from an area with many roots and the other from an area with few or no roots. Soil samples representing each root density and horizon for the 8 replicate trees of each species in a soil type were pooled, making a total of 60 samples for analysis. Garin computed separate analyses of variance for five mechanical and five chemical characteristics of each sample. Overall contrasts between the two soils and the three horizons were very marked, and in some cases the differences between the five tree species as well.

The mean difference between zones with many and with few roots or one or more interactions of this difference with soils, horizons, or species of tree proved statistically significant in seven measures, designated in units of the difference between zones of many and of few roots as:

y_1 = percent loss on ignition.
y_2 = percent base saturation ($\times 0.1$).
y_3 = pH ($\times 10$).
y_4 = total exchange capacity, in mg equivalents per 100g of soil.
y_5 = percent total nitrogen ($\times 10$).
y_6 = percent moisture equivalent.
y_7 = percent clay.

This study is concerned with the relative importance of these seven criteria in determining comparative root densities as indicators of the choice shown by these two panels of young trees. From their combined response could some criteria be omitted without loss? Did the ratings differ with soil type, horizon, and species of tree?

2. Test of Computing Units

The problem will be recognized as one suitable for discriminant analysis (Fisher, 1936). Because of the disturbing effect of nonnormality or of an outlier or wild observation, the data for each criterion were first examined for normality. As a working hypothesis the soil on each plantation was assumed to vary randomly between species and between neighboring zones of many and of few roots. Because of sharp differences between the two soils and the three soil horizons, each criterion for a given soil and horizon was tested separately with a rankit diagram (Bliss, 1967), on the assumption that its ten measurements could be considered a normal sample. The 42 plots of rankits for a series of ten against the ordered

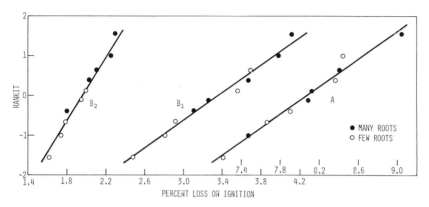

Fig. 3.1 Graphic test of normality for percent loss on ignition, from rankits plotted against ordered initial measurements for each of three horizons on Charlton soil, representing two root concentrations in five species of tree.

measurements were fitted by eye with straight lines passing through their means. As in the examples in Figure 3.1 none showed a consistent curvature about these lines that would suggest the transformation of any criterion to other units.

In a few cases, however, the end measurement in a series fell markedly to the right (or left) of a line which agreed well with the other nine. When checked by the Dixon (1951) gap test, four apparent outliers had a small enough probability of occurrence to be suspect, as in the example in Figure 3.2. The remaining nine measurements in these graphs were then

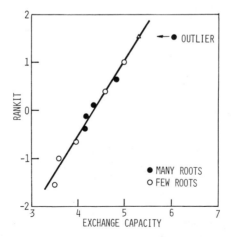

Fig. 3.2 Rankit diagram showing replacement Δ of an apparent outlier for exchange capacity, from horizon B_1 in Charlton soil.

assumed to be censored normal samples, each with a missing end measurement. A replacement was computed for each from its estimated mean $\tilde{y}$ (Bliss and Griminger, 1969) as $10\tilde{y} - \Sigma\, y$.

Some measurements differed by tenfold or more between soils and horizons. Over so wide a range the size of a difference y between two paired measurements could depend upon their sum u. To test this possibility, each difference was plotted against the sum of the two values. Of the seven soil measurements considered here, only three showed a significant ($P < 0.05$ for y_5 and y_7) or near significant ($P = 0.07$ for y_2) trend of the difference upon the sum of the two readings. To correct this dependence, the differences y_2, y_5, and y_7 have been adjusted to residuals around their respective regressions, as $y_i + b_i(\bar{u}_i - u_i)$.

Three additional outliers were first replaced, each a single difference falling well above the band defined by the remaining 29 measurements. In one (for y_6) with no trend of the difference upon the sum, the outlier was significant by the gap test (at $P = 0.01$) and replaced as the missing upper reading in a censored normal sample. In the other two criteria the remaining 29 differences increased y_5 or decreased y_7 with the sum, the apparent outlier differing from its predicted value on the regression with a probability of $P = 0.03$ for y_5 and $P = 0.09$ for y_7. Each was replaced by an average of two estimates of the missing end of a censored sample, one based upon other differences within a limited range of u and the second based upon its

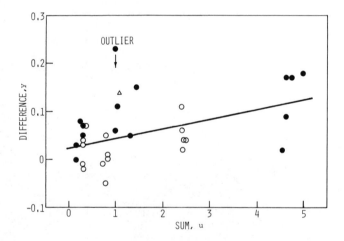

Fig. 3.3 Regression of the difference y in root density for percent ($\times 10$) of total nitrogen against the sum u of the initial measurements, computed with the replacement Δ for the outlier. Differences in Charlton soil are plotted as solid circles; residuals from the regression are the y_5's in Table 3.1.

expectation from the standard deviation about the regression. Each regression was then recomputed with the replacement and is shown graphically for y_5 in Figure 3.3.

With apparent outliers reduced to their normal limits and any dependence of the difference upon the sum removed, 30 differences in root density were available for each soil criterion. The 7 differences in soil characteristics have been listed in Table 3.1 for Merrimac loamy sand and for Charlton fine sandy loam for each tree species and soil horizon. The

Table 3.1

Basic Differences y_i in Seven Soil Characteristics for (many–few) Roots, after Reduction of Outlying Components to Their Normal Limits and Correction for Regression of the Differences y_2, y_5, and y_7 upon Their Respective Sums (u_i).

Soil	Tree	Hori-zon	y_1	y_2	y_3	y_4	y_5	y_6	y_7
Merri-mac	WP	A	0.30	−0.059	−0.3	−0.08	0.04	−0.38	0.36
		B_1	0.34	−0.100	−0.7	0.39	0.16	1.04	1.03
		B_2	0.20	0.349	0.4	0.15	0.06	0.43	0.28
	RP	A	0.09	−0.110	−0.1	0.09	0.22	−0.48	−0.54
		B_1	0.26	0.026	−0.1	0.00	0.26	0.23	2.41
		B_2	0.48	−0.244	0.1	0.00	0.95	0.34	0.10
	NS	A	0.14	−0.108	0.1	0.00	0.44	0.04	−0.78
		B_1	−0.11	−0.539*	−0.8	−0.46	0.08	−0.98	−0.93
		B_2	0.07	0.399	−0.2	0.07	0.57	−0.35	−0.08
	WA	A	0.15	0.269	−0.3	0.17	0.94	0.15	−1.11
		B_1	0.08	0.517	0.0	0.15	−0.33	−0.35	−1.05
		B_2	0.53	0.241	−0.4	0.36	0.67	0.47	−0.08
	RO	A	0.00	0.675	0.4*	0.50	−0.27	0.05	0.47
		B_1	0.32	0.677	0.5	0.61	0.66	0.88	−0.64
		B_2	−0.15	1.132	0.3	0.22	0.17	−0.25	−1.19
Charl-ton	WP	A	0.18	0.038	−0.2	0.32	1.06	0.46	2.37*
		B_1	0.42	0.299	−0.4	0.25	0.56	0.98	1.01
		B_2	0.31	−0.640	−0.1	−0.30	0.97	0.55	−0.66
	RP	A	0.26	−0.130	−0.5	−0.63	−0.40	−0.37	−0.01
		B_1	0.30	−0.776	−0.1	0.59	0.72	1.08	2.71
		B_2	0.24	−0.426	−0.5	0.31	0.77	0.91	−0.36
	NS	A	0.46	0.226	−0.6	0.27	0.29	−0.11	0.48
		B_1	0.75	−0.207	−0.1	0.21	1.51*	1.33	0.00
		B_2	0.11	−0.699	0.1	−0.77	0.29	0.11	0.87
*Replacements for				−0.742	0.9		2.30		3.6

Table 3.1 (*continued*)

Soil	Tree	Hori-zon	Difference in Soil Characteristic						
			y_1	y_2	y_3	y_4	y_5	y_6	y_7
Charl-ton	WA	A	0.61	−0.238	0.0	−0.18	1.12	0.06	1.45
		B_1	0.42	0.895†	−0.1	0.30†	1.54	1.59†	−1.00
		B_2	0.56	−0.722	−0.6	0.30	1.08	0.78	−1.20
	RO	A	0.04	−0.219	−0.1	−0.45	1.09	−0.42	−2.06
		B_1	0.78	0.823	0.5	0.83	1.22	0.92	−1.32
		B_2	0.16	0.991	−0.2	0.62	0.60	−0.40	−0.47
†Replacements for				1.400		1.25		2.76	
Merrimac total			2.70	3.125	−1.1	2.17	4.62	0.84	−1.75
Charlton			5.60	−0.785	−2.9	1.67	12.42	7.47	1.81
Charlton–Merrimac			2.90	−3.910	−1.8	−0.50	7.80	6.63	3.56
Total	WP	total	1.75	−0.113	−1.3	0.73	2.85	3.08	4.39
	RP		1.63	−1.660	−1.2	0.36	2.52	1.71	4.31
	NS		1.42	−0.928	−1.5	−0.68	3.18	0.04	−0.44
	WA		2.35	0.962	−1.4	1.10	5.02	2.70	−2.99
	RO		1.15	4.079	1.4	2.33	3.47	0.78	−5.21
Dec–con.			0.90	20.525	8.0	9.47	8.37	0.78	−41.12
Total		A	2.23	0.344	−1.6	0.01	4.53	−1.00	0.63
		B_1	3.56	1.615	−1.3	2.87	6.38	6.72	2.22
		B_2	2.51	0.381	−1.1	0.96	6.13	2.59	−2.79
Overall total			8.30	2.340	−4.0	3.84	17.04	8.31	0.06

Soils: Merrimac loamy sand, Charlton fine sandy loam.

Soil measurements: $y_1 = \%$ loss on ignition, $y_2 = \%$ base saturation ($\times 0.1$), $y_3 = $ pH ($\times 10$), $y_4 = $ total exchange capacity, $y_5 = \%$ total nitrogen ($\times 10$), $y_6 = \%$ moisture equivalent, $y_7 = \%$ clay.

Trees: conifers (C)—WP white pine, RP red pine, NS Norway spruce; deciduous (D) —WA white ash, RO red oak.

total nitrogen y_5 and also the exchange capacity y_4 would be expected to increase with the percent of organic matter in the soil as measured by the percent loss on ignition y_1. Similarly, some relation would be anticipated between base saturation y_2 and pH y_3. Of these measures pH is the only one which might have been affected by a high root concentration. The others were presumably independent of the number of roots at the point of sampling.

3. Analysis of the Response to Root Density

A first step was to compute an analysis of variance for each soil measurement from the individual differences y_i in Table 3.1 of many minus few or no roots and their totals. The mean squares from these analyses in Table 3.2 list in the first row the overall effect of root density with one degree of freedom. Rows 2–4 give the interaction of y_i with the three independent variables: two soil types, three soil horizons, and five species of tree. The remaining interactions of three or more factors in row 5

Table 3.2

Mean Squares from Analyses of Variance of the Differences (y_i) in Soil Characteristics in Table 3.1.

Row	Term	d.f.	(y_1^2)	(y_2^2)
1	Root density (RD)	1	2.29633*	0.18252
2	RD × soils	1	0.28033‡	0.50960
3	RD × horizons	2	0.04916	0.05233
4	RD × trees	4	0.03345	0.83742†
5	Other interactions	22	0.04628	0.18570
6	RD × soils × horizons	2	0.05922	0.37332
7	RD × soils × trees	4	0.04822	0.01512
8	RD × horizons × trees	8	0.06219	0.23572
9	RD × 3-way interactions	8	0.02617	0.17407
10	RD × deciduous-conifers	1	0.00450	2.34042†
11	RD × trees of same type	3	0.04310	0.33642

Row	(y_3^2)	(y_4^2)	(y_5^2)	(y_6^2)	(y_7^2)
1	0.53333‡	0.49152§	9.67872*	2.30187†	0.00012
2	0.10800	0.00833	2.02800†	1.46523‡	0.42245
3	0.00633	0.21217	0.10075	1.49239‡	0.65541
4	0.25417§	0.20061	0.15643	0.27080	3.08855§
5	0.10497	0.12839	0.22480	0.27211	1.15643
6	0.17100	0.21844	0.34333	0.64639	0.57516
7	0.06050	0.00854	0.24794	0.25942	0.70768
8	0.14092	0.11464	0.19370	0.21482	1.72961
9	0.07475	0.17955	0.21471	0.24217	0.95293
10	0.35556§	0.49823§	0.38920	0.00338	9.39364‡
11	0.22037	0.10141	0.07884	0.35994	0.98686

*P < 0.001.   ‡P < 0.05.
†P < 0.01.   §P < 0.10.

form a potential error. To test its homogeneity, its constituents have been isolated in the next four rows. Apart from two exceptionally small interactions in row 7, for y_2 and y_4, no interaction mean square was enough larger than the other three to warrant omitting it from the composite error in row 5. This error has been allotted 22 degrees of freedom, even though an average of one outlier for each criterion has been made less extreme.

An inverse matrix c_{ij} was then computed from the sums of squares and products for root density plus the composite error, each with 23 degrees of freedom, corresponding to rows 1 + 5 in Table 3.2. This matrix and the sums of the differences $\sum y_i$ led directly to the discriminant functions L_i. Because the matrix was so computed (Cochran and Bliss, 1948; Bliss, 1970), the discriminators could be compared by an analysis of variance in units of a convenient dummy variable, the divisor for each overall difference in root concentration, $[x^2] = 2 \times 3 \times 5 = 30$, isolating the contribution of each criterion as in the analysis of a multiple regression.

The combined effect in Table 3.3 of all seven measures accounted for 82% of $[x^2]$ after allowing for the correlation between them, but individually they contributed very unequally to this total. The measure that contributed least ($F = 0.02$) was percent moisture equivalent y_6, despite its significance in Table 3.2. With the inverse matrix adjusted for its omission, the discriminant equation and analysis of variance were recomputed. Similar omissions in successive steps were percent clay y_7, exchange capacity y_4, and percent base saturation y_2, none of them

Table 3.3

Analysis of Variance of the Discriminant Function for Soil Measurements that Maximized Root Response

Row	Term		d.f.	SS	F	d.f.	SS	F
1	Effect of all L_i		7	24.64844	10.50	3	23.83968	25.80
2	Test of $L_1 =$	1.44981	1	1.98296	5.91	1	2.17126	7.05
3	Test of $L_2 =$	0.37213	1	0.45765	1.36		...	...
4	Test of $L_3 =$	−0.81414	1	1.38083	4.12	1	1.42016	4.61
5	Test of $L_4 =$	−0.51517	1	0.38092	1.14		...	...
6	Test of $L_5 =$	0.63513	1	1.62794	4.85	1	1.44973	4.71
7	Test of $L_6 =$	−0.04361	1	0.00500	0.01		...	...
8	Test of $L_7 =$	0.09509	1	0.19750	0.59		...	...
9	Error		16	5.35156	...	20	6.16032	...
10	Total		23	30.00000	...	23	30.00000	...
11	Error mean square, s^2		16	0.33537	...	20	0.30802	...

Final estimates: $L_1 = 1.40289$, $L_3 = 0.73161$, $L_5 = 0.54397$.

contributing significantly to discrimination by the trees in these two panels. The coefficients for the remaining three criteria—percent loss on ignition y_1, pH y_3, and percent nitrogen y_5 in the equation

$$Z_* = 1.40289 \, y_1 - 0.73161 \, y_3 + 0.54397 \, y_5,$$

determined their relative sizes in characterizing the location preferred by these tree roots. From the sums of squares in rows 1 and 10 in the right side of Table 3.3 they accounted for 79% of the dummy variable. The sums of squares for individual L_i's in rows 2, 4, and 6, however, were of very different relative magnitudes than the L_i's themselves.

Since discriminant coefficients are in arbitrary units, they may be replaced by their relative values. When shortened to one-digit weights of $w_1 = 1.0$, $w_3 = -0.5$, and $w_5 = 0.4$, their combined effect in row 1 of Table 3.4 was reduced from that for the L_i's by less than 0.02%, increasing the error variance in row 5 from $s^2 = 0.30802$ to 0.30819. These weights for the three measures have been based upon the overall sum of the differences $\sum y_i$ plus an error based upon their higher order interactions with soils, horizons, and species of tree—all major sources of variation in one or more soil characteristics. Was the weighted root density response the same in both soils, in the three horizons, and for the five species?

Table 3.4

Analysis of the Weights, $w_1 = 1.0$, $w_3 = -0.5$ and $w_5 = 0.4$, Based upon the L_i's in Table 3.3, With Test of First-Order Interactions.

Row	Term	d.f.	SS	MS	F
1	Combined effect of w_i's	3	23.83615	7.94538	25.78
2	RD × soils	1	3.92494	3.92494	12.74
3	RD × horizons	2	0.46700	0.23350	0.76
4	RD × trees	4	2.14489	0.53622	1.74
5	Error	20	6.16385	0.30819	...
6	Total	23	30.00000	...	...

This has been tested in rows 2–4 of Table 3.4, which showed a significant difference in the response to the two soils, but not to horizons or species of tree. Optimal weights for these three soil measurements would not be the same in Merrimac loamy sand and in Charlton fine sandy loam. Their relative values have been estimated alternatively in Table 3.5 from the inverse matrix of the error sums of squares and products for y_1, y_3, and y_5 and the sums of the differences $\sum y_i$ for each soil and their total. Two-digit relative weights covering both soils were the same as before—$w_1 = 1.00$

Table 3.5

Inverse Matrix From the Error Sums of Squares and Products for y_1, y_3 and y_5 ($n = 22$), and Discriminant Functions for Each and Both Soils.

		y_1	y_3	y_5
	c_{1j}	1.225930	−0.107774	−0.221475
	c_{3j}	...	0.463796	−0.047743
	c_{5j}	...	...	0.252152
Merrimac	$\sum y_i$	2.70	−1.1	4.62
	L_i	2.40535	−1.02174	0.61948
	w_i	1.00	−0.42	0.26
Charlton	$\sum y_i$	5.60	−2.9	12.42
	L_i	4.42703	−2.54151	2.02992
	w_i	1.00	−0.57	0.46
Total	$\sum y_i$	8.30	−4.0	17.04
	L_i	6.83238	−3.56325	2.64940
	w_i	1.00	−0.52	0.39

for percent loss on ignition, $w_3 = -0.52$ for pH, and $w_5 = 0.39$ for percent nitrogen—but for Merrimac they were 1.00, −0.42, and 0.26; for Charlton 1.00, −0.57, and 0.46.

4. Root Density Interactions

The combination of soil measurements that would best separate the root density (RD) response to the two soils changed considerably when recomputed from the sum of the differences $\sum y_{\text{Charlton}} - \sum y_{\text{Merrimac}}$ (Table 3.1). A parallel analysis starting with the sums of squares and products for the interaction RD × soils (row 2 in Table 3.2) plus error (row 5) for all seven measurements led to an inverse matrix and analysis of variance with the same dummy variable, $[x^2] = 30$. The discriminant function shortened progressively to a different selection of four criteria, each with $F > 2$, in the analysis of variance. Their L_i's accounted for 50% of the variation in the dummy variate and had relative weights of $w_3 = -1.00$ for pH, of $w_4 = -0.88$ for exchange capacity, of $w_5 = 0.61$ for percent nitrogen, and of $w_6 = 0.78$ for percent moisture equivalent, with an error variance of $s^2 = 0.8530$.

Although the RD response of the five tree species in Table 3.4 (row 4) did not differ significantly, inspection of the sums of the y_i's in Table 3.1

suggested a possibly significant contrast of RD × (deciduous − coniferous) species, with one degree of freedom. For each criterion the differences from the totals of the 6 y_i's for each species were

$$\text{dec} - \text{con} = 3(\sum y_{WA} + \sum y_{RO}) - 2(\sum y_{WP} + \sum y_{RP} + \sum y_{NS}).$$

When squared and divided by $6(2 \times 3^2 + 3 \times 2^2) = 180$, the resulting sums of squares in row 10 of Table 3.2 proved significant for y_2 and for y_7 and near significant ($P < 0.10$) for y_3 and y_4. The remaining differences between species (row 11) were well within the range of random error. The sums of squares and products for dec–con were then added to the corresponding error terms (with 22 degrees of freedom) for computing the inverse matrix and discriminant function.

To convert its analysis of variance to units of the dummy variate in Tables 3.3–3.5, $[x^2] = 30$, each sum of squares, computed as before, was divided by 6 in a final additional step. The analysis of variance of the contributions of each criterion, starting again with all seven measurements and omitting the smallest contributor in successive reductions, gave the smallest error variance for five L_i's, ($s^2 = 0.67450$ with 18 degrees of freedom), having relative weights of $w_2 = 1.00$ for percent base saturation, $w_3 = 1.95$ for pH, $w_4 = 1.94$ for exchange capacity, $w_6 = -0.98$ for percent moisture equivalent, and $w_7 = -0.77$ for percent clay. Further omissions of $L_6(F = 1.30)$ and then of $L_4(F = 1.11)$ reduced the discriminant weights for the remaining three criteria to $w_2 = 1.00$, $w_3 = 0.78$, and $w_7 = -0.40$ ($s^2 = 0.6892$ with 20 degrees of freedom). These last three accounted for 54% of the variation in the dummy variate.

5. Analysis Omitting pH y_3

Of the seven soil measurements, the only one which might have been influenced by the concentration of roots was pH, where root exudates could have increased soil acidity at the points of collection. Although pH ranged from 5.13 to 5.68 in the initial measurements, differences associated with root density varied only from $y_3 = -0.08$ to 0.05, with one outlier of 0.09 reduced to 0.04. The trees were too young for dead and decaying roots to have contributed appreciably to the organic matter measured by y_1 or by y_5. Since the differences in pH could have been in part a response to root density, the analysis has been recomputed with the remaining six criteria.

Without pH the percent moisture equivalent y_6 remained in the discriminant function when the error mean square was minimal ($s^2 = 0.3501$) along with percent loss on ignition y_1 and percent nitrogen y_5, although in the initial analysis y_6 was the first to be omitted. Relative weights were $w_1 = 1.00$, $w_5 = 0.41$, and $w_6 = -0.22$—the three accounting for 77% of the dummy variate, or only 2% less than when based upon y_1, y_3, and y_5. The interaction of RD × soils again was significant ($P < 0.02$), and the interactions of RD × horizons and of RD × trees were less than the error.

The combination of measurements that would maximize the difference between soils, when recomputed without pH from the error + the sum of the differences between Merrimac and Charlton, reduced to three criteria with a minimal error variance of $s^2 = 0.9351$. Both y_1 and y_5 remained in the equation, but y_4 replaced y_6, with relative weights of $w_1 = 1.00$ for percent loss on ignition, $w_4 = -0.51$ for exchange capacity, and $w_5 = 0.57$ for percent nitrogen—accounting for 38% of the dummy variate. A similar comparison without y_3 of deciduous trees and conifers needed only two criteria, base saturation y_2 and percent clay y_7, which together accounted for 50% of its dummy variate, with relative weights of $w_2 = 1.00$ and $w_7 = -0.33$.

6. Summary

In separate analyses of the seven soil measurements in the present study, four differed significantly between zones of high root concentration and parallel zones with few or no roots. When the correlation between criteria was removed, one of these, percent moisture equivalent y_6, proved unnecessary. It might then be supposed that only the remaining three would be needed in estimating the most favorable soil combination for the roots of these trees, two of them related factors, percent loss on ignition y_1 and percent nitrogen y_5, and one that could be modified by contact with roots, pH y_3. The discriminant function based upon these three measures could be reduced to one- or two-digit relative weights with a negligible loss in effectiveness. In tests of the interaction of root density with the three independent variables—for soils, horizons, and species of tree—only the contrast between the two soils was significant.

A new discriminant function was then computed, based upon the differences between the two soils in all seven measures. Two were retained from the first discriminant, y_3 and y_5; but y_1 was replaced by exchange capacity y_4 and moisture equivalent y_6. Although trees did not differ significantly in terms of the initial relative weights, the contrasts between deciduous and coniferous trees for all seven criteria when added to the error reduced to still another combination—this time of base saturation y_2, pH y_3, and percent clay y_7. Thus the significant contrasts in the root density response required different combinations of all seven soil measurements.

Because pH was known to be modified by root exudates, this measure could be in part a response to root density rather than a cause, and hence a debatable criterion in predicting a favored soil environment for these tree species. Accordingly, the analysis was repeated, omitting y_3. The discriminant for the overall response reduced again to loss on ignition y_1 and percent nitrogen y_5, both significant, plus moisture equivalent y_6—the three accounting for 77% of the dummy variate. The interaction of RD $\times$ soils was significant as before, and both RD $\times$ horizons and RD $\times$ trees were less than its error. Reanalysis with the initial error plus RD $\times$ soils also reduced to y_1 and y_5, but with exchange capacity y_4 replacing y_6. For discriminating between deciduous trees and conifers, two criteria accounted for half of $[x^2]$, base saturation y_2 and percent clay y_7. Hence all clearly independent variables were needed, whether pH was included or omitted. Among the measures tested, no one set alone discriminated between the conditions favoring many over few roots. The results imply instead that different criteria must be considered, depending on whether individual species are compared in a given soil, or several species on different soils.

References

Bliss, C. I. 1967, 1970. *Statistics in biology*, vols. 1 and 2. New York: McGraw-Hill.

Bliss, C. I., and Paul Griminger. 1969. Response criteria for the bioassay of vitamin K. *Biometrics* 25: 735–45.

Cochran, W. G., and C. I. Bliss. 1948. Discriminant functions with covariance. *Ann. Math. Stat.* 19: 151–76.

Dixon, W. J. 1951. Ratios involving extreme values. *Ann. Math. Stat.* 22: 68–78.

Fisher, R. A. 1936. The use of multiple measurements in taxonomic problems. *Ann. Eugenics* 7: 179–88.

Garin, G. I. 1942. Distribution of roots of certain tree species in two Connecticut soils. Conn. Agr. Expt. Sta. Bull. 454.

4

Binomial Sequential Design of Experiments

J. D. BORWANKER, H. T. DAVID,
and CHARLES D. INGWELL

1. Introduction

IN SECTIONS 2 AND 3 we consider how to select sequentially, under a certain general loss structure, one of two available binomial sources of information, differing both in reliability and sampling cost. Our main result for this general situation (Theorem 3.1) is that the Bayes strategies have the expected characterization under our condition (3.4). We do not know whether condition (3.4) would arise naturally in adapting the theory of [7] to general-loss sequential design. Note that (3.4) requires only a certain asymptotic relative equivalence of terminal decisions with respect to total terminal loss, without requiring—e.g., in [11]—asymptotically unbounded sampling cost.

The case of equal sampling cost under the usual additive loss structure— (2.1) below— and models related to that case have been studied extensively, e.g., in [3], [4], [5], [6], and [11]. Unequal sampling cost under additive loss is treated in [7] as part of a general theory of optimal

J. D. BORWANKER, formerly of the University of Minnesota, is now at the Indian Institute of Technology.

The research of H. T. DAVID, Professor of Statistics, Iowa State University, was done in part while on leave at the Statistical Center, Department of Statistics, University of Minnesota.

CHARLES D. INGWELL, formerly of Iowa State University, is now at the Western Electric Company.

This research was sponsored in part by the Statistical Center, Department of Statistics, University of Minnesota; the Statistical Laboratory, Iowa State University; and the Systems and Research Center, Honeywell, Inc., St. Paul, Minnesota.

Arnold Kanarick and Jane Huntington of Honeywell, Inc., asked us to look at a general loss and unequal costs. Elaine Frankowski of the University of Minnesota Computing Center, and Al Peterson of Honeywell, Inc., carried out computations reported in Table 4.1.

45

stopping and also in [2] and [12]. Nonetheless, we address ourselves to some specific questions in the area of additive loss as opposed to general loss. Specifically in section 4 we consider the inadmissibility of one of the two sources, OC and ASN computations, and characterization and computation of the Bayes procedures. We note in this connection that the investigations in [2] and [12] stress the possibility, when there are two states of nature, that posterior probability regions corresponding to particular sources are intervals. However, in view of our basic symmetry assumption in section 2, which associates a single information parameter π with each source, these regions in our case inevitably are composed of several intervals, for all but one of the sources.

2. The Model

There are two states of nature θ_1, θ_2, two terminal decisions a_1, a_2, two sources of information A, B, and two signals 0, 1. The cost of one observation of type A is C_A, and $P(1|A, \theta_1) = P(0|A, \theta_2) = \pi_A \geq 1/2$; and similarly for B, so that there is for each source a cost parameter C and an information parameter π. For these we shall assume $C_A > C_B, \pi_A > \pi_B$. Losses due to terminal decision and to sampling are jointly given by four functions $L_{ij}(y) = L(\theta_i, a_j, y)$, $i, j = 1, 2$, which are nonnegative for $y \geq 0$ and are positive for $y > 0$, where y denotes the cumulative cost of sampling "so far." A traditional additive form of the four functions L_{ij} is

$$L_{ij}(y) = y + (1 - \delta_{ij})L_i. \tag{2.1}$$

On the other hand, a "doomsday" situation where a treasure M is successively depleted by sampling and ultimately lost by a wrong terminal decision, may be represented by

$$\begin{aligned} L_{ij}(y) &= y\delta_{ij} + (1 - \delta_{ij})M, &\quad y \leq M \\ &= M + \varepsilon. &\quad y > M \end{aligned} \tag{2.2}$$

Indeed this is the nonadditive loss situation originally presented to us by the authors of [9].

3. Bayes Risks and Strategies

The setting is as in section 5.2.3 of [11], with $p_1, p_1^* = \pi_A, \pi_B$ and $p_2, p_2^* = 1 - \pi_A, 1 - \pi_B$; in other words, there is a sequence $X_1, X_2, \ldots$ of independently distributed chance variables, each of which takes on the value 0 or 1, of which the odd ones are of type A and the even ones are of type B. This determines unique distributions P_{θ_1} and P_{θ_2} on the space X of sequences $x = \{x_1, x_2, \ldots\}$.

The randomized strategies δ are as defined in section 5.2.3 of [11]; denoting the four functions $L_{ij}(y)$ by $\mathscr{L}$, we define $R(\xi, \mathscr{L}, \delta)$ as the risk under strategy δ, when the prior is ξ. Similarly define $R(\xi, \mathscr{L}, \delta|y)$ as the conditional risk, under δ and ξ, given the cumulative sampling cost Y is equal to y.

It is readily checked that the conclusion of Wald's Theorem 3.2 in [11] relating regular to weak intrinsic convergence holds for the general loss functions $L_{ij}(y)$:

$$\delta^n \rightsquigarrow \delta^0 \Rightarrow \lim_n R(\xi, \mathscr{L}, \delta^n) \geq R(\xi, \mathscr{L}, \delta^0), \tag{3.1}$$

where $\rightsquigarrow$ denotes regular convergence.

Next define

$$f(\xi, C, \mathscr{L}) = \inf_{\delta \in \mathscr{S}_C} R(\xi, \mathscr{L}, \delta), \tag{3.2}$$

where $\mathscr{S}_C$ is the set of all decision rules for which the probability of a total sampling expense greater than or equal to C is zero.

Similarly define

$$f(\xi, \mathscr{L}) = \inf_{\mathscr{S}} R(\xi, \mathscr{L}, \delta), \tag{3.3}$$

where $\mathscr{S}$ is the set of all decision rules δ. A condition for the convergence of $f(\xi, C, \mathscr{L})$ to $f(\xi, \mathscr{L})$ is given in the following lemma.

LEMMA 3.1. If

$$\lim_{C \to \infty} \frac{\max\limits_{C \leq y < C + C_A} \min\limits_{j} \max\limits_{i} L_{ij}(y)}{\min\limits_{y \geq C} \min\limits_{j} \min\limits_{i} L_{ij}(y)} = 1, \tag{3.4}$$

then

$$f(\xi, C, \mathscr{L}) \xrightarrow{C} f(\xi, \mathscr{L}). \tag{3.5}$$

PROOF. As in section 4.1.2 of [11], let $\{\delta_m\}$ be a sequence of rules such that $R(\xi, \mathscr{L}, \delta_m) \xrightarrow{m} f(\xi, \mathscr{L})$. Defining $R(\xi, \mathscr{L}, \delta_m|y) = 0$ if $P_{\xi, \delta_m}(y) = 0$, we have

$$R(\xi, \mathscr{L}, \delta_m) \geq \sum_{y \geq C} R(\xi, \mathscr{L}, \delta_m|y) P_{\xi, \delta_m}(y)$$

$$\geq \min_{y \geq C} \min_{j} \min_{i} L_{ij}(y) \cdot P_{\xi, \delta_m}(Y \geq C),$$

or

$$P_{\xi, \delta_m}(Y \geq C) \leq R(\xi, \mathscr{L}, \delta_m) / \min_{y \geq C} \min_{j} \min_{i} L_{ij}(y).$$

Let δ_m^C be δ_m truncated at C. That is, δ_m^C proceeds as δ_m as long as $Y < C$ and takes the minimum risk terminal action otherwise. Then

$$f(\xi, \mathcal{L}) \leq f(\xi, C + C_A, \mathcal{L}) \leq R(\xi, \mathcal{L}, \delta_m^C)$$

$$\leq R(\xi, \mathcal{L}, \delta_m) + \big[\max_{C \leq y < C + C_A} \min_j \max_i L_{ij}(y)$$

$$- \min_{y \geq c} \min_j \min_i L_{ij}(y)\big] \cdot P_{\xi, \delta_m}(Y \geq C)$$

$$\leq R(\xi, \mathcal{L}, \delta_m) \cdot \left[\frac{\displaystyle\max_{C \leq y < C + C_A} \min_j \max_i L_{ij}(y)}{\displaystyle\min_{y \geq C} \min_j \min_i L_{ij}(y)} \right].$$

Letting first m and then C tend to infinity yields (3.5) in view of (3.4).

Note that the loss structure in (2.1) satisfies (3.4); on the other hand, (3.4) is far from necessary for (3.5). For example, any loss structure as that of (2.2), such that for some $\bar{C}$,

$$\frac{\displaystyle\max_{\bar{C} \leq y < \bar{C} + C_A} \min_j \max_i L_{ij}(y)}{\displaystyle\min_{y \geq \bar{C} + C_A} \min_j \min_i L_{ij}(y)} \leq 1$$

will lead to $f(\xi, C, \mathcal{L}) = f(\xi, \mathcal{L})$ for $C \geq \bar{C} + C_A$.

The conditional version of Lemma 3.1 is given by the following corollary.

COROLLARY 3.1. Let $R(\xi, \mathcal{L}, \delta | x_{i_1}, \ldots, x_{i_j})$ be the evident conditional analogue of $R(\xi, \mathcal{L}, \delta)$ for every δ for which there is positive probability of observing $x_{i_1}, \ldots, x_{i_j}$ (where an x_{i_k} is either 0, type A; 1, type A; 0, type B; or 1, type B). Similarly for $C > (j)(C_A)$, let $f(\xi, C, \mathcal{L} | x_{i_1}, \ldots, x_{i_j})$ be the evident conditional analogue of $f(\xi, C, \mathcal{L})$, and $f(\xi, \mathcal{L} | x_{i_1}, \ldots, x_{i_j})$ the analogue of $f(\xi, \mathcal{L})$. Then under (3.2),

$$f(\xi, C, \mathcal{L} | x_{i_1}, \ldots, x_{i_j}) \to f(\xi, \mathcal{L} | x_{i_1}, \ldots, x_{i_j}) \quad \text{for all} \quad x_{i_1}, \ldots, x_{i_j}.$$

PROOF. Let $\mathcal{L}_y(t) = \mathcal{L}(t + y)$. Then the functions $L_y(t)$ are easily seen to satisfy condition (3.4), so that from the definition of the conditional risk, we have for increasing C,

$$f(\xi, C, \mathcal{L} | x_{i_1}, \ldots, x_{i_j}) = f(\xi_j, C - y_j, \mathcal{L}_{y_j}) \overset{C}{\to} f(\xi_j, \mathcal{L}_{y_j}) \quad \text{by (3.5)}$$

$$= f(\xi, \mathcal{L} | x_{i_1}, \ldots, x_{i_j}), \tag{3.6}$$

where in the above, y_j is the cumulative cost of $x_{i_1}, \ldots, x_{i_j}$, and ξ_j is the posterior probability resulting from ξ and $x_{i_1}, \ldots, x_{i_j}$.

Next define

$$D_1(\xi, \mathcal{L}) = \xi L_{11}(0) + (1 - \xi) L_{21}(0)$$

$$D_2(\xi, \mathcal{L}) = \xi L_{12}(0) + (1 - \xi) L_{22}(0)$$

$$D_3(\xi, C, \mathscr{L}) = [\xi\pi_A + (1 - \xi)(1 - \pi_A)]f\left[\frac{\xi\pi_A}{\xi\pi_A + (1 - \xi)(1 - \pi_A)},\right.$$

$$\left. C - C_A, \mathscr{L}_{C_A}\right]$$

$$+[(1 - \xi)\pi_A + \xi(1 - \pi_A)]f\left[\frac{\xi(1 - \pi_A)}{\xi(1 - \pi_A) + (1 - \xi)\pi_A},\right.$$

$$\left. C - C_A, \mathscr{L}_{C_A}\right]$$

$$D_4(\xi, C, \mathscr{L}) = [\xi\pi_B + (1 - \xi)(1 - \pi_B)]f\left[\frac{\xi\pi_B}{\xi\pi_B + (1 - \xi)(1 - \pi_B)},\right.$$

$$\left. C - C_B, \mathscr{L}_{C_B}\right]$$

$$+[(1 - \xi)\pi_B + \xi(1 - \pi_B)]f\left[\frac{\xi(1 - \pi_B)}{\xi(1 - \pi_B) + (1 - \xi)\pi_B},\right.$$

$$\left. C - C_B, \mathscr{L}_{C_B}\right]. \tag{3.7}$$

The following recursion formula for $f(\xi, C, \mathscr{L})$ is easy to verify. For $C \geq C_A$,

$$f(\xi, C, \mathscr{L}) = \min[D_1(\xi, \mathscr{L}), D_2(\xi, \mathscr{L}), D_3(\xi, C, \mathscr{L}), D_4(\xi, C, \mathscr{L})]. \tag{3.8}$$

In view of the first equality in (3.6) we therefore have for $y_j \leq C$

$$f(\xi, C, \mathscr{L}|x_{i_1}, \ldots, x_{i_j}) = f(\xi_j, C - y_j, \mathscr{L}_{y_j})$$
$$= \min[D_1(\xi_j, \mathscr{L}_{y_j}), D_2(\xi_j, \mathscr{L}_{y_j}), \tag{3.9}$$
$$D_3(\xi_j, C - y_j, \mathscr{L}_{y_j}), D_4(\xi_j, C - y_j, \mathscr{L}_{y_j})].$$

Defining $D_3(\xi, \mathscr{L})$ and $D_4(\xi, \mathscr{L})$ in terms of (3.3), as $D_3(\xi, C, \mathscr{L})$ and $D_4(\xi, C, \mathscr{L})$ are defined in terms of (3.2), (3.6) also yields

$$f(\xi_j, \mathscr{L}_{y_j}) = f(\xi, \mathscr{L}|x_{i_1}, \ldots, x_{i_j})$$
$$= \min[D_1(\xi_j, \mathscr{L}_{y_j}), D_2(\xi_j, \mathscr{L}_{y_j}),$$
$$D_3(\xi_j, \mathscr{L}_{y_j}), D_4(\xi_j, \mathscr{L}_{y_j})]. \tag{3.10}$$

Note that all the functions f and D above, by virtue of being risk infemums over suitable sets of strategies, are concave in their first argument.

Analogously to section 4.1.3 of [11], we now characterize the Bayes strategies.

THEOREM 3.1. Under condition (3.4) the strategies $\tilde{\delta}$ which proceed at every stage in accordance with the minimum of $D_1(\xi_j, \mathscr{L}_{y_j})$, $D_2(\xi_j, \mathscr{L}_{y_j})$, $D_3(\xi_j, \mathscr{L}_{y_j})$, and $D_4(\xi_j, \mathscr{L}_{y_j})$ are the Bayes strategies.

PROOF. $\tilde{\delta}$ *are Bayes strategies*: Assume without loss of generality that there are no ties. Given $\varepsilon > 0$, let C be such that for all $\bar{C} \geq C$,

$$\frac{\min_j \max_i L_{ij}(\bar{C})}{\min_{i,\,j,\,y > \bar{C}} L_{ij}(y)} < 1 + \varepsilon. \tag{3.11}$$

Since

$$D_1(\xi_j, \mathscr{L}_{y_j}) = \xi_j L_{11}(y_j) + (1 - \xi_j)L_{21}(y_j)$$
$$\leq \max[L_{11}(y_j), L_{21}(y_j)]$$

and similarly

$$D_2(\xi_j, \mathscr{L}_{y_j}) \leq \max[L_{12}(y_j), L_{22}(y_j)],$$

we have

$$\min\left[D_1(\xi_j, \mathscr{L}_{y_j}), D_2(\xi_j, \mathscr{L}_{y_j})\right] \leq \min_j \max_i L_{ij}(y_j). \tag{3.12}$$

Again

$$\min\left[D_3(\xi_j, \mathscr{L}_{y_j}), D_4(\xi_j, \mathscr{L}_{y_j})\right] \geq \min_{i,\,jy > y_j} L_{ij}(y). \tag{3.13}$$

Hence for $y_j \geq C$ we have from (3.11), (3.12), and (3.13)

$$1 \leq \frac{\min[D_1(\xi_j, \mathscr{L}_{y_j}), D_2(\xi_j, \mathscr{L}_{y_j})]}{\min[D_1(\xi_j, \mathscr{L}_{y_j}), D_2(\xi_j, \mathscr{L}_{y_j}), D_3(\xi_j, \mathscr{L}_{y_j}), D_4(\xi_j, \mathscr{L}_{y_j})]} \leq 1 + \varepsilon,$$

so that for $y_j \geq C$ we get by (3.10)

$$\min[D_1(\xi_j, \mathscr{L}_{y_j}), D_2(\xi_j, \mathscr{L}_{y_j})] \leq (1 + \varepsilon)f(\xi_j, \mathscr{L}_{y_j}).$$

Let $\tilde{\delta}^C$ be $\tilde{\delta}$ truncated at C (see Lemma 3.1). Then for any $x_{i_1}, \ldots, x_{i_j}$, observed with positive probability by $\tilde{\delta}$ such that $y_j \geq C > y_{j-1}$,

$$R(\xi, \mathscr{L}, \tilde{\delta}^C | x_{i_1}, \ldots, x_{i_j}) = \min[D_1(\xi_j, \mathscr{L}_{y_j}), D_2(\xi_j, \mathscr{L}_{y_j})]$$
$$\leq (1 + \varepsilon)f(\xi_j, \mathscr{L}_{y_j})$$
$$= (1 + \varepsilon)f(\xi, \mathscr{L} | x_{i_1}, \ldots, x_{i_j}). \tag{3.14}$$

Now if $\tilde{\delta}$ calls for a terminal action after observing $x_{i_1}, \ldots, x_{i_{j-1}}$, then

$$R(\xi, \mathscr{L}, \tilde{\delta}^C | x_{i_1}, \ldots, x_{i_{j-1}}) = \min[D_1(\xi_{j-1}, \mathscr{L}_{y_{j-1}}), D_2(\xi_{j-1}, \mathscr{L}_{y_{j-1}})],$$
$$= f(\xi, \mathscr{L} | x_{i_1}, \ldots, x_{i_{j-1}}). \tag{3.15}$$

On the other hand if $\tilde{\delta}$ calls for sampling A, which means that

$$D_3(\xi, \mathscr{L} | x_{i_1}, \ldots, x_{i_{j-1}})$$

is the minimum D, then

$$R(\xi, \mathcal{L}, \tilde{\delta}^C | x_{i_1}, \ldots, x_{i_{j-1}})$$

$$= \sum_{x_{ij}=0, 1} R(\xi, \mathcal{L}, \tilde{\delta}^C | x_{i_1}, \ldots, x_{i_j}) P(x_{i_j} | \xi)$$

$$\leq (1 + \varepsilon) \sum_{x_{ij}=0, 1} f(\xi, \mathcal{L} | x_{i_1}, \ldots, x_{i_j}) P(x_{i_j} | \xi)$$

$$= (1 + \varepsilon) D_3(\xi, \mathcal{L} | x_{i_1}, \ldots, x_{i_{j-1}})$$

$$= (1 + \varepsilon) f\xi, \mathcal{L} | x_{i_1}, \ldots, x_{j-1}) \tag{3.16}$$

and similarly for the case when the minimum $D = D_4$. Hence for any of the four eventualities

$$R(\xi, \mathcal{L}, \tilde{\delta}^C | x_{i_1}, \ldots, x_{i_{j-1}}) \leq (1 + \varepsilon) f(\xi, \mathcal{L} | x_{i_1}, \ldots, x_{i_{j-1}}).$$

Induction backwards now yields

$$R(\xi, \mathcal{L}, \tilde{\delta}^C) \leq (1 + \varepsilon) f(\xi, \mathcal{L}). \tag{3.17}$$

Letting $C \to \infty$, we note, by (3.1), that

$$R(\xi, \mathcal{L}, \tilde{\delta}) \leq \lim_C R(\xi, \mathcal{L}, \tilde{\delta}^C) \leq (1 + \varepsilon) f(\xi, \mathcal{L}).$$

The result follows since ε was arbitrary and $R(\xi, \mathcal{L}, \tilde{\delta}) \geq f(\xi, \mathcal{L})$.

Under the condition (3.4) of Lemma 3.1 one also has that $\tilde{\delta}$ *are the only Bayes strategies*. Let δ be any strategy differing from the $\tilde{\delta}$ for ξ. Then, again as in [11], the set M^* of sequences on which δ differs from $\tilde{\delta}$ is not empty. For a given sequence x in M^* let δ agree up to $x_{i_1}, \ldots, x_{i_j}$. The conditional analogue of the first part of the proof then guarantees that

$$R(\xi, \mathcal{L}, \tilde{\delta} | x_{i_1}, \ldots, x_{i_j}) = f(\xi, \mathcal{L} | x_{i_1}, \ldots, x_{i_j}) < R(\xi, \mathcal{L}, \delta | x_{i_1}, \ldots, x_{i_j})$$

for each sequence in M^*. Thus

$$R(\xi, \mathcal{L}, \tilde{\delta}) = f(\xi, \mathcal{L}) < R(\xi, \mathcal{L}, \delta).$$

4. Additive Loss

For the traditional loss structure (2.1) define $f(\xi) \equiv f(\xi, \mathcal{L}_0)$. Then

$$f(\xi_j, \mathcal{L}_{y_j}) = f(\xi_j) + y_j \tag{4.1}$$

and

$$D_1(\xi_j, \mathcal{L}_{y_j}) = (1 - \xi_j) L_2 + y_j \tag{4.2}$$

$$D_2(\xi_j, \mathcal{L}_{y_j}) = \xi_j L_1 + y_j,$$

so that the recursion relation (3.10) becomes the functional equation,

$$f(\xi_j) = \min\{(1 - \xi_j)L_2, \xi_j L_1, [\pi_A \xi_j + (1 - \pi_A)(1 - \xi_j)]$$
$$f[\pi_A \xi_j/\pi_A \xi_j + (1 - \pi_A)(1 - \xi_j)] + [(1 - \pi_A)\xi_j + \pi_A(1 - \xi_j)]$$
$$f[\xi_j(1 - \pi_A)/\pi_A(1 - \xi_j) + \xi(1 - \pi_A)] + C_A,$$
$$[\pi_B \xi_j + (1 - \pi_B)(1 - \xi_j)]f[\pi_B \xi_j/\pi_B \xi_j + (1 - \pi_B)(1 - \xi_j)]$$
$$+[(1 - \pi_B)\xi_j + \pi_B(1 - \xi_j)]$$
$$f[\pi_B(1 - \xi_j)/\pi_B(1 - \xi_j) + \xi_j(1 - \pi_B)] + C_B\}$$
$$\equiv \min[D_1(\xi_j), D_2(\xi_j), D_3(\xi_j), D_4(\xi_j)] \tag{4.3}$$

for $f(\xi)$, which, except for the differing costs C_A and C_B, is the relation (5.119) in [11]. Further, the Bayes strategies $\tilde{\delta}$ now involve the comparison of the four quantities on the right-hand side of (4.3). The following properties of $D_3(\xi)$ and $D_4(\xi)$ are readily verifiable.

1. $D_3(0) = D_3(1) = C_A$;
2. $D_4(0) = D_4(1) = C_B$;
3. $D_3(\xi)$ and $D_4(\xi)$ are concave, and hence continuous on $(0, 1)$;
4. under the additional symmetry assumption $L_1 = L_2$, $D_3(\xi)$ and $D_4(\xi)$ both are symmetric about $1/2$; and
5. under $L_1 = L_2$, $D_3(\xi)$ is "shallower" than $D_4(\xi)$, in the sense that $D_3(1/2) - C_A < D_4(1/2) - C_B$. (4.4)

As indicated in [7], the class $\tilde{\delta}$ of Bayes strategies corresponds to the partitioning of the closed ξ-interval I: $[0, 1]$ as follows:

1. An interval $I_2 = [0, a)$ associated with a_2;
2. an interval $I_1 = (b, 1]$ associated with a_1;
3. a set I_3 consisting of the union of an at most countable number of open intervals lying in $I - I_1 - I_2$, each associated with the further sampling of A;
4. similarly for I_4 and B; and
5. $I - I_1 - I_2 - I_3 - I_4$, a set on which at least two of the four D functions are tied, with arbitrary randomization called for between the corresponding alternatives. (4.5)

Whether I_3 and I_4 can consist of an infinite number of intervals is an open question. The few cases that have been investigated by computing the functions D_3 and D_4 by iterating (4.3) have revealed the data in Table 4.1:

Table 4.1

Examples of Regions I_1, I_2, I_3, I_4

$L_1 = L_2$	C_A	C_B	π_A	π_B	I_1	I_2	I_3	I_4
20	1.75	1.0	.75	.70	(.815, 1]	[0, .185)	(.185, .815)	
20	2.00	1.0	.81	.70	(.765, 1]	[0, .235)	(.285, .715)	(.235, .285)∪ (.715, .765)
20	2.00	1.0	.79	.70	(.755, 1]	[0, .245)	(.345, .655)	(.245, .345)∪ (.655, .755)
20	2.00	1.0	.77	.70	(.735, 1]	[0, .265)		(.265, .735)
1	.0023	.0007	.8808	.7311	(.9975, 1]	[0, .0025)	(.0180, .9820)	(.0025, .0181)∪ (.9820, .9975)

Approximate OC computations can be carried out for such procedures in essentially the manner given by Wald in [11] or in [10], yielding the following lemma.

LEMMA 4.1. Given ξ, let δ_ξ be the Bayes plan against ξ. Let $\bar\xi$ be the inf of the posterior probabilities of θ_1 leading to a_1 under δ_ξ, and let $\underline\xi$ be the sup of the posterior probabilities of θ_1 leading to a_2 under δ_ξ. For parameters C_A, C_B, π_A, π_B, L_1, and L_2 leading to large $\bar\xi$ and small $\underline\xi$ and for ξ not too far from 1/2, the Bayes procedure δ_ξ has the following OC behavior.

$$\text{Prob}[a_1|\theta^{(h)}] \doteq \frac{\mathscr{A}^h - 1}{\mathscr{A}^h - \mathscr{B}^h},$$

where $\theta^{(h)}$ represents a one-parameter family of pairs $(\pi_A, \pi_B)_h$:

$$\theta^{(h)} = (\pi_A, \pi_B)_h = \left[\frac{\pi_A^h}{\pi_A^h + (1 - \pi_A)^h}, \frac{\pi_B^h}{\pi_B^h + (1 - \pi_B)^h} \right],$$

and also

$$\mathscr{A} = \left[\frac{(1 - \bar\xi)\xi}{(1 - \xi)\bar\xi} \right] \quad \mathscr{B} = \left[\frac{(1 - \underline\xi)\xi}{(1 - \xi)\underline\xi} \right].$$

ASN computations must be expanded to computations of expected numbers of observations of both types, say $E_A(\pi_A, \pi_B)$ and $E_B(\pi_A, \pi_B)$. It seems helpful here to recognize the general class of random walks in fact involved. These walks traverse several adjacent intervals before ultimate absorption at the outer endpoint of either of the two extreme intervals. A particular interval corresponds to a particular posterior probability range for which the Bayes procedures call for consulting a particular source. The character of the walk changes as it leaves one interval and enters an adjacent one. In other words, the transition matrix or kernel of the walk does not have the typical translation structure $f(y - x)$, but rather the structure $f[y - x, I(x)]$, where $I(x)$ indicates the interval containing x.

For certain step-size ratios the resulting difference equations are especially easy to solve and provide a simple alternative to the approach in [2]. To take a simple example, consider the case of a walk beginning at the origin which takes steps $+1$ and -1 with probabilities respectively π_B and $1 - \pi_B$, as long as the cumulative step sum is $+1, 0$ or -1; steps $+2$ and -2 with probabilities respectively π_A and $1 - \pi_A$, as long as the cumulative step sum is $+2, +4, -2$, or -4 and which is absorbed at ± 6. Writing $E_B(z)$ for the expected number of remaining B-consultations when the walk is at z and writing ρ_A for $1 - \pi_A$ and ρ_B for $1 - \pi_B$, we have

$$E_B(6) = E_B(-6) = 0,$$

$$E_B(4) = \pi_A E_B(6) + \rho_A E_B(2) = \rho_A E_B(2),$$

$$E_B(-4) = \pi_A E_B(-2),$$

$$E_B(2) = \pi_A E_B(4) + \rho_A E_B(0) = \frac{\rho_A E_B(0)}{1 - \pi_A \rho_A},$$

$$E_B(-2) = \frac{\pi_A E_B(0)}{1 - \pi_A \rho_A},$$

$$E_B(1) = \pi_B[E_B(2) + 1] + \rho_B[E_B(0) + 1]$$

$$= 1 + \rho_B E_B(0) + \pi_B \left[\frac{\rho_A E_B(0)}{1 - \pi_A \rho_A} \right]$$

$$= \frac{1 - \pi_A \rho_A - [(1 - \pi_A \rho_A)\rho_B + \pi_B \rho_A]E_B(0)}{1 - \pi_A \rho_A},$$

$$E_B(-1) = \frac{1 - \pi_A \rho_A + [(1 - \pi_A \rho_A)\pi_B + \pi_A \rho_B]E_B(0)}{1 - \pi_A \rho_A},$$

$$E_B(0) = \pi_B[E_B(1) + 1] + \rho_A[E_B(-1) + 1],$$

and, solving for $E_B(0)$,

$$E_B(0) = \frac{2(1 - \pi_B \rho_B)}{1 - \pi_A \rho_A - 2\pi_B \rho_B + 2\pi_B \pi_A \rho_B \rho_A - \pi_B^2 \rho_A - \pi_A \rho_B^2}.$$

Similarly,

$$E_A(6) = E_A(-6) = 0,$$

$$E_A(4) = \pi_A[E_A(6) + 1] + \rho_A[E_A(2) + 1] = 1 + \rho_A E_A(2),$$

$$E_A(-4) = 1 + \pi_A E_A(-2),$$

$$E_A(2) = \pi_A[E_A(4) + 1] + \rho_A[E_A(0) + 1] = \frac{1 + \pi_A + \rho_A E_A(0)}{1 - \pi_A \rho_A},$$

$$E_A(-2) = \frac{1 + \rho_A + \pi_A E_A(0)}{1 - \pi_A \rho_A},$$

$$E_A(1) = \pi_B E_A(2) + \rho_B E_A(0)$$

$$= \frac{\pi_B + \pi_B \pi_A + [\pi_B \rho_A + (1 - \pi_A \rho_A)\rho_B]E_A(0)}{1 - \pi_A \rho_A},$$

$$E_A(-1) = \frac{\rho_B + \rho_B \rho_A + [\pi_A \rho_B + (1 - \pi_A \rho_A)\pi_B]E_A(0)}{1 - \pi_A \rho_A},$$

$$E_A(0) = \pi_B E_A(1) + \rho_B E_A(-1),$$

and solving for $E_A(0)$,

$$E_A(0) = \frac{\pi_B^2(1 + \pi_A) + \rho_B^2(1 + \rho_A)}{1 - \pi_A \rho_A - 2\pi_B \rho_B + 2\pi_B \pi_A \rho_B \rho_A - \pi_B^2 \rho_A - \pi_A \rho_B^2}.$$

As a matter of fact, absorption probabilities (i.e., OC ordinates) are computable in similar fashion:

$$L(6) = 1,$$

$$L(-6) = 0,$$

$$L(4) = \pi_A L(6) + \rho_A L(2) = \pi_A + \rho_A L(2),$$

$$L(-4) = \pi_A L(-2) + \rho_A L(-6) = \pi_A L(-2),$$

$$L(2) = \pi_A L(4) + \rho_A L(0) = \frac{\pi_A^2 + \rho_A L(0)}{1 - \pi_A \rho_A},$$

$$L(-2) = \pi_A L(0) + \rho_A L(-4) = \frac{\pi_A L(0)}{1 - \pi_A \rho_A},$$

$$L(1) = \pi_B L(2) + \rho_B L(0)$$

$$= \frac{\pi_B \pi_A^2 + [\pi_B \rho_A + (1 - \pi_A \rho_A)\rho_B]L(0)}{1 - \pi_A \rho_A},$$

$$L(-1) = \pi_B L(0) + \rho_B L(-2) = \frac{[\pi_A \rho_B + \pi_B(1 - \pi_A \rho_A)]L(0)}{1 - \pi_A \rho_A},$$

$$L(0) = \pi_B L(1) + \rho_B L(-1)$$

$$= \frac{\pi_B^2 \pi_A^2}{1 - \pi_A \rho_A - 2\pi_B \rho_B + 2\pi_B \pi_A \rho_B \rho_A - \pi_B^2 \rho_A - \pi_A \rho_B^2}.$$

Our next concern is identification of parametric cases for which no Bayes strategy will call for consulting source A (respectively B). This implies identifying parametric cases for which one of the two sources is inadmissible, (i.e., is not called for by any admissible strategy). For example, sufficient conditions for the inadmissibility of source A are contained in the following lemma.

LEMMA 4.2. If

1. $L_1 = L_2 = L$,

2. $1 - \pi_B \leq \dfrac{C_B}{L}$,

either

3. $C_A + (1 - \pi_A)L > \dfrac{L}{2}$,   (4.6)

or

4. $\dfrac{\pi_A(1 - \pi_A)}{(2\pi_A - 1)L}[(2\pi_B - 1)L - 2C_B]^2 < C_A - L\pi_A - (C_B - L\pi_B)$,

(4.7)

then A is inadmissible.

PROOF. In view of concavity, source A will be inadmissible if $D_3(1/2) > L/2$, and condition (4.6) reflects this. Lacking this, source A will be inadmissible if $D_3(\xi)$ lies above $D_4(\xi)$ for all ξ in $I - I_1 - I_2$, and we propose to show that (4.7) ensures this.

To begin with our conditions imply that $D_3(1/2) > D_4(1/2)$, which permits us to consider Figure 4.1.

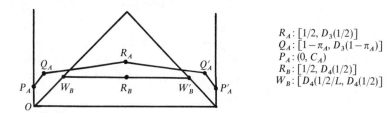

$R_A: [1/2, D_3(1/2)]$
$Q_A: [1 - \pi_A, D_3(1 - \pi_A)]$
$P_A: (0, C_A)$
$R_B: [1/2, D_4(1/2)]$
$W_B: [D_4(1/2)/L, D_4(1/2)]$

Fig. 4.1 Risk bounds.

It is clear that concavity ensures that the curve $P_A Q_A R_A Q'_A P'_A$ forms a lower bound for $D_3(\xi)$. Moreover, the line $W_B R_B W'_B$ bounds $D_4(\xi)$ from above in the interval $I - I_1 - I_2$. Hence it is sufficient, for ensuring that D_3 lie above D_4 in $I - I_1 - I_2$, that the point W_B lie below the line segment $Q_A R_A$, and condition (4.7) ensures this.

Finally, we have noted that Table 4.1 suggests a certain aspect of the Bayes solutions. Namely that when the parameters are such that both source A and source B are consulted first under some δ_ξ, source A is consulted at $\xi = 1/2$. We are rather inclined to this conjecture; i.e., that whenever the Bayes procedures do not call exclusively for one source, the "better" source is called for when information is less sharp.

To date, our attack on this problem has been limited to showing that for certain specific plans calling for consulting B on a central interval there do not exist parameters L, C_A, C_B such that D_3 equals D_4 at the two end points ξ^* of that interval, and such that D_3 equals the appropriate one of D_1 and D_2 at the end points ξ' of the two extreme "action" intervals. The technique is to utilize the supposed equality of D_3 and D_4 at ξ^* and the supposed equality of D_3 and D_1 or D_2 at ξ' to generate several procedures, all of which supposedly are Bayes procedures δ_{ξ^*}. One then observes that there are no parameters L, C_A, C_B endowing these several procedures δ_{ξ^*} with the same risk at ξ^*. The explicit ASN and OC expressions derived above turn out to be of use in the risk computations required.

References

[1] Borwanker, J. D., and H. T. David. 1968. Binomial sequential design of experiments with general loss and unequal sampling costs. (Abstr.) *Ann. Math. Stat.* 39: 1771–72.

[2] Bohrer, Robert. 1968. Operating characteristics of some sequential design rules. *Ann. Math. Stat.* 39: 1176–85.

[3] Bradt, Russell N., and Samuel Karlin. 1956. On the design and comparison of certain dichotomous experiments. *Ann. Math. Stat.* 27: 390–409.

[4] Chernoff, Herman. 1959. Sequential design of experiments. *Ann. Math. Stat.* 30: 755–70.

[5] DeGroot, M. H. 1962. Uncertainty, information and sequential experiments. *Ann. Math. Stat.* 33: 404–19.

[6] Feldman, Dorian. 1962. Contributions to the two-armed bandit problem. *Ann. Math. Stat.* 33: 847–56.

[7] Haggstrom, Gus W. 1966. Optimal stopping and experimental design. *Ann. Math. Stat.* 37: 7–29.

[8] Ingwell, Charles D. 1969. A symmetric binomial sequential design. Unpubl. M.S. thesis, Iowa State Univ., Ames.

[9] Kanarick, Arnold F., Jane M. Huntington, and Ronald C. Petersen. 1969. Multi-source information acquisition with optional stopping. *Human Factors* 11: 379–86.

[10] Lawing, William D., and H. T. David. 1966. Likelihood ratio computations of operating characteristics. *Ann. Math. Stat.* 37: 1704–16.

[11] Wald, Abraham. 1950. *Statistical decision functions.* New York: Wiley.

[12] Whittle, P. 1965. Some general results in sequential design (with discussion). *J. Roy. Statist. Soc.* Ser. B, 27: 371–94.

5

Recursive Sets of Rules in Statistical Decision Processes

GEORGE W. BROWN

1. Introduction

THIS PAPER is concerned with an approach to the problem of calculating approximate solutions for recurrent decision problems, including sequential analysis as an important special case. The approach is based on a particular choice of models and an appropriate method of representing approximate solutions suitable for treatment by a variety of algorithms. Attention is directed mainly at the model and at some important properties of the representation method, rather than at particular algorithms.

The model adopted is essentially that of a stationary, infinite-horizon, discrete Markovian process, with underlying states not necessarily directly observable, with nonnegative current costs, and with the objective of minimizing total expected costs under the assumption that finite total cost is attainable. All variables, corresponding to states, observables, and actions, take on finitely many values. Although the model corresponds to undiscounted costs, the case of discounted costs, with costs of arbitrary sign and discount factor less than unity, may be covered by adding a constant to make all costs nonnegative, then postulating a constant probability of termination instead of discounting the costs. Cases not covered by the model directly are those in which the total costs are infinite where the objective is usually to minimize a cost rate or are those with costs of arbitrary sign, nondiscounted. Appropriate modeling can cover finite horizons, terminating processes, and other kinds of time-dependencies

GEORGE W. BROWN is Professor and Dean, Graduate School of Administration, University of California at Irvine.

within the stationary model. The choice of the undiscounted form was dictated largely by the desire to cover the commonest examples of sequential analysis.

In representing approximate solutions to the decision problem set by the model, decision rules are represented by recursive definitions in which an initial action is specified, followed by application of a rule selected as a function of the observation obtained. Recursive sets of rules are those in which each rule is recursively definable in terms of member rules of the set. Operating characteristics are the expected total costs associated with initial state values, computable for each rule. A recursive set of rules, taken together with the operating characteristics, may be viewed as an "approximate" solution to the decision problem by providing a repertoire from which an initial rule may be selected. Given a priori probabilities for the initial states, an approximate Bayes solution corresponding to the repertoire would be obtained by choosing the initial rule with least expected cost.

A good algorithm for solving the decision problem should lead to repertoires whose minimum Bayesian expected costs approach the minimum attainable with any decision rule. An important result of this paper is that under certain restrictions a repertoire based on a recursive set of rules and satisfying the familiar optimality condition must itself be optimal. Another important result is that any repertoire based on a recursive set of rules may be used in an alternative fashion by updating the a priori probabilities, choosing each time the initial action of the currently "optimal" rule, with results at least as good as those attainable by direct use of the best initial rule.

This paper draws heavily upon ideas advanced and developed by others under a variety of headings, largely during the past 25 years. Sequential analysis, statistical decision theory, dynamic programming, Markovian decision processes, and list-processing languages for computer programming have all contributed directly and indirectly to the stock of ideas upon which the paper depends. With apologies to those authors not cited here who have made important contributions to the various pertinent literatures, who have influenced this author, and whose ideas may either have been shamelessly adopted or inadvertently reinvented, a brief set of references is appended. Among the debts specifically acknowledged are those to Wald [1] for sequential analysis and [2] for statistical decision theory, to Blackwell and Girshick [3] for their extensions of Wald's ideas, to Bellman [4] for dynamic programming, and to Blackwell [5] and [6] and Howard [7] for Markovian decision processes. An excellent bibliography on dynamic programming may be found in Wagner [8]. A different direction for generalization of Markovian decision processes is presented in Martin [9], in which transition probabilities are subject to uncertainty, although states are observed. The intersection of these generalizations is almost certainly nonvoid, but they appear to be far from equivalent.

2. The Model

At time $t = 0, 1, 2, \ldots$ denote the underlying state by s_t, the selected action by a_t, and the resulting observation by x_t. The cycle is imagined to start with s_t (not necessarily known), followed by the selection of a_t, followed by the observation x_t; the state s_{t+1} then starts the next cycle. Assume given the function

$$\text{Prob}(s_{t+1} = s', x_t = x/s_t = s, a_t = a) = P(s', x/s, a) \qquad (2.1)$$

which defines the joint transition and observation probabilities, independent of t, for transition to the state next time and for the value of the observed current variable conditioned on the current state and action. The function is defined for all combinations of values, with the domain of each variable a finite set. Assume given also the function $c(s, a) \geq 0$, to be interpreted as the expected current cost c_t if $s_t = s$ and $a_t = a$. The criterion to be minimized is $C = E[\sum_t c_t]$.

Let a decision rule be defined as any function which assigns a_t as a function of all $\{x_{t'}, a_{t'}\}$, with $t' < t$ for all t. It is assumed that there exists at least one rule r with finite expected total costs, i.e., such that

$$E[C/r, s_0 = s] = E[\sum_t c_t] = F_r(s) < \infty \text{ for all } s. \qquad (2.2)$$

For fixed r the function $F_r(s)$ is referred to as the *operating characteristic* of r. The notation makes explicit the dependence of the expected total cost on the initial state assumed.

To see that the model covers the discounted cost case, assume that $P(s', x/s, a)$ is given as in (2.1) but that the current cost function is $d(s, a)$, with signs arbitrary, and that the criterion to be minimized is $\sum_t \beta^t d_t$, with $0 < \beta < 1$. Let $c(s, a) = d(s, a) + K$, where K is sufficiently large to make $c(s, a) \geq 0$ for all s, a. Then for all rules, discounted total costs based on $c(s, a)$ will differ by the constant $K/(1 - \beta)$ from total costs based on $d(s, a)$. The decision problem is thus unchanged by converting it to one with nonnegative costs. It suffices now to add a trapping state in which all costs are to be null—replacing $P(s', x/s, a)$ by $\beta P(s', x/s, a)$ for the original states— filling out the extra values required in a manner consistent with the condition that the probability of entering the trapping state from each of the original states shall be $1 - \beta$, whatever action a

is chosen. For any rule r the new undiscounted current cost at time t is β^t (the probability of survival to time t) times the original current cost c_t. Thus the transformation yields expected costs, without discounting, identical to those associated with the original costs discounted according to the factor β for each time period.

To see how one variety of time dependence may be modeled within the stationarity assumption, consider, for example, a cyclic situation in which actions at even values of t have different consequences than at odd values of t. Simply double the number of states, with half corresponding to even values and half corresponding to odd values of t, and construct the appropriate probabilities for transitions and observations. The effect is to make the time dependence implicit while preserving the stationary form. Within the restriction that the total number of states must remain finite there exists considerable freedom to model time dependencies implicitly.

3. An Example for the Basic Model

A simple concrete example for the basic model is furnished by considering a sequential experiment for testing a simple hypothesis against a single alternative. In constructing the example, it is assumed that the experimenter has one of the following options: (1) he may make a single observation at unit cost, (2) he may receive a pair of independent observations at a cost of $1\frac{1}{2}$ units, (3) he may accept the null hypothesis without further experimentation, or (4) he may reject the null hypothesis without further experimentation. Costs of wrong decisions are placed at 10 units and 20 units respectively, when the null hypothesis is assumed true or false. The null hypothesis corresponds to a probability of 0.2 for a particular event, while the alternative hypothesis corresponds to probability 0.8 for the same event; the observable is the number of occurrences of the event, whether in single or double trials. The decision problem is modeled as follows:

$$P(s', x/s, a)$$

$s=1, a=1$				$s=1, a=2$				$s=1,2; a=3,4$		
$s' \backslash x = 0$	1	2		$s' \backslash x = 0$	1	2		$s' \backslash x = 0$	1	2
1	.8	.2	0	1	.64	.32	.04	1	0	0	0
2	0	0	0	2	0	0	0	2	0	0	0
3	0	0	0	3	0	0	0	3	1	0	0

$s = 2, a = 1$			
$s' \backslash x = $	0	1	2
1	0	0	0
2	.2	.8	0
3	0	0	0

$s = 2, a = 2$			
$s' \backslash x = $	0	1	2
1	0	0	0
2	.04	.32	.64
3	0	0	0

$s = 3, a = 1, 2, 3, 4$			
$s' \backslash x = $	0	1	2
1	0	0	0
2	0	0	0
3	1	0	0

$$(3.1)$$

and

$c(s, a)$				
$s \backslash a = $	1	2	3	4
1	1	$1\frac{1}{2}$	0	10
2	1	$1\frac{1}{2}$	20	0
3	0	0	0	0

$$(3.2)$$

Identification is readily made with the preceding description: $s = 1$ corresponds to the null hypothesis, $s = 2$ corresponds to the alternative hypothesis being true, and for both $s = 1$ and $s = 2$ no final action has been taken, while $s = 3$ corresponds to termination of the experiment; $a = 1$ corresponds to the choice of a single trial, $a = 2$ corresponds to the choice of a pair of trials, $a = 3$ corresponds to acceptance of the null hypothesis, while $a = 4$ corresponds to rejection of the null hypothesis; and $x = 0, 1, 2$ corresponds to the number of observed occurrences of the event. The table of current costs (3.2) reflects both the costs of experimentation and the costs of erroneous final choices.

The example of this section, while not entirely devoid of interest for its own sake, exercises very little of the generality possible within the model. It is apparent that the model may be used to represent situations in which the underlying statistical parameters are themselves subject to Markovian transition or in which the costs are subject to transition, or some actions may be taken for information and some to reduce costs. The model is therefore appropriate not only for the context of sequential experimentation but also for a large number of contexts like repair and replacement models, inventory management subject to variation in demand parameters, and other procurement and allocation problems which can be represented as recurrent decision problems. The example, in spite of its special nature, will serve to illustrate various aspects of the development which follows.

4. Recursive Sets of Rules

The specification of a decision rule involves the formulation of the dependence of the action to be taken at time t upon all earlier actions and observations. By specifying the initial action of a rule (i.e., the action to be taken at $t = 0$), the definition of a rule may be reduced to the definitions of the various procedures to be followed, starting at $t = 1$ and depending on the outcome of x_0. This suggests that rules may be recursively definable in terms of other rules by specifying the initial action a and a function $f(x)$ which maps the observable into the set of all rules, so that after action a is taken and x is observed, the rule $f(x)$ is applied as if $t = 1$ were its time origin. The notation

$$r:[a;f(x)] \tag{4.1}$$

will stand for such a recursive specification of the rule r. Note that nothing prevents the rule r from occurring as a value of $f(x)$; r is said to be *recursively defined* in terms of the rules $\{f(x)\}$.

By viewing the total expected cost as the sum of the current cost at $t = 0$ and the expected total future cost beginning with $t = 1$ and assuming (4.1), the recursion

$$F_r(s) = c(s, a) + \sum_{s', x} P(s', x/s, a) F_{f(x)}(s'); \text{ all } s \tag{4.2}$$

yields the operating characteristic of r in terms of those of the rules $\{f(x)\}$, whenever the latter are all finite.

Now let R be a subset of the set of all rules $\bar{R}$. R is said to be a *recursive set* of rules if each rule in R is recursively definable in terms of members of R, i.e., if

$$r:[a_r; f_r(x)],$$

with

$$f_r(x) \in R$$

whenever

$$r \in R. \tag{4.3}$$

$\bar{R}$ itself clearly qualifies as a recursive set of rules. Most recursive sets of rules with which we will be concerned will be finite sets. For any recursive set of rules, the recursion relations (4.2), when appropriately subscripted for each r, yield a simultaneous set of equations in the quantities $F_r(s)$, as follows:

$$F_r(s) = c(s, a_r) + \sum_{s', x} P(s', x/s, a_r) F_{f_r(x)}(s'); r \in R, \text{ all } s. \tag{4.4}$$

Unfortunately, the simultaneous equations (4.4) cannot by themselves completely determine the operating characteristics, since solutions are ambiguous at least to the extent of an additive constant by virtue of the fact that $\sum_{s', x} P(s', x/s, a) = 1$. Some additional information will be needed to determine the operating characteristics of the rules. It should be noted, however, that the joint set of definitions (4.3) completely defines each of the member rules of a recursive set R in an operational manner, with the result that the operating characteristics are uniquely determined by (4.3), provided they are finite.

Table 5.1 provides an example of a simple recursive set of rules pertaining to the model example of section 3. Dashed entries in the table correspond to entries which may be filled in arbitrarily, either because all future costs will be necessarily null whatever specification is filled in or because the entry corresponds to an observable with null probability. The table is somewhat easier to interpret when presented in this fashion; the reader is encouraged to fill out the specification of the open entries if he feels cheated.

Table 5.1

Recursive Set R_1

r	a_r	$f_r(0)$	$f_r(1)$	$f_r(2)$	$F_r(1)$	$F_r(2)$	$F_r(3)$
1	3	...	...	...	0	20	0
2	4	...	...	...	10	0	0
3	1	1	2	...	3	5	0
4	2	1	3	2	2.86	3.90	0

In Table 5.1 rules 1 and 2 generate only current costs, with the result that their operating characteristics can be written down directly from (3.2). The recursion equations then yield the operating characteristic for rule 3, then finally for rule 4. Rules 1 and 2 correspond to outright acceptance or rejection of the null hypothesis without further experimentation. Rule 3 corresponds to making a single trial, then terminates appropriately according to whether $x = 0$ or $x = 1$. Rule 4 corresponds to the choice of a pair of trials leading to appropriate termination if $x = 0$ or $x = 2$ but appealing to rule 3 in case $x = 1$ is observed. Rule 3 happens to be dominated by rule 4 in the ordinary sense of the term. Consequences of this dominance will be considered next.

5. Recursive Sets of Rules as Repertoires

The fundamental problem in attempting to deal computationally with recurrent decision problems is to exhibit a set of rules which will serve or nearly so, (instead of the set $\bar{R}$ of all rules) as the basic repertoire from which to choose. The familiar Bayesian construction seeks to minimize the expected cost $\sum_s \pi_s F_r(s)$ over $\bar{R}$, where $\{\pi_s\}$ is a given set of a priori probabilities for the initial state s_0. Letting $\{\pi_s\}$ vary generates a family of Bayes solutions (providing $\sum_s \pi_s F_r(s)$ actually attains a minimum) which taken together can characterize an optimal solution to the decision problem. In any case, the envelope of all the functions $\sum_s \pi_s F_r(s)$ for $r \in \bar{R}$ generates the concave function

$$\psi(\pi) = \inf_{\bar{R}} \sum_s \pi_s F_r(s) \tag{5.1}$$

which is the greatest lower bound to the expected Bayesian cost, given initial $\pi = \{\pi_s\}$. The most obvious way, therefore, to use a finite repertoire R is to choose from R given π that rule which minimizes the same Bayesian cost. Confining the repertoire to R leads to a minimum expected cost function

$$\psi_R(\pi) = \min_R \sum_s \pi_s F_r(s). \tag{5.2}$$

A good repertoire should yield $\psi_R(\pi)$ close to $\psi(\pi)$.

The fact that non-Bayesian use of a set of rules R as a repertoire requires some other basis for picking an admissible rule or may require a randomized rule does not invalidate computational methods which are restricted to Bayes solutions within R. With the model chosen here, it is a direct consequence of the duality theorem of linear programming that probability mixtures of Bayes solutions in R will permit dominating any rule in R which, while being admissible in R, is not itself a Bayes solution in R. In this sense the Bayes solutions in R provide a basis for admissible randomized rules in R.

If R is a finite recursive set of rules with known operating characteristics, then R itself may be viewed directly as an approximate solution to the decision problem. It is easy to find a minimizing r given initial π, to use that r as an entry point to the table, to apply the recursion as indicated after taking the indicated action and observing x, and thus to determine the next entry point in the table. Such an application of the table may be termed *literal application* of the repertoire R. As will be shown below, other possibilities may exist, either for improvement of R or for improved uses of R.

Referring back to recursive set R_1 (Table 5.1), recall that rule 3 is dominated by rule 4. Although rule 3 would not be selected initially by the procedure described in the preceding paragraph, rule 4 might be, and literal application of rule 4 might lead to rule 3 next time. It is thus suggested that there may be more intelligent ways to use R_1 than by literal application of the repertoire. For example, taking account of the fact that rule 4 is better than rule 3, what would be the consequence of choosing to follow rule 4 whenever the observed x points to the use of rule 3? Such departure from the literal use of R_1 can only decrease the expected cost of the procedure. The same observed dominance can be used to modify R_1 by redefining, instead of rule 4, the rule 4′ which is similar to 4 but replaces the reference to rule 3 by reference to 4′ itself. The recursion equation again determines the operating characteristic, and the modified table becomes Table 5.2. Comparison with Table 5.1 shows a substantial improvement for rule 4′ over rule 4.

Table 5.2

Recursive Set R_2

r	a_r	$f_r(0)$	$f_r(1)$	$f_r(2)$	$F_r(1)$	$F_r(2)$	$F_r(3)$
1	3	...	...	...	0	20	0
2	4	...	...	...	10	0	0
4′	2	1	4′	2	2.794	3.382	0

Another alternative use for a repertoire R may be the method of Bayesian updating.

6. Bayesian Updating

If π is an initial set of a priori probabilities, the choice of a and observation of x lead to the possibility of using posterior probabilities of the states s' as new a priori probabilities next time. Routine use of Bayes' formula leads to the result

$$\text{Prob}(s'/\pi, a, x) = \frac{\sum_s \pi_s P(s', x/s, a)}{\sum_{s', s} \pi_s P(s', x/s, a)},$$

if

$$P(s', x/s, a) \neq 0 \text{ for all } s', s. \tag{6.1}$$

Given π, a, and x, the new set of probabilities might appropriately be denoted by $\pi'(\pi/a, x)$.

Given the possibility of calculating, after each step of the decision process, an updated set of a priori probabilities reflecting the past history, a repertoire for which operating characteristics are known can be used to select a rule which minimizes the expected cost for the purpose of adopting the initial action of that rule. Continued use of this process may result in a procedure which corresponds to none of the rules of the repertoire itself since each updating causes the rule selection process to be repeated again. It seems a reasonable guess that use of a repertoire for Bayesian updating and choice of action a cannot for an initial π lead to a higher cost than would the use of the lowest cost rule of the repertoire in literal fashion. In the case of R_1 (the repertoire of Table 5.1) Bayesian updating would yield results equivalent to the improved repertoire R_2 (Table 5.2) by virtue of the fact that if $a = 2$ and $x = 1$, $\pi' = \pi$, as may be seen by applying (6.1) to the probabilities specified by the example in (3.1). Thus if rule 4 is the least-cost Bayes solution within R_1 and if $x = 1$, rule 4 will again be selected by Bayesian updating. While Bayesian updating is practically more complicated than literal use of a repertoire, it not only suppresses references to dominated rules but also suppresses references to rules which are not least cost for any π, whether dominated or not.

The Bayesian updating process corresponds to the construction (implicitly) of a function $a_R(\pi)$, the initial action of the least-cost rule in R given π. No problem is caused by failure of $a_R(\pi)$ to be always single valued. Note that when R is $\bar{R}$, the corresponding function $a(\pi)$ can be determined whether minimum costs are attained or not, since there are only a finite number of values of a and some one of them must occur infinitely often as initial action in any infinite sequence of rules. If $a_R(\pi)$ were explicitly available for all π, it would be a *policy* prescription based on R. By contrast if $\psi_R(\pi)$ were explicitly available for all π, it would be a cost function corresponding to the *value* system associated with R. Traditional algorithmic approaches to dynamic decision problems deal with one or both of these functions. When the underlying states are not necessarily known, direct representation will not be practical if the number of states exceeds a very few. Thus it appears that representation of approximate solutions by repertoires of rules may keep the representation in bounds, while retaining the ability to obtain either $a_R(\pi)$ or $\psi_R(\pi)$ for particular values of π.

7. The Optimality Condition

The purpose of this section is to derive the familiar optimality condition in the form appropriate to the model. Setting

$$P(x/\pi, a) = \sum_{s', s} \pi_s P(s', x/s, a),$$

the marginal probability of x given initial a priori probabilities π and initial action a, the optimality condition can be written

$$\psi(\pi) = \min_a [\sum_s \pi_s c(s, a) + \sum_x P(x/\pi, a)\psi(\pi'(\pi/a, x))], \qquad (7.1)$$

expressing the intuitive notion that the best overall result is the best combination of an action in the present followed by optimum results in the future, whatever x may be observed following the initial action. Derivation of the optimality condition proceeds easily from the recursive representation of rules in $\bar{R}$ as in (4.2). Multiplying each of the equations (4.2) by π_s and summing,

$$\sum_s \pi_s F_r(s) = \sum_s \pi_s c(s, a) + \sum_{s, s', x} \pi_s P(s', x/s, a) F_{f(x)}(s'), \qquad (7.2)$$

for all $r \in \bar{R}$. Ordering the triple summation in the sequence x, s', s from the outside to the inside and recalling the definition of $P(x/\pi, a)$ above and of $\pi'(\pi/a, x)$ from (6.1), (7.2) becomes

$$\sum_s \pi_s F_r(s) = \sum_s \pi_s c(s, a) + \sum_x P(x/\pi, a) \sum_{s'} \pi'_{s'}(\pi/a, x) F_{f(x)}(s'). \quad (7.3)$$

Consider now the problem of calculating the infimum on both sides over all rules r. On the left the result would be $\psi(\pi)$, while on the right we may consider all initial actions a followed by any $f(x)$. Corresponding to a fixed a, the last term on the right-hand side is a sum of nonnegative terms, each of which may be taken individually to the infimum, with a and x fixed. (7.1) is therefore the result of proceeding to the infimum on both sides of (7.3).

Examining the process of deriving the optimality condition, it is seen that if $R \subseteq \bar{R}$ and the infimum process is confined to $r \in R$, the right-hand side may go lower than the left, since R may not contain all the candidate rules appealed to on the right-hand side. Thus, for any set of rules R we have the inequality

$$\psi_R(\pi) \geq \min_a \left[\sum_s \pi_s c(s, a) + \sum_x P(x/\pi, a) \psi_R(\pi'(\pi/a, x)) \right]. \quad (7.4)$$

Failure to satisfy the optimality condition with $\psi_R(\pi)$ at a given π suggests new possibilities, either for the use of R as a repertoire or for the improvement of R by augmentation. Thus, using a particular initial value of π, finding the minimizing a for the right-hand side of (7.4) may yield a different action than the a corresponding directly to $\psi_R(\pi)$. The alternate process supplies a best action based on a one-period "look-ahead," assuming subsequent use of R. Here again it seems reasonable that no loss can be sustained as compared with literal use of R or even as compared with updated Bayesian use of R. Furthermore, the process of optimizing on the right-hand side of (7.4) may serve to define a new rule recursively defined on R, which does better at the chosen π than any member of R, thus permitting an improvement to be made by adjoining the new rule to R. If R is already a recursive set of rules, the augmented set will of course also be a recursive set. The process described is identical to the process of constructing the best rule, recursively definable on R at the value π.

Referring back to the set R_2 (Table 5.2), choices of $\pi = .8, .2, 0$ and $\pi = .3, .7, 0$ each led to new rules through use of the "look-ahead" procedure just described. Relabeling rule 4' as rule 3 and inserting the additional rules, Table 5.3 is obtained.

Table 5.3
Recursive Set R_3

r	a_r	$f_r(0)$	$f_r(1)$	$f_r(2)$	$F_r(1)$	$F_r(2)$	$F_r(3)$
1	3	...	...	...	0	20	0
2	4	...	...	...	10	0	0
3	2	1	3	2	2.794	3.382	0
4	1	3	2	...	5.235	1.676	0
5	1	1	3	...	1.559	7.706	0

Further attempts to modify and improve Table 5.3 were unsuccessful. The table is in fact the complete solution to the example of section 3, subject to the hazards of the author's arithmetic. If the construction described above never yields a new rule whatever π value is chosen, then of course the optimality condition is necessarily satisfied. It remains to be seen under what conditions the optimality condition itself implies that the set R of rules for which it is satisfied will be an optimal set. Given the undiscounted model adopted here, it will be necessary to adopt some additional condition to ensure sufficiency of the optimality condition.

8. Effectiveness of Bayesian Updating

Section 6 described Bayesian updating as an alternative to literal application of a repertoire R of rules. This section will show that the expected cost, given an initial value of π, is no greater with Bayesian updating than that of the least-cost rule in R at the same π. The proof is valid for any recursive set R, finite or not.

As in the preceding sections let $\psi_R(\pi)$ be the least cost and let $a_R(\pi)$ be the initial action corresponding to least cost. Since R is not necessarily finite, the following definitions will serve:

$$\psi_R(\pi) = \inf_R \sum_s \pi_s F_r(s) \tag{8.1}$$

and

$a_R(\pi)$ is the initial action of each member of a sequence r_n, with

$$\sum_s \pi_s F_{r_n}(s) \to \psi_R(\pi). \tag{8.2}$$

Now define a derived recursive set of rules R' as follows:

$$r_\pi \in R' \text{ if } r_\pi \text{ is of the form } [a_R(\pi); r_{\pi'}], \text{ where } \pi' = \pi'(\pi/a_R(\pi), x).$$
(8.3)

The function π' corresponds to the function defined by (6.1), the updated state probability vector; and it may be seen that (8.3) defines (for each initial π) r_π as the decision rule which executes the Bayesian updated use of the repertoire R. The theorem of this section may be formulated as

Theorem: For all π, $\sum_s \pi_s F_{r_\pi}(s) \leq \psi_R(\pi)$.   (8.4)

Proof of the theorem is accomplished by constructing a sequence $R_n(\pi)$ of sets of rules for each n and for each π such that members of $R_n(\pi)$ will agree with r_π for $t = 0, 1, 2, \ldots n - 1$, appealing to some member of R at $t = n$. The $\{R_n(\pi)\}$ are defined inductively:

For all π, $R_0(\pi)$ is R, and $r \in R_{n+1}(\pi)$ if r is of the form

$$r : [a_R(\pi); r'], \text{ with } r' \in R_n(\pi') \text{ and } \pi' = \pi'(\pi/a_R(\pi), x).$$ (8.5)

Thus members of $R_1(\pi)$ begin with the action $a_R(\pi)$ and appeal to any member of R at $t = 1$; members of $R_2(\pi)$ begin with $a_R(\pi)$, form π', choose $a_R(\pi')$ at $t = 1$, etc.

Associated with each $R_n(\pi)$ is the function

$$\phi_n(\pi) = \inf_{R_n(\pi)} \sum_s \pi_s F_r(s),$$ (8.6)

the greatest lower bound of Bayesian expected costs at π, using a member of $R_n(\pi)$ chosen with the same π. Since $R_0(\pi)$ is R itself,

$$\phi_0(\pi) = \psi_R(\pi).$$ (8.7)

Now, for any rule of the form $[a; f(x)]$ the Bayesian cost may be written

$$\sum_s \pi_s F_r(s) = \sum_s \pi_s c(s, a) + \sum_x P(x/\pi, a) \sum_s \pi'_s F_{f(x)}(s),$$ (8.8)

where

$$\pi' = \pi'(\pi/a, x),$$

following (7.3). Applying (8.8) to rules of $R_{n+1}(\pi)$ on the left and recalling (8.5), $f(x)$ will be a member of $R_n(\pi')$ for fixed π and x, with $a = a_R(\pi)$. Taking the infimum on the left and taking the infimum for each term (in which x is fixed) yields the recursion

$$\phi_{n+1}(\pi) = \sum_s \pi_s c(s, a_R(\pi)) + \sum_x P(x/\pi, a_R(\pi))\phi_n(\pi'),$$ (8.9)

with

$$\pi' = \pi'(\pi/a_R(\pi), x).$$

Happily, (8.9) implies that if $\phi_{n+1}(\pi) \leqq \phi_n(\pi)$ for all π, then the same inequality holds for the next value of n. To see that $\phi_1(\pi) \leqq \phi_0(\pi)$, return to (8.8) and note that the infimum will not be increased by choosing $a = a_R(\pi)$ by virtue of (8.2) and that on the right-hand side each term $\sum_s \pi'_s F_{f(x)}(s)$ is certainly at least $\psi_R(\pi') = \phi_0(\pi')$ by virtue of the fact that $f(x) \in R$. Setting $n = 0$ in (8.9) shows that $\phi_1(\pi)$ is equal to an expression which has just been shown to be not greater than $\phi_0(\pi)$. Summarizing,

$$\phi_{n+1}(\pi) \leqq \phi_n(\pi) \leqq \ldots \phi_0(\pi) = \psi_R(\pi). \tag{8.10}$$

Recall now that every member of $R_n(\pi)$ agrees as a decision rule with r_π for the first n time periods, $t = 0, 1, 2, \ldots n - 1$. For a member of $R_n(\pi)$ the expected Bayesian cost of the first n periods is not greater than $\phi_n(\pi)$. For r_π the expected Bayesian cost of the first n periods is the same as for a member of $R_n(\pi)$, hence is not greater than $\phi_n(\pi)$. The fact that as n increases one nondescending series has its members bounded by the terms of a nonascending sequence establishes that the total Bayesian cost of r_π is finite and satisfies (8.4).

It may not be obvious where the recursive property of the set R was used in the proof of the theorem. The crucial point is when the first inequality is established, at which time, after (8.9), the fact that $f(x) \in R$ if $r \in R$ was required. It is not difficult to construct a counterexample, if R is not recursive.

The recursion (8.9), used in the proof, is related to the functional equation

$$\phi(\pi) = \sum_s \pi_s c(s, a_R(\pi)) + \sum_x P(x/\pi, a_R(\pi)) \phi(\pi'), \tag{8.11}$$

with

$$\pi' = \pi'(\pi/a_R(\pi), x).$$

This functional equation will be satisfied both by $\sum_s \pi_s F_{r_\pi}(s)$ and by $\lim \phi_n(\pi)$. In the case of the former, (8.8) applied to the definition of r_π establishes the functional equation. Again, it is not difficult to establish by a counterexample that the two functions need not be identical.

An important corollary to the theorem, obtained by letting R be $\bar{R}$, the set of all rules, is that $\psi(\pi)$ is attainable. In this case r_π, a perfectly good decision rule, has Bayesian expected cost $\psi(\pi)$.

9. Sufficiency of the Optimality Condition

In section 7 it was shown that $\psi(\pi)$ satisfies the optimality condition (7.1). It was further suggested that algorithmic methods of approximating solutions to the decision problem formulated in section 2 might be based on trying to satisfy the optimality condition. The theorem presented in this section proposes a simple condition, one for which it should not be difficult to provide, to ensure that solutions to (7.1) will agree with $\psi(\pi)$.

Before proceeding with the formal statement and proof of the sufficiency of the optimality condition, the additional assumption introduced is essentially that the function satisfying (7.1) vanishes wherever $\psi(\pi)$ itself vanishes. Given the model used in this paper, $\psi(\pi) = 0$ corresponds to "perfect information" from the point of view of the decision problem. At such a value of π there must exist a rule which guarantees null costs forever. In practical applications it would appear that such π values are so special there will be no difficulty isolating them together with the rules which yield null costs. If this is so, there will be no difficulty including those rules in any set of rules proposed as an approximating repertoire, thus assuring that the assumption will be satisfied. Reference to the example of section 3 and to the cost table (3.2) shows immediately that $\psi(\pi) = 0$ implies that either π_1 or π_2 must vanish. Furthermore, inclusion of rules beginning with a_3 and with a_4 will guarantee the assumption.

The formal statement of sufficiency is as follows:

Theorem: Let $\psi(\pi)$ be the minimum Bayesian expected cost for initial state probability vector π corresponding to the model of section 2; let $\phi(\pi)$ be a bounded function which satisfies (7.1), with $\phi(\pi) \geq \psi(\pi)$ and with $\phi(\pi) = 0$ and continuous wherever $\psi(\pi) = 0$. Then $\phi(\pi) = \psi(\pi)$, for all π. (9.1)

Proof of (9.1) proceeds from the fact (7.1) may be rewritten for $\psi(\pi)$, to reflect that $a(\pi)$ is a minimizing choice of a, as

$$\psi(\pi) = \sum_s \pi_s c(s, a(\pi)) + \sum_x P(x/\pi, a(\pi))\psi(\pi'), \qquad (9.2)$$

with

$$\pi' = \pi'(\pi/a(\pi), x).$$

At the same time since $\phi(\pi)$ satisfies (7.1) and $a(\pi)$ may fail to minimize the right-hand side of (7.1) when applied to $\phi(\pi)$, we obtain

$$\phi(\pi) \leq \sum_s \pi_s c(s, a(\pi)) + \sum_x P(x/\pi, a(\pi))\phi(\pi'), \qquad (9.3)$$

with π' as in (9.2). Subtracting (9.2) from (9.3),

$$\phi(\pi) - \psi(\pi) \leq \sum_x P(x/\pi, a(\pi))[\phi(\pi') - \psi(\pi')], \qquad (9.4)$$

which may be reinterpreted as

$$\phi(\pi) - \psi(\pi) \leq E[\phi(\pi') - \psi(\pi')], \qquad (9.5)$$

with
$$\pi' = \pi'(\pi/a(\pi), x).$$

In (9.5) π' is viewed as the random successor to π under $a(\pi)$. If $\pi^{(t)}$ is defined as the random successor at time t under repetition of this process, (9.5) implies

$$\phi(\pi) - \psi(\pi) \leq E[\phi(\pi^{(t)}) - \psi(\pi^{(t)})]. \tag{9.6}$$

Returning now to (9.2) and imagining repeated application to successors of π, it can be seen that $E[\psi(\pi^{(t)})]$ is the remainder of the infinite series, the sum of which is $\psi(\pi)$. Hence

$$\lim E[\psi(\pi^{(t)})] = 0, \text{ as } t \to \infty. \tag{9.7}$$

Simply sketching the rest of the proof, if the initial π is chosen so that $\phi(\pi) - \psi(\pi)$ is sufficiently near to its supremum, assumed positive, then successor values can be found for sufficiently large t at which ψ is as small as desired while ϕ is bounded above a positive constant by virtue of (9.6) and (9.7). The assumptions on ϕ and the fact that ψ is continuous yield a value of π at which ψ vanishes while ϕ is positive, thus contradicting the hypotheses of the theorem. Continuity of $\psi(\pi)$ or of any $\psi_R(\pi)$ is easily established as a consequence of the definition of $\psi_R(\pi)$, and the proof is omitted here.

10. Concluding Remarks

An interesting class of algorithmic methods is associated with random generation of initial π values, at each of which can be constructed the best rule, recursively definable in terms of the repertoire currently available at the time that π value is considered. A number of variants have been tested empirically with good results. Variant approaches differ in the manner in which successive π values are obtained. Consider the fact that sampling of π values may be uniform over the simplex of all π values or may be associated with sequences of successor values derived from Monte Carlo application of the repertoire currently available. Questions of convergence rates, of bounds on the error, or other efficiency measures are entirely open. It is expected that future efforts will be devoted to such questions.

Also open are questions of extensibility of the approach to cases excluded by the model adopted here—cases mentioned in the introduction as well as cases in which the domains of the variables are not necessarily finite. A particular extension of interest would be the case in which domains are finite, costs are allowed to be negative as well as positive or zero, and expected total costs are bounded from below as well as from above.

A point which should be emphasized was alluded to at the end of section 6 and provided the main motivation for the development of this paper: the repertoire of rules permits effective representation of an approximating system for the recurrent decision problem without involving explicit representation of either policies or values as functions of π. Thus it becomes practical to consider problems in which the dimensionality of the state-probability space would prohibit tabular representation of the functions ordinarily used. Finally, what more appropriate representation for a concave function could exist than the coefficients of the linear functions whose envelope defines the function?

References

[1] Abraham Wald. 1947. *Sequential analysis*. New York: Wiley.
[2] ———. 1950. *Statistical decision functions*. New York: Wiley.
[3] David Blackwell and M. A. Girshick. 1954. *Theory of games and statistical decisions*. New York: Wiley.
[4] Richard Bellman. 1957. *Dynamic programming*. Princeton: Princeton Univ. Press.
[5] David Blackwell. 1962. Discrete dynamic programming. *Ann. Math. Stat.* 33: 719–26.
[6] ———. 1965. Discounted dynamic programming. *Ann. Math. Stat.* 36: 226–35.
[7] Ronald A. Howard. 1960. *Dynamic programming and Markov processes*. Cambridge: M.I.T. Press.
[8] Harvey M. Wagner. 1969. *Principles of operations research*. Englewood Cliffs, N.J., Prentice-Hall.
[9] J. J. Martin. 1967. *Bayesian decision problems and Markov chains*. New York: Wiley.

6

Observational Studies

WILLIAM G. COCHRAN

1. Introduction

OBSERVATIONAL STUDIES are a class of statistical studies that have increased in frequency and importance during the past 20 years. In an observational study the investigator is restricted to taking selected observations or measurements on the process under study. For one reason or another he cannot interfere in the process in the way that one does in a controlled laboratory type of experiment.

Observational studies fall roughly into two broad types. The first is often given the name of "analytical surveys." The investigator takes a sample survey of a population of interest and proceeds to conduct statistical analyses of the relations between variables of interest to him. An early example was Kinsey's study (1948) of the relation between the frequencies of certain types of sexual behavior and variables like the age, sex, social level, religious affiliation, rural-urban background, and direction of social mobility of the person involved. Dr. Kinsey gave much thought to the methodological problems that he would face in planning his study. More recently, in what is called the "midtown Manhattan study" (Srole et al., 1962), a team of psychiatrists studied the relation in Manhattan, New York, between age, sex, parental and own social level, ethnic origin, generation in the United States, and religion and nonhospitalized mental illness.

WILLIAM G. COCHRAN is Professor of Statistics, Harvard University, Cambridge, Massachusetts.
This work was assisted by a contract with the Office of Naval Research, Navy Department.

The second type of observational study is narrower in scope. The investigator has in mind some agents, procedures, or experiences that may produce certain causal effects (good or bad) on people. These agents are like those the statistician would call *treatments* in a controlled experiment, except that a controlled experiment is not feasible. Examples of this type abound. A simple one structurally is a Cornell study of the effect of wearing a lap seat belt on the amount and type of injury sustained in an automobile collision. This study was done from police and medical records of injuries in automobile accidents. The prospective smoking and health studies (1964) are also a well-known example. These are comparisons of the death rates and causes of death of men and women with different smoking patterns in regard to type and amount. An example known as the "national halothane study" (Bunker et al., 1969) attempted to make a fair comparison of the death rates due to the five leading anesthetics used in hospital operations.

Several factors are probably responsible for the growth in the number of studies of this kind. One is a general increase in funds for research in the social sciences and medicine. A related reason is the growing awareness of social problems. A study known as the "Coleman report" (1966) has attracted much discussion. This was begun because Congress gave the U.S. Office of Education a substantial sum and asked it to conduct a nationwide survey of elementary schools and high schools to discover to what extent minority-group children in the United States (Blacks, Indians, Puerto Ricans, Mexican-Americans, and Orientals) receive a poorer education than the majority whites. A third reason is the growing area of program evaluation. All over the world, administrative bodies—central, regional, and local—spend the taxpayers' money on new programs intended to benefit some or all of the population or to combat social evils. Similarly, a business organization may institute changes in its operations in the hope of improving the running of the business. The idea is spreading that it might be wise to devote some resources to trying to measure both the intended and the unintended effects of these programs. Such evaluations are difficult to do well, and they make much use of observational studies. Finally, some studies are undertaken to investigate stray reports of unexpected effects that appear from time to time. The halothane study is an example; others are studies of side effects of the contraceptive pill and studies of health effects of air pollution.

This paper is confined mainly to the second, narrower class of observational studies, although some of the problems to be considered are also met in the broader analytical ones.

For this paper I naturally sought a topic that would reflect the outlook and research interests of George Snedecor. In his career activity of helping investigators, he developed a strong interest in the design of experiments, a subject on which numerous texts are now available. The planning of observational studies, in which we would like to do an experiment but cannot, is a closely related topic which cries aloud for George's mature wisdom and the methodological truths that he expounded so clearly.

Succeeding sections will consider some of the common issues that arise in planning.

2. The Statement of Objectives

Early in the planning it is helpful to construct and discuss as clear and specific a written statement of the objectives as can be made at that stage. Otherwise it is easy in a study of any complexity to take later decisions that are contrary to the objectives or to find that different team members have conflicting ideas about the purpose of the study. Some investigators prefer a statement in the form of hypotheses to be tested, others in the form of quantities to be estimated or comparisons to be made. An example of the *hypothesis* type comes from a study (Buck et al., 1968), by a Johns Hopkins team, of the effects of coca-chewing by Peruvian Indians. Their hypotheses were stated as follows.

1. Coca, by diminishing the sensation of hunger, has an unfavorable effect on the nutritional state of the habitual chewer. Malnutrition and conditions in which nutritional deficiencies are important disease determinants occur more frequently among chewers than among control subjects.
2. Coca chewing leads to a state of relative indifference which can result in inferior personal hygiene.
3. The work performance of coca chewers is lower than that of comparable non-chewers.

One objection sometimes made to this form of statement is its suggestion that the answers are already known, and thus it hints at personal bias. However, these statements could easily have been put in a neutral form, and the three specific hypotheses about coca were suggested by a previous League of Nations commission. The statements perform the valuable purpose of directing attention to the comparisons and measurements that will be needed.

3. The Comparative Structure of the Plan

The statement of objectives should have suggested the type of comparisons on which logical judgments about the effects of treatment would be based. Some of the most common structures are outlined below. First, the study may be restricted to a single group of people, all subject to the same treatment. The timing of the measurements may take several forms.

1. After Only (i.e., after a period during which the treatment should have had time to produce its effects).
2. Before and After (planned comparable measurements both before and after the period of exposure to the agent or treatment).
3. Repeated Before and Repeated After.

In both (1) and (2) there may be a series of After measurements if there is interest in the long-term effects of treatment.

Single-group studies are so weak logically that they should be avoided whenever possible, but in the case of a compulsory law or change in business practice, a comparable group not subject to the treatment may not be available. In an After Only study we can perhaps judge whether or not the situation after the period of treatment was satisfactory but have no basis for judging to what extent, if any, the treatment was a cause, except perhaps by an opinion derived from a subjective impression as to the situation before exposure. Supplementary observations might of course teach something useful about the operation of a law—e.g., that it was widely disobeyed through ignorance or unpopularity with the public or that it was unworkable as too complex for the administrative staff.

In the single-group Before and After study we at least have estimates of the changes that took place during the period of treatment. The problem is to judge the role of the treatment in producing these changes. For this step it is helpful to list and judge any other contributors to the change that can be envisaged. Campbell and Stanley (1966) have provided a useful list with particular reference to the field of education.

Consider a Before-After rise. This might be due to what I vaguely call "external" causes. In an economic study a Before-After rise might accompany a wide variety of "treatments", good or bad, during a period of increasing national employment and prosperity. In educational examinations contributors might be the increasing maturity of the students or familiarity with the tests. In a study of an apparently low group on some variable (e.g., poor at some task) a rise might be due to what is called the regression effect. If a person's score fluctuates from time to time through human variability or measurement error, the "low" group selected is likely to contain persons who were having an unusually bad day or had a negative error of measurement on that day. In the subsequent After measurement, such persons are likely to show a rise in score even under no treatment—either they are having one of their "up" days or the error of measurement is positive on that day. After World War I the French government instituted a wage bonus for civil servants with large families to stimulate an increase in the birthrate and the population of France. I have been told the primary effect was an influx of men with large families into French civil service jobs, creating a Before-After rise that might be interpreted as a success of the "treatment." An English Before-After evaluation of a publicity campaign to encourage people to come into London clinics for needed protective shots obtained a Before-After drop in number of shots given. The clinics, who were asked to keep the records, had persuaded patrons to come in at once if they were known to be intending to have shots (Before), so that these people would be out of the way when the presumed big rush from the campaign started.

A time-series study with repeated measurements Before and After presents interesting problems—that of appraising whether the Before-After change during the period of treatment is real in relation to changes that occur from external causes in the Before and After periods and that of deciding what is suggested about the time-response curve to the treatment. Campbell and Ross (1968) give an excellent account of the types of analysis and judgment needed in connection with a study of the Connecticut state law imposing a crackdown on speeding, and Campbell (1969) has discussed the role of this and other techniques in a highly interesting paper on program evaluation.

Single-group studies emphasize a characteristic that is prominent in the analysis of nearly all observational studies—the role of judgment. No matter how well-constructed a mathematical model we have, we cannot expect to plan a statistical analysis that will provide an almost automatic verdict. The statistician who intends to operate in this field must cultivate an ability to judge and weigh the relative importance of different factors whose effects cannot be measured at all accurately.

Reverting to types of structure, we come now to those with more than one group. The simplest is a two-group study of treated and untreated groups (seat-belt wearers and nonwearers). We may also have various treatments or forms of treatment, as in the smoking and health studies (pipes, cigars, cigarettes, different amounts smoked, and ex-smokers who had stopped for different lengths of time and had previously smoked different amounts). Both After Only and Before and After measurements are common. Sometimes both an After Only and a Before-After measurement are recommended for each comparison group if there is interest in studying whether the taking of the Before measurement influenced the After measurement.

Comparison groups bring a great increase in analytical insight. The influence of external causes on both groups will be similar in many types of study and will cancel or be minimized when we compare treatment with no treatment. But such studies raise a new problem—How do we ensure that the groups are comparable? Some relevant statistical techniques are outlined in section 6. In regard to incomparability of the groups the Before and After study is less vulnerable than the After Only since we should be able to judge comparability of the treated and untreated groups on the response variable at a time when they have not been subjected to the difference in treatment. Occasionally, we might even be able to select the two groups by randomization, having a randomized experiment instead of an observational study; but this is not feasible when the groups are self-selected (as in smokers) or selected by some administrative fiat or outside agent (e.g., illness).

4. Measurements

The statement of objectives will also have suggested the types of measurements needed; their relevance is obviously important. For instance, early British studies by aerial photographs in World War II were reported to show great damage to German industry. Knowing that early British policy was to bomb the town center and that German factories were often concentrated mainly on the outskirts, Yates (1968) confined his study to the factory areas, with quite a different conclusion which was confirmed when postwar studies could be made. The question of what is considered relevant is particularly important in program evaluation. A program may succeed in its main objectives but have undesirable side effects. The verdict on the program may differ depending on whether or not these side effects are counted in the evaluation.

It is also worth reviewing what is known about the accuracy and precision of proposed measurements. This is especially true in social studies, which often deal with people's attitudes, motivations, opinions, and behavior—factors that are difficult to measure accurately. Since we may have to manage with very imperfect measurements, statisticians need more technical research on the effects of errors of measurement. Three aspects are: (1) more study of the actual distribution of errors of measurement, particularly in multivariate problems, so that we work with realistic models; (2) investigation, from these models, of the effects on the standard types of analysis; (3) study of methods of remedying the situation by different analyses with or without supplementary study of the error distributions. To judge by work to date on the problem of estimating a structural regression, this last problem is formidable.

It is also important to check comparability of measurement in the comparison groups. In a medical study a trained nurse who has worked with one group for years but is a stranger to the other group might elicit different amounts of trustworthy information on sensitive questions. Cancer patients might be better informed about cases of cancer among blood relatives than controls free from cancer.

The scale of the operation may also influence the measuring process. The midtown Manhattan study, for instance, at first planned to use trained psychiatrists for obtaining the key measurements, but they found that only enough psychiatrists could be provided to measure a sample of 100. The analytical aims of the study needed a sample of at least 1,000. In numerous instances the choice seems to lie between doing a study much smaller and narrower in scope than desired but with high quality of measurement, or an extensive study with measurements of dubious quality. I am seldom sure what to advise.

In large studies one occasionally sees a mistake in plans for measurement that is perhaps due to inattention. If two laboratories or judges are needed to measure the responses, an administrator sends all the treatment group to laboratory 1 and the untreated to laboratory 2—it is at least a tidy decision. But any systematic difference between laboratories or judges becomes part of the estimated treatment effect. In such studies there is usually no difficulty in sending half of each group, selected at random, to each judge.

5. Observations and Experiments

In the search for techniques that help to ensure comparability in observational studies, it is worth recalling the techniques used in controlled experiments, where the investigator faces similar problems but has more resources to command. In simple terms these techniques might be described as follows.

Identify the major sources of variation (other than the treatments) that affect the response variable. Conduct the experiment and analysis so that the effects of such sources are removed or balanced out. The two principal devices for this purpose are blocking and the analysis of covariance. Blocking is employed at the planning stage of the experiment. With two treatments, for example, the subjects are first grouped into pairs (blocks of size 2) such that the members of a pair are similar with respect to the major anticipated sources of variation. Covariance is used primarily when the response variable y is quantitative and some of the major extraneous sources of variation can also be represented by quantitative variables $x_1, x_2, \ldots$. From a mathematical model expressing y in terms of the treatment effects and the values of the x_i, estimates of the treatment effects are obtained that have been adjusted to remove the effects of the x_i.

Covariance and blocking may be combined.

For minor and unknown sources of variation, use randomization. Roughly speaking, randomization makes such sources of error equally likely to favor either treatment and ensures that their contribution is included in the standard error of the estimated treatment effect if properly calculated for the plan used.

In general, extraneous sources of variation may influence the estimated treatment effect τ in two ways. They may create a bias B. Instead of estimating the true treatment effect τ, the expected value of $\hat{\tau}$ is $(\tau + B)$, where B is usually unknown. They also increase the variance of $\hat{\tau}$. In experiments a result of randomization and other precautions (e.g., blindness in measurement) is that the investigator usually has little worry about bias. Discussions of the effectiveness of blocking and covariance (e.g., Cox, 1957) are confined to their effect on $V(\hat{\tau})$ and on the power of tests of significance.

In observational studies we cannot use random assignment of subjects, but we can try to use techniques like blocking and covariance. However, in the absence of randomization these techniques have a double task—to remove or reduce bias and to increase precision by decreasing $V(\hat{\tau})$. The reduction of bias should, I think, be regarded as the primary objective—a highly precise estimate of the wrong quantity is not much help.

6. Matching and Adjustments

In observational studies as in experiments we start with a list of the most important extraneous sources of variation that affect the response variable. The Cornell study, based on automobile accidents involving seat-belt wearers and nonwearers, listed 12 major variables. The most important was the intensity and direction of the physical force at impact. A head-on collision at 60 mph is a very different matter from a sideswipe at 25 mph. In the smoking–death-rate studies age gradually becomes a predominating variable for men over 55. In the raw data supplied to the Surgeon General's committee by the British and Canadian studies and in a U.S. study cigarette smokers and nonsmokers had about the same death rates. The high death rates occurred among the cigar and pipe smokers. If these data had been believed, television warnings might now be advising cigar and pipe smokers to switch to cigarettes. However, cigar and pipe smokers in these studies were found to be markedly older than nonsmokers, while cigarette smokers were, on the whole, younger. All studies regarded age as a major extraneous variable in the analysis. After adjustment for age differences, death rates for cigar and pipe smokers were close to those for nonsmokers; those for cigarette smokers were consistently higher.

In observational studies three methods are in common use in an attempt to remove bias due to extraneous variables.

Blocking, usually known as matching in observational studies. Each member of the treated group has a match or partner in the untreated group. If the x variables are classified, we form the cells created by the multiple classification (e.g., x_1 with 3 classes and x_2 with 4 classes create 12 cells). A match means a member of the same cell. If x is quantitative (discrete or continuous), a common method is to turn it into a classified variate (e.g., age in 10-year classes). Another method, caliper matching, is to call x_{11i} (in group 1) and x_{12j} (in group 2) matches with respect to x_1 if $|x_{11i} - x_{12j}| \leq a$.

Standardization (adjustment by subclassification). This is the analogue of covariance when the x's are classified and we do not match. Arrange the data from the treated and untreated samples in cells, the ith cell containing say n_{1i}, n_{2i} observations with response means $\bar{y}_{1i}$, $\bar{y}_{2i}$. If the effect τ of the treatment is the same in every cell, this method depends on the result that for any set of weights w_i with $\Sigma w_i = 1$, the quantity $\hat{\tau} = \Sigma w_i(\bar{y}_{1i} - \bar{y}_{2i})$ is an unbiased estimate of τ (apart from any within-cell biases). The weights can therefore be chosen to minimize $V(\hat{\tau})$. If it is clear that τ varies from cell to cell as often happens, the choice of weights becomes more critical, since it determines the quantity $\Sigma w_i \tau_i$ that is being estimated. In vital statistics a common practice is to take the weights from some standard population to which we wish the comparison to apply.

Covariance (with x's quantitative), used just as in experiments. The idea of matching is easy to grasp, and the statistical analysis is simple. On the operational side, matching requires a large reservoir in at least one group (treated or untreated) in which to look for matches. The hunt for matches (particularly with caliper matching) may be slow and frustrating, although computers should be able to help if data about the x's can be fed into them. Matching is avoided when the planned sample size is large, there are numerous treatments, subjects become available only slowly through time, and it is not feasible to measure the x's until the samples have already been chosen and y is also being measured.

There has been relatively little study of the effects of these devices on bias and precision, although particular aspects have been discussed by Billewicz (1965), Cochran (1968) and Rubin (1970). If x is classified and two members of the same class are identical in regard to the effect of x on y, matching and standardization remove all the bias, while matching should be somewhat superior in regard to precision. I am not sure, however, how often such ideal classifications actually exist. Many classified variables, especially ordered classifications, have an underlying quantitative x—e.g., for sex with certain types of response there is a whole gradation from very manly men to very womanly women. This is obviously true for quantitative x's that are deliberately made classified in order to use within-cell matching. In such cases, matching and standardization remove between-cell bias but not within-cell bias. Of an initial bias in means $\mu_{1x} - \mu_{2x}$ they remove about 64%, 80%, 87%, 91%, and 93% with 2, 3, 4, 5, and 6 classes, the actual amount varying a little with the choice of class boundaries and the nature of the x distribution (Cochran, 1968). Caliper matching removes about 76%, 84%, 90%, 95%, and 99% with $a/\sigma_x = 1, 0.8, 0.6, 0.4$, and 0.2. These percentages also apply to y under a linear or nearly linear regression of y on x.

With a quantitative x, covariance adjustments remove all the initial bias if the correct model is fitted, and they are superior to within-class matching of x when this assumption holds. In practice, covariance nearly always means linear covariance to most users, and some bias remains after covariance adjustment if the y, x relation is nonlinear and a linear covariance is fitted. If nonlinearity is of the type that can be approximated by a quadratic curve, results by Rubin (1970) suggest that the residual bias should be small if $\sigma_{1x}^2 = \sigma_{2x}^2$ and x is symmetrical or nearly so in distribution. When $\sigma_{1x}^2/\sigma_{2x}^2$ is 1/2 or 2, the adjustment can either overcorrect or undercorrect to a material extent.

Caliper matching, on the other hand, and even within-class matching do not lean on an assumed linear relation between y and x. If $\sigma_{1x}^2/\sigma_{2x}^2$ is near 1 (perhaps between 0.8 and 1.2), the evidence to date suggests, however, that linear covariance is superior to within-class matching in removing bias under a moderately curved y, x relation, although more study of this point is needed. Linear covariance applied to even loosely caliper-matched samples should remove nearly all the initial bias in this situation. Billewicz (1965) compared linear covariance and within-class matching (3 or 4 classes) in regard to precision in a model in which x was distributed as $N(0, 1)$ in both populations. For the curved relations $y = 0.4x - 0.1x^2$, $y = 0.8x - 0.14x^2$, and $y = \tanh x$ he found covariance superior in precision on samples of size 40.

Larger studies in which matching becomes impractical present difficult problems in analysis. Protection against bias from numerous x variables is not easy. Further, if there are say four x variables, the treatment effect may change with the levels of x_2 and x_3. For applications of the conclusions it may be important to find this out. The obvious recourse is to model construction and analysis based on the model, which has been greatly developed, particularly in regression. Nevertheless the Coleman report on education (1966) and the national halothane study (Bunker et al., 1969) illustrate difficulties that remain.

7. Further Points on Planning

Sample Size

Statisticians have developed formulas that provide guidance on the sample size needed in a study. The formulas tend to be harder to use in observational studies than in experiments because less may be known about the likely values of population parameters that appear in the formulas and the formulas assume that bias is negligible. Nevertheless there is frequently something useful to be learned—for instance, that the proposed size looks adequate for estimating a single overall effect of the treatment, but does not if the variation in effect with an x is of major interest.

Nonresponse

Certain administrative bodies may refuse to cooperate in a study; certain people may be unwilling or unable to answer the questions asked or may not be found at home. In modern studies, standards with regard to the nonresponse problem seem to me to be lax. In both the smoking and Coleman studies nonresponse rates of over 30% were common. The main difficulty with nonresponse is not the reduction in sample size but that nonrespondents may be to some extent different types of people from respondents and give different types of answers, so that results from respondents are biased in this sense. Fortunately, nonresponse can often be reduced materially by hard work during the study, but definite plans for this need to be made in advance.

Pilot Study

The case for starting with a small pilot study should be considered—for instance, to work out the field procedures and check the understanding and acceptability of the questions and the interviewing methods and time taken. When information is wanted on a new problem, the cheapest and quickest method is to base a study on routine records that already exist. However, such records are often incomplete and have numerous gross errors. A law or administrative rule specifying that records shall be kept does not ensure that the records are usable for research purposes. A good pilot study of the records should reveal the state of affairs. It is worth looking at variances; a suspiciously low variance has sometimes led to detection of the practice of copying previous values instead of making an independent determination.

Critique

When the draft of plans for a study is prepared, it helps to find a colleague willing to play the role of devil's advocate— to read the plan and to point out any methodological weaknesses that he sees. Since observational studies are vulnerable to such defects, the investigator should of course also be doing this, but it is easy to get in a rut and overlook some aspect. It helps even more if the colleague can suggest ways of removing or reducing these faults.

In the end, however, the best plan that investigator and colleague can devise may still be subject to known weaknesses. In the report of the results these should be discussed in a clearly labeled section, with the investigator's judgment about their impact.

Sampled and Target Populations

Ideally, the statistician would recommend that a study start with a probability sample of the target population about which the investigator wishes to obtain information. But both in experiments and in observational surveys many factors—feasibility, costs, geography, supply of subjects, opportunity—influence the choice of samples. The population actually sampled may therefore differ in several respects from the target population. In his report the investigator should try to describe the sampled population and relevant target populations and give his opinion as to how any differences might affect the results, although this is admittedly difficult.

One reason why this step is useful is that an administrator in California, say, may want to see the results of a good study on some social issue for policy guidance and may find that the only relevant study was done in Philadelphia or Sweden. He will appreciate help in judging whether to expect the same results in California.

8. Judgment about Causality

Techniques of statistical analysis of observational studies have in general employed standard methods and will not be discussed here. When the analysis is completed, there remains the problem of reaching a judgment about causality. On this point I have little to add to a previous discussion (Cochran, 1965). It is well known that evidence of a relationship between x and y is no proof that x causes y. The scientific philosophers to whom we might turn for expert guidance on this tricky issue are a disappointment. Almost unanimously and with evident delight they throw the idea of cause and effect overboard. As the statistical study of relationships has become more sophisticated, the statistician might admit, however, that his point of view is not very different, even if he wishes to retain the terms cause and effect.

The probabilistic approach enables us to discard oversimplified deterministic notions that make the idea look ridiculous. We can conceive of a response y having numerous contributory causes, not just one. To say that x is a cause of y does not imply that x is the only cause. With 0, 1 variables we may merely mean that if x is present, the probability that y happens is increased—but not necessarily by much. If x and y are continuous, a causal relation may imply that as x increases, the average value of y increases, or some other feature of its distribution changes. The relation may be affected by the levels of other variables; it may be strengthened or weakened or entirely disappear, depending on these levels. One can see why the idea becomes tortuous. For successful prediction, however, a knowledge of the nature and stability of these relationships is an essential step and this is something that we can try to learn in observational studies.

A claim of proof of cause and effect must carry with it an explanation of the mechanism by which the effect is produced. Except in cases where the mechanism is obvious and undisputed, this may require a completely different type of research from the observational study that is being summarized. Thus in most cases the study ends with an opinion or judgment about causality, not a claim of proof.

Given a specific causal hypothesis that is under investigation, the investigator should think of as many consequences of the hypothesis as he can and in the study try to include response measurements that will verify whether these consequences follow. The cigarette-smoking and death-rate studies are a good example. For causes of death to which smoking is thought to be a leading contributor, we can compare death rates for nonsmokers and for smokers of different amounts, for ex-smokers who have stopped for different lengths of time but used to smoke the same amount, for ex-smokers who have stopped for the same length of time but used to smoke different amounts, and (in later studies) for smokers of filter and nonfilter cigarettes. We can do this separately for men and women and also for causes of death to which, for physiological reasons, smoking should not be a contributor. In each comparison the direction of the difference in death rates and a very rough guess at the relative size can be made from a causal hypothesis and can be put to the test.

The same can be done for any alternative hypotheses that occur to the investigator. It might be possible to include in the study response measurements or supplementary observations for which alternative hypotheses give different predictions. In this way, ingenuity and hard work can produce further relevant data to assist the final judgment. The final report should contain a discussion of the status of the evidence about these alternatives as well as about the main hypothesis under study.

In conclusion, observational studies are an interesting and challenging field which demands a good deal of humility, since we can claim only to be groping toward the truth.

References

Buck, A. A. et al. 1968. Coca chewing and health. *Am. J. Epidemiol.* 88: 159–77.

Bunker, J. P. et al., eds., 1969. The national halothane study. Washington, D.C.: USGPO.

Billewicz, W. Z. 1965. The efficiency of matched samples. *Biometrics* 21: 623–44.

Campbell, D. T. 1969. Reforms as experiments. *Am. Psychologist* 24: 409–29.

Campbell, D. T., and H. L. Ross. 1968. The Connecticut crackdown on speeding: Time series data in quasi-experimental analysis. *Law and Society Rev.* 3: 33–53.

Campbell, D. T., and J. C. Stanley. 1966. *Experimental and quasi-experimental designs in research.* Chicago: Rand McNally.

Cochran, W. G. 1965. The planning of observational studies. *J. Roy. Statist. Soc.* Ser. A, 128: 234–66.

———. 1968. The effectiveness of adjustment by classification in removing bias in observational studies. *Biometrics* 24: 295–314.

Coleman, J. S. 1966. Equality of educational opportunity. Washington, D.C.: USGPO.

Cox, D. R. 1957. The use of a concomitant variable in selecting an experimental design. *Biometrika* 44: 150–58.

Kinsey, A. C., W. B. Pomeroy, and C. E. Martin. 1948. Sexual behavior in the human male. Philadelphia: Saunders.

Rubin, D. B. 1970. The use of matched sampling and regression adjustment in observational studies. Ph.D. thesis, Harvard Univ., Cambridge.

Srole, L., T. S. Langner, S. T. Michael, M. K. Opler, and T. A. C. Rennie. 1962. Mental health in the metropolis. (The midtown Manhattan study.) New York: McGraw-Hill.

U.S. Surgeon-General's committee. 1964. Smoking and health. Washington, D.C.: USGPO.

Yates, F. 1968. Theory and practice in statistics. *J. Roy. Statist. Soc.* Ser. A, 131: 463–77.

7

Construction of Classes of Experimental Designs Using Transversals in Latin Squares and Hedayat's Sum Composition Method

WALTER T. FEDERER

1. Introduction

A NEW METHOD of constructing a pair of mutually orthogonal Latin squares of order n has been discovered by A. Hedayat (see Keller, 1969; Federer et al., 1969; Hedayat and Seiden, 1970) and is called the *sum composition method*, since it makes use of pairs of mutually orthogonal Latin squares of orders n_1 and n_2, such that $n_1 + n_2 = n$ to produce the larger pair of mutually orthogonal Latin squares of order n. The method makes use of parallel and common-parallel transversals in Latin squares of order n_1, for $n_1 > n_2$, and of the projection of the transversals into the last n_2 rows and n_2 columns of a square of side n.

For the sum composition method the k parallel transversals may be selected in many ways, but we shall restrict selection to be among the k parallel transversals which when projected produce a balanced incomplete block design arrangement. Within this restriction it is shown how to construct plans for balanced incomplete block designs, Youden square designs, partially balanced Latin rectangle designs, $T:TT$ designs, $P:PP$ designs, supplemented balanced and partially balanced designs of the $O:SS$ and $O:S_pS_p$ types, augmented designs of various types, and several types of designs for successive (or simultaneous) experiments on the same set of experimental material. Illustrative examples of designs in each of the classes are presented.

WALTER T. FEDERER is Professor of Biological Statistics, Cornell University, Ithaca, New York.

The statistical analysis for most of the designs is relatively simple computationally. The analyses are not presented in this paper, nor is the randomization procedure for the various classes. However, for a single experiment a permutation of the rows, columns, and treatments in any given plan should suffice for most purposes.

2. Definition and Notation

In the following the symbol $O(n, t)$ denotes a set of t mutually orthogonal Latin squares of order n. A *transversal*, or *directrix*, of a Latin square $L(n)$ of order n on an n set Σ is a collection of n cells such that the entries of these cells exhaust the set Σ and every row and every column of $L(n)$ is represented. Two transversals in $L(n)$ are said to be *parallel* if they have no cell in common. A collection of n cells is said to form a *common transversal* for an $O(n, t)$ set if the collection is a transversal for each of the t Latin squares. Similarly, two common transversals are said to be *common-parallel transversals* if they have no cell of the squares in common. In the following $O(4, 2)$ set (Table 7.1) two common-parallel transversals are presented. One of the common transversals is indicated by underlining the elements in the two squares, and the other common transversal is indicated by the parentheses around the elements.

Table 7.1

1	2	(3)	<u>4</u>		1	2	(3)	<u>4</u>
(2)	<u>1</u>	4	3		(4)	<u>3</u>	2	1
<u>3</u>	(4)	1	2		2	(1)	4	3
4	3	<u>2</u>	(1)		3	4	<u>1</u>	(2)

Not all Latin squares of order n have a transversal, and still fewer have n transversals. For any given Latin square of order n it may be extremely difficult to determine whether or not one or more transversals exist, even though a relatively simple procedure has been developed by Hedayat and Federer (1970) for constructing a Latin square with a transversal. However, if an $O(n, 2)$ set of Latin squares is available, it is a simple matter to find the n transversals in each of the two squares. One superimposes one square on the other and notes the n positions for a single element of one of the two squares. These n positions form a transversal for the second square. Continuing this process for all n elements in the first square, the n transversals of the second square are obtained. The n transversals of the first square may be obtained similarly. This method of finding a square with n transversals is available for all n except $n = 2$ and 6. $O(n, 2)$ sets for all odd n are obtained by a cyclical permutation of the elements in the set (see Hedayat and Federer, 1969). A new method of constructing an $O(n, 2)$ set, Hedayat's *sum composition method* is especially useful for even n and for $n = 4q + 2$ where $q = 2, 3, \ldots$. The group construction method (see Federer et al., 1969) is useful for constructing $O(n, 2)$ sets when n is not of the form $4q + 2$. Any $O(n, 2)$ set suffices to determine the n parallel transversals in the two squares.

If an $O(n, 3)$ set exists, it is a relatively simple matter to find the n common-parallel transversals for any pair of the Latin squares in the set by using the third square. One of the Latin squares is superimposed upon each of the other two. The n positions of a specified element in the first square determine a common transversal in the other two. Continuing this process for the remaining $n - 1$ elements of the first square, the n common-parallel transversals of the other two are obtained. To illustrate, consider the $O(4, 3)$ set, Table 7.2.

Table 7.2

$L_1(4)$			
1^*	$\underline{2}$	3^+	4^-
$\underline{2}$	1^*	4^-	3^+
3^+	4^-	1^*	$\underline{2}$
4^-	3^+	$\underline{2}$	1^*

$L_2(4)$			
1^*	$\underline{2}$	3^+	4^-
$\underline{4}$	3^*	2^-	1^+
2^+	1^-	4^*	$\underline{3}$
3^-	4^+	$\underline{1}$	2^*

$L_3(4)$			
1^*	$\underline{2}$	3^+	4^-
$\underline{3}$	4^*	1^-	2^+
4^+	3^-	2^*	$\underline{1}$
2^-	1^+	$\underline{4}$	3^*

The four common-parallel transversals in $L_2(4)$ and $L_3(4)$ are those indicated with the symbols $*$, $^+$, $^-$, $_-$, which correspond respectively to elements 1, 3, 4, and 2 in $L_1(4)$.

An $O(n, t)$ set suffices to obtain the n common-parallel transversals in $t - 1$ of the Latin squares of order n. The procedure for doing this is a straightforward extension of the procedure described for an $O(n, 3)$ set.

3. The Sum Composition Method of Constructing $O(n, 2)$ Sets

The method of sum composition of $O(n, 2)$ sets is described in detail by Hedayat and Seiden (1970). An example will suffice for the purposes of this paper. Among other results, Hedayat and Seiden (1970) proved that a sufficient condition for the construction of an $O(n, 2)$ set by the method of sum composition is that an $O(n_1, 2)$ set and an $O(n_2, 2)$ set be available, that $n_1 + n_2 = n$, that $n_1 \geq 2n_2$, that the $O(n_1, 2)$ set contains at least $2n_2$ common-parallel transversals, and that either (1) $7 \leq n_1 = p^\alpha$, p an odd prime, or $8 \leq n_1 = 2^\alpha$, α a positive integer, and $n_2 = (n_1 - 1)/2$ or $n_2 = n_1/2$, respectively; (2) $7 \leq n_1 = p^\alpha$, p a prime having any of the following forms: $3m + 1$, $8m + 1$, $8m + 3$, $24m + 11$, $60m + 23$, or $60m + 47$ and $n_2 = 3$; (3) $n_1 = p^\alpha$, p a prime of the form $8m + 1$, or $8m + 3$, $m \neq 0$ and $n_2 = 4$.

Let $n_1 = 7$ and $n_2 = 3$, and let the $O(7, 2)$ and $O(3, 2)$ sets be as in Table 7.3.

Table 7.3

$C_1(10)$

0	1	2*	3	4+	5	6			
2	3*	4	5+	6	0	1			
4*	5	6+	0	1	2	3			
6	0+	1	2	3	4	5*			
1+	2	3	4	5	6*	0			
3	4	5	6	0*	1	2+			
5	6	0	1*	2	3+	4			
							7	8	9
							8	9	7
							9	7	8

$C_2(10)$

0	1	2	(3	4	5)	6⁻			
4	5	(6	0	1)	2⁻	3			
1	(2	3	4)	5⁻	6	0			
(5	6	0)	1⁻	2	3	4			
2	3)	4⁻	5	6	0	(1			
6)	0⁻	1	2	3	(4	5			
3⁻	4	5	6	(0	1	2)			
							7	8	9
							9	7	8
							8	9	7

To demonstrate that in Table 7.3 the marked left diagonals are common-parallel transversals, we need to consider the $O(7, 3)$ set of which the two squares in Table 7.3 and the square in Table 7.4 are members.

Table 7.4

$$L_1(7) = \begin{array}{|c|c|c|c|c|c|c|}
\hline
0 & 1 & 2 & 3 & 4 & 5 & 6 \\
\hline
1 & 2 & 3 & 4 & 5 & 6 & 0 \\
\hline
2 & 3 & 4 & 5 & 6 & 0 & 1 \\
\hline
3 & 4 & 5 & 6 & 0 & 1 & 2 \\
\hline
4 & 5 & 6 & 0 & 1 & 2 & 3 \\
\hline
5 & 6 & 0 & 1 & 2 & 3 & 4 \\
\hline
6 & 0 & 1 & 2 & 3 & 4 & 5 \\
\hline
\end{array}$$

The left diagonal in $C_1(10)$ containing the elements with an underline corresponds to the left diagonal containing the element 1 in $L_1(7)$; the left diagonal in $C_1(10)$ marked with an asterisk corresponds to the left diagonal containing the element 2 in $L_1(7)$; the left diagonal in $C_1(10)$ marked with a plus sign superscript corresponds to the left diagonal containing the element 4 in $L_1(7)$; the left diagonal in $C_2(10)$ marked with a left parenthesis corresponds to the left diagonal containing the element 3 in $L_1(7)$; the left diagonals marked with a right parenthesis and with a minus superscript in $C_2(10)$ correspond to those in $L_1(7)$ containing elements 5 and 6 respectively. The remaining common transversal is unmarked, and it corresponds to the diagonal in $L_1(7)$ containing the element 0.

The next step after selecting the $n_2 = 3$ parallel transversals in each of the two Latin squares of order $n_1 = 7$ in $C_1(n_1 + n_2) = C_1(10)$ and $C_2(n_1 + n_2) = C_2(10)$ is to project the elements from each selected transversal into an empty row and into an empty column of both $C_1(n_1 + n_2)$ and $C_2(n_1 + n_2)$. For $n_1 + n_2 = 10$, Table 7.5 results.

The last step in constructing an $O(10, 2)$ set is to insert one element from the Latin square of order $n_2 = 3$ in each of the $n_2 = 3$ transversals to obtain the $O(10, 2)$ set shown in Table 7.6.

Latin squares of order $n_1 = 11$ and $n_2 = 5$ may be used to construct an $O(16, 2)$ set by the method of sum composition. Let the Latin square of order 11 in Table 7.7 be used to determine the common-parallel transversals in the construction of $C_1(16)$ and $C_2(16)$ (Table 7.8) in the same manner as for $n_1 = 7$.

CONSTRUCTION OF CLASSES OF EXPERIMENTAL DESIGNS

Table 7.5

$C_3(10)$

0			3		5	6	<u>1</u>	2*	4⁺
		4		6	0	1	<u>2</u>	3*	5⁺
	5		0	1	2		<u>3</u>	4*	6⁺
6		1	2	3			<u>4</u>	5*	0⁺
	2	3	4			0	<u>5</u>	6*	1⁺
3	4	5			1		<u>6</u>	0*	2⁺
5	6			2		4	<u>0</u>	1*	3⁺
<u>2</u>	<u>1</u>	<u>0</u>	<u>6</u>	<u>5</u>	<u>4</u>	<u>3</u>	7	8	9
4*	3*	2*	1*	0*	6*	5*	8	9	7
1⁺	0⁺	6⁺	5⁺	4⁺	3⁺	2⁺	9	7	8

$C_4(10)$

0	1	2		4			6⁻	5)	(3
4	5		0			3	2⁻	1)	(6
1		3			6	0	5⁻	4)	(2
	6			2	3	4	1⁻	0)	(5
2			5	6	0		4⁻	3)	(1
		1	2	3		5	0⁻	6)	(4
	4	5	6		1		3⁻	2)	(0
3⁻	0⁻	4⁻	1⁻	5⁻	2⁻	6⁻	7	8	9
6)	3)	0)	4)	1)	5)	2)	9	7	8
(5	(2	(6	(3	(0	(4	(1	8	9	7

The right diagonals from $C_1(16)$ to be projected into the last $n_2 = 5$ rows and $n_2 = 5$ columns are those corresponding to the right diagonals in $L_1(11)$ containing the elements of 0, 4, 7, 9, and 10. Here, as in the cases of $C_1(10)$ and $C_2(10)$, diagonals were selected such that the projected elements into rows would form a balanced incomplete block design arrangement within columns, and the projected elements into columns would form a balanced incomplete block design arrangement within rows.

Table 7.6

$L_1(10)$

0	7	8	3	9	5	6	1	2	4
7	8	4	9	6	0	1	2	3	5
8	5	9	0	1	2	7	3	4	6
6	9	1	2	3	7	8	4	5	0
9	2	3	4	7	8	0	5	6	1
3	4	5	7	8	1	9	6	0	2
5	6	7	8	2	9	4	0	1	3
2	1	0	6	5	4	3	7	8	9
4	3	2	1	0	6	5	8	9	7
1	0	6	5	4	3	2	9	7	8

$L_2(10)$

0	1	2	7	4	8	9	6	5	3
4	5	7	0	8	9	3	2	1	6
1	7	3	8	9	6	0	5	4	2
7	6	8	9	2	3	4	1	0	5
2	8	9	5	6	0	7	4	3	1
8	9	1	2	3	7	5	0	6	4
9	4	5	6	7	1	8	3	2	0
3	0	4	1	5	2	6	7	8	9
6	3	0	4	1	5	2	9	7	8
5	2	6	3	0	4	1	8	9	7

Table 7.7

$L_1(11) =$

0	1	2	3	4	5	6	7	8	9	10
10	0	1	2	3	4	5	6	7	8	9
9	10	0	1	2	3	4	5	6	7	8
8	9	10	0	1	2	3	4	5	6	7
7	8	9	10	0	1	2	3	4	5	6
6	7	8	9	10	0	1	2	3	4	5
5	6	7	8	9	10	0	1	2	3	4
4	5	6	7	8	9	10	0	1	2	3
3	4	5	6	7	8	9	10	0	1	2
2	3	4	5	6	7	8	9	10	0	1
1	2	3	4	5	6	7	8	9	10	0

In selecting the particular transversals giving this arrangement, one need only check to ascertain that *one* element occurs equally frequent with all other elements. The nature of projection of transversals makes it unnecessary to check frequency of occurrence for all pairs of elements.

Also, from any balanced incomplete block design plan for $v = n_1 = b$, $k = r$, and $\lambda = k(k - 1)/(n_1 - 1)$ the elements in any single block may be used to denote the transversals to be projected; then, from a *different*

Table 7.8

$C_1(16)$

0	1	2	3	4	5	6	7	8	9	10					
1	2	3	4	5	6	7	8	9	10	0					
2	3	4	5	6	7	8	9	10	0	1					
3	4	5	6	7	8	9	10	0	1	2					
4	5	6	7	8	9	10	0	1	2	3					
5	6	7	8	9	10	0	1	2	3	4					
6	7	8	9	10	0	1	2	3	4	5					
7	8	9	10	0	1	2	3	4	5	6					
8	9	10	0	1	2	3	4	5	6	7					
9	10	0	1	2	3	4	5	6	7	8					
10	0	1	2	3	4	5	6	7	8	9					
											11	12	13	14	15
											12	13	14	15	11
											13	14	15	11	12
											14	15	11	12	13
											15	11	12	13	14

$C_2(16)$

0	1	2	3	4	5	6	7	8	9	10					
2	3	4	5	6	7	8	9	10	0	1					
4	5	6	7	8	9	10	0	1	2	3					
6	7	8	9	10	0	1	2	3	4	5					
8	9	10	0	1	2	3	4	5	6	7					
10	0	1	2	3	4	5	6	7	8	9					
1	2	3	4	5	6	7	8	9	10	0					
3	4	5	6	7	8	9	10	0	1	2					
5	6	7	8	9	10	0	1	2	3	4					
7	8	9	10	0	1	2	3	4	5	6					
9	10	0	1	2	3	4	5	6	7	8					
											11	12	13	14	15
											13	14	15	11	12
											15	11	12	13	14
											12	13	14	15	11
											14	15	11	12	13

Table 7.9

$C_3(16)$

1	2	3		5	6		8				0	4	7	9	10
	3	4	5		7	8		10			2	6	9	0	1
		5	6	7		9	10		1		4	8	0	2	3
3			7	8	9		0	1			6	10	2	4	5
	5			9	10	0		2	3		8	1	4	6	7
5		7			0	1	2			4	10	3	6	8	9
6	7		9			2	3	4			1	5	8	10	0
	8	9		0			4	5	6		3	7	10	1	2
8		10	0		2			6	7		5	9	1	3	4
9	10		1	2		4				8	7	0	3	5	6
10	0	1		3	4		6				9	2	5	7	8
0	2	4	6	8	10	1	3	5	7	9	11	12	13	14	15
7	9	0	2	4	6	8	10	1	3	5	12	13	14	15	11
4	6	8	10	1	3	5	7	9	0	2	13	14	15	11	12
2	4	6	8	10	1	3	5	7	9	0	14	15	11	12	13
1	3	5	7	9	0	2	4	6	8	10	15	11	12	13	14

$C_4(16)$

0				4		6	7		9	10	8	5	3	2	1
2	3				7		9	10		1	0	8	6	5	4
4	5	6			10		1	2			3	0	9	8	7
	7	8	9			2		4	5		6	3	1	0	10
8		10	0	1			5		7		9	6	4	3	2
10	0		2	3	4			8			1	9	7	6	5
	2	3		5	6	7				0	4	1	10	9	8
3		5	6		8	9	10				7	4	2	1	0
	6		8	9		0	1	2			10	7	5	4	3
		9		0	1		3	4	5		2	10	8	7	6
			1		3	4		6	7	8	5	2	0	10	9
6	9	1	4	7	10	2	5	8	0	3	11	12	13	14	15
1	4	7	10	2	5	8	0	3	6	9	13	14	15	11	12
9	1	4	7	10	2	5	8	0	3	6	15	11	12	13	14
5	8	0	3	6	9	1	4	7	10	2	12	13	14	15	11
7	10	2	5	8	0	3	6	9	1	4	14	15	11	12	13

Table 7.10

$L_1(16)$

11	1	2	3	12	5	6	13	8	14	15	0	4	7	9	10
15	11	3	4	5	12	7	8	13	10	14	2	6	9	0	1
14	15	11	5	6	7	12	9	10	13	1	4	8	0	2	3
3	14	15	11	7	8	9	12	0	1	13	6	10	2	4	5
13	5	14	15	11	9	10	0	12	2	3	8	1	4	6	7
5	13	7	14	15	11	0	1	2	12	4	10	3	6	8	9
6	7	13	9	14	15	11	2	3	4	12	1	5	8	10	0
12	8	9	13	0	14	15	11	4	5	6	3	7	10	1	2
8	12	10	0	13	2	14	15	11	6	7	5	9	1	3	4
9	10	12	1	2	13	4	14	15	11	8	7	0	3	5	6
10	0	1	12	3	4	13	6	14	15	11	9	2	5	7	8
0	2	4	6	8	10	1	3	5	7	9	11	12	13	14	15
7	9	0	2	4	6	8	10	1	3	5	12	13	14	15	11
4	6	8	10	1	3	5	7	9	0	2	13	14	15	11	12
2	4	6	8	10	1	3	5	7	9	0	14	15	11	12	13
1	3	5	7	9	0	2	4	6	8	10	15	11	12	13	14

$L_2(16)$

0	11	12	13	4	14	6	7	15	9	10	8	5	3	2	1
2	3	11	12	13	7	14	9	10	15	1	0	8	6	5	4
4	5	6	11	12	13	10	14	1	2	15	3	0	9	8	7
15	7	8	9	11	12	13	2	14	4	5	6	3	1	0	10
8	15	10	0	1	11	12	13	5	14	7	9	6	4	3	2
10	0	15	2	3	4	11	12	13	8	14	1	9	7	6	5
14	2	3	15	5	6	7	11	12	13	0	4	1	10	9	8
3	14	5	6	15	8	9	10	11	12	13	7	4	2	1	0
13	6	14	8	9	15	0	1	2	11	12	10	7	5	4	3
12	13	9	14	0	1	15	3	4	5	11	2	10	8	7	6
11	12	13	1	14	3	4	15	6	7	8	5	2	0	10	9
6	9	1	4	7	10	2	5	8	0	3	11	12	13	14	15
1	4	7	10	2	5	8	0	3	6	9	13	14	15	11	12
9	1	4	7	10	2	5	8	0	3	6	15	11	12	13	14
5	8	0	3	6	9	1	4	7	10	2	12	13	14	15	11
7	10	2	5	8	0	3	6	9	1	4	14	15	11	12	13

balanced incomplete block design plan the elements of any block containing k of the remaining $n_1 - k$ symbols may be used to denote the transversals to be projected in the second square. Not all balanced incomplete block design plans can be utilized to determine the transversals to be projected in the second square. A permutation of the transversals to be projected and the columns (or rows) into which they are projected may be necessary to obtain an $O(n, 2)$ set and a balanced incomplete block design arrangement in the last k rows and columns.

The parallel transversals selected for projection in $C_2(16)$ are those corresponding to right diagonals in $L_1(11)$ containing the elements 1, 2, 3, 5, and 8. The resulting $C_3(16)$ and $C_4(16)$ plans are given in Table 7.9.

Inserting an element from the Latin square of order $n_2 = 5$ into a transversal which was projected produces a Latin square of order 16. The two squares resulting in an $O(16, 2)$ set are shown in Table 7.10.

This discussion will not be carried further here since the tables shown are sufficient to illustrate a number of types of experimental design plans that may be constructed using the procedures and concepts given.

4. Youden Square and Balanced Incomplete Block Design Plans Constructed from Common-Parallel Transversals in an $O(n_1 + n_2, 2)$ Set

In a balanced incomplete block design with v treatments (or elements) replicated r times in the $v = n$ blocks of size $r = k$, every pair of treatments occurs together $\lambda = r(k - 1)/(v - 1) = k(k - 1)/(n - 1)$ times, where λ is an integer. A Youden square design has balanced incomplete block design properties in columns and has randomized complete block properties in rows. (The randomized complete block property in rows means the treatment parameters are orthogonal to the row parameters.) If T denotes the (totally) balanced property of one set of parameters within a second set; if O denotes orthogonality of two sets of parameters; and if rows, columns, and treatments are always referred to in this order, then a Latin square design is of the $O:OO$ type, and a Youden square design is of an $O:OT$ type. In the latter type, the O to the left of the colon denotes the fact that columns are orthogonal to rows; the OT designation to the right of the colon denotes the fact that treatments are orthogonal to rows and totally balanced with respect to columns. This symbolism was introduced and used to designate various types of designs for successive experiments by Hoblyn, Pearce, and Freeman (1954), by Pearce (1960, 1963), by Clarke (1963), and by Freeman (1964).

In the second step of the method of sum composition of a pair of orthogonal Latin squares of side $n = n_1 + n_2$, k transversals were projected into the last k rows and the last k columns of $C_1(n_1 + n_2)$ and $C_2(n_1 + n_2)$ to form $C_3(n_1 + n_2)$ and $C_4(n_1 + n_2)$ plans. When $\lambda = k(k - 1)/(n - 1) =$ an integer, a Youden square plan results from an appropriate selection of the k transversals. The Youden square plans in Table 7.11 were obtained from the projected sets in $C_3(10)$ and $C_4(10)$ (Table 7.5).

Table 7.11

From rows 8, 9, and 10 of $C_3(10)$

2	1	0	6	5	4	3
4	3	2	1	0	6	5
1	0	6	5	4	3	2

From columns 8, 9, and 10 of $C_3(10)$

1	2	3	4	5	6	0
2	3	4	5	6	0	1
4	5	6	0	1	2	3

From rows 8, 9, and 10 of $C_4(10)$

3	0	4	1	5	2	6
6	3	0	4	1	5	2
5	2	6	3	0	4	1

From columns 8, 9, and 10 of $C_4(10)$

6	2	5	1	4	0	3
5	1	4	0	3	6	2
3	6	2	5	1	4	0

In each of the four plans of Table 7.11 the elements (or treatments) are in a balanced incomplete block design arrangement within columns with $\lambda = 1$ and $k = r = 3$. The row and element parameters are orthogonal to each other.

If the remaining four transversals in $C_3(10)$ and $C_4(10)$ are projected into four rows and four columns, four more Youden square plans result. The parameters of these plans are $v = n_1 = 7$, $k = r = 4$, and $\lambda = k(k - 1)/(n - 1) = 2$.

Let us now turn to plans $C_3(16)$ and $C_4(16)$ (Table 7.9). The last five rows and first eleven columns and the first eleven rows and the last five columns form four different Youden square plans with the parameters $v = n = 11$, $r = k = 5$, and $\lambda = k(k - 1)/(n - 1) = 2$. Projection of the remaining six transversals in the first eleven rows and eleven columns of $C_3(16)$ and $C_4(16)$ into rows and columns produces a quartet of Youden square plans with the parameters $v = b = 11$, $k = r = 6$, and $\lambda = 3$.

It should be noted here that not all $O(n, 2)$ sets constructed by the sum composition method and the projection of $k = n_2$ transversals into the last k rows and the last k columns of $C_1(n_1 + n_2)$ and $C_2(n_1 + n_2)$ produce a Youden square plan. It is necessary to find k transversals which produce blocks for which *one* of the elements occurs λ times with every other element. It is not necessary to check occurrences for all pairs of elements because of the nature of parallel transversals. Then, from among the remaining $n - k = n_1$ transversals it is necessary to find k different transversals such that the projected transversals into the last k rows and last k columns produce a Youden square plan.

Table 7.12 is a list of values of $n_1 = v \le 50$ and $k = r < n_1 - 1$ for which $\lambda = k(k - 1)/(n_1 - 1)$ is an integer, where v is the number of treatments or elements in the set and k is block size.

Table 7.12

$v = n_1$	$k = r$	λ	$v = n_1$	$k = r$	λ	$v = n_1$	$k = r$	λ
7	3	1	27	13	6	39	19	9
7	4	2	27	14	7	39	20	10
11	5	2	29	8	2	41	16	6
11	6	3	29	21	15	41	25	15
13	4	1	31	6	1	43	7	1
13	9	6	31	25	20	43	36	30
15	7	3	31	10	3	43	15	5
15	8	4	31	21	14	43	28	18
16	6	2	31	15	7	43	21	10
16	10	6	31	16	8	43	22	11
19	9	4	34	12	4	45	12	3
19	10	5	34	22	14	45	33	24
21	5	1	35	17	8	46	10	2
21	16	12	35	18	9	46	36	28
23	11	5	36	15	6	47	23	11
23	12	6	36	21	12	47	24	12
25	9	3	37	9	2	49	16	5
25	16	10	37	28	21	49	33	22

Youden square plans may also be formed by deleting one row (or column) from any Latin square plan. This procedure does not involve the use of projection of parallel transversals. These plans are not listed in Table 7.12.

Table 7.12 has some interesting features in connection with the sum composition method. So far, this method has not produced an $O(n, 3)$ set. The numbers $n_1 = 31$ and 43 appear to be interesting candidates to produce an $O(31 + 5, 3)$ set and an $O(43 + 7, 3)$ set. When projected, the parallel transversals should form rows of different Youden squares. That is, to produce an $O(36, 3)$ set, take five transversals of the six which produce a 6-row × 31-column Youden square, five from the ten which form a 10-row × 31-column Youden square, and five from the fifteen which form a 15-row × 31-column Youden square. Perhaps an $O(36, 5)$ set could be formed in this manner. Also, it is interesting to note that whenever n_1 is of the form $4q + 3$, $q = 1, 2, \ldots,$ Youden square plans are formed for $k = (n_1 - 1)/2$ and $k = (n_1 + 1)/2$. The corresponding λ values are $\lambda = (n_1 - 3)/4 = (k - 1)/2$ and $\lambda = (n_1 + 1)/4 = k/2$. The numbers 31 and 43 are of the form $4q + 3$, but additional partitions producing Youden square plans are available for these numbers. Extension of Table 7.12 to values of 100 or higher may throw additional light on these features.

5. $T{:}TT$ Type Plans for n_1 Rows, Columns, and Treatments

In the upper left 7 × 7 partitioning of $C_3(10)$ and $C_4(10)$ (Table 7.5), the treatments form a balanced incomplete block design arrangement within rows and within columns with $\lambda = 2$. With the $21 = kn_1$ entries omitted from the k transversals from this part of $C_3(10)$ and $C_4(10)$, the columns are in a balanced arrangement within rows, and *vice versa*. In this design $\lambda = 2$ for columns within rows and for treatments within rows and within columns.

Likewise, in the upper left 11 × 11 partitioning of $C_3(16)$ and $C_4(16)$ (Table 7.9) with $6(11)$ entries remaining, the treatments form a balanced incomplete block design arrangement within rows and within columns with $\lambda = 3$; the columns are in a balanced incomplete block design arrangement within rows, and *vice versa*, with $\lambda = 3$ also.

Both of the above designs are of the $T{:}TT$ type. This appears to be a new class of designs for situations where the kn_1 deleted entries may be left blank, where the kn_1 deleted entries may be replaced by a dummy treatment, or where additional treatments are to be inserted in the kn_1 spaces. This last class of plans will be considered later. Thus, from the upper left $n_1 \times n_1$ part of the $C_3(n_1 + n_2)$ and $C_4(n_1 + n_2)$ plans, a class of designs of the $T{:}TT$ type may be easily constructed from one of the steps in the sum composition method of constructing an $O(n_1 + n_2, 2)$ set. Since the block size and the value of λ for treatments are identical within rows and within columns and a balanced arrangement is obtained, this results in a relatively simple statistical analysis.

6. Partially Balanced Latin Rectangle and $P:PP$ Plans

Instead of projecting the k transversals which result in a Youden square plan, fewer than the k transversals resulting in a balanced incomplete block design arrangement may be projected to produce a partially balanced Latin rectangle arrangement. For example, in $C_3(10)$ and $C_4(10)$ (Table 7.5) the use of only two of the three selected projected diagonals results in a 2-row $\times$ 7-column Latin rectangle; the set of three projected transversals considered are those resulting in a Youden square plan. In the $C_3(16)$ and $C_4(16)$ (Table 7.9) plans two, three, or four of the five selected transversals producing a Youden square plan may be used to produce a partially balanced 2-row, 3-row, or 4-row $\times$ 11-column Latin rectangle. Since the selection of fewer than k transversals is from the k transversals producing a balanced arrangement, the resulting Latin rectangles will be as nearly balanced as possible.

If the k parallel transversals producing a Youden square plan and $1, 2, \ldots$ additional parallel transversals are projected, the resulting plans will be in the class of partially balanced Latin rectangle plans, except in those instances where sufficient parallel transversals are projected to result in a balanced arrangement. In $C_3(10)$ and $C_4(10)$ the projection of any pair of parallel transversals from among the remaining four results in a 5-row $\times$ 7-column partially balanced Latin rectangle plan; the projection of any triplet of parallel transversals from among the remaining four results in a 6-row $\times$ 7-column Youden square plan. In $C_3(16)$ and $C_4(16)$ the projection of an additional one, two, three, or four additional parallel transversals from among the remaining six results in a partially balanced Latin rectangle. The projection of any five additional parallel transversals to the five already projected results in a Youden square plan.

In the $n_1 \times n_1$ part of plans $C_3(n_1 + n_2)$ and $C_4(n_1 + n_2)$ which remains after the projection of fewer than (or more than) the k transversals resulting in a balanced arrangement, the elements or the treatments in the resulting plan have the partially balanced incomplete block design property within rows and within columns. Likewise, the columns have the partially balanced incomplete block design property within rows, and vice versa. Thus, the remaining entries in the $n_1 \times n_1$ part, after projecting $k - q$, $q = 1, 2, \ldots, k - 2$ parallel transversals, form a $P:PP$ type design, where P denotes partially balanced. Likewise, after the projection of $k + 1, k + 2, \ldots, n_1 - 2$ parallel transversals where the first k form a Youden square plan, the remaining entries in the $n_1 \times n_1$ parts of $C_3(n_1 + n_2)$ and $C_4(n_1 + n_2)$ plans form a $P:PP$ type plan. The $P:PP$ class of designs is useful in the same situations discussed for the $T:TT$ class of designs.

7. Supplemented Balanced and Partially Balanced Plans

A supplemented balanced plan is one in which one element or treatment appears once in each of the n_1 rows and once in each of the n_1 columns; n_1 elements or treatments appear $n_1 - 1$ times in the n_1 rows and in the n_1 columns (see Pearce, 1960). The treatment appearing n_1 times is inserted in a transversal of a Latin square of order n_1; the treatments which appear $n_1 - 1$ times form a balanced arrangement within rows and within columns. This type of arrangement of the $n_1 + 1$ elements or treatments in the n_1-row $\times$ n_1-column arrangement is denoted as supplemented balance and symbolized as S. Since rows and columns are orthogonal to each other, this plan is denoted as an $O:SS$ type of plan.

We may extend this idea by inserting k additional treatments in the places occupied by the projected k parallel transversals, such that one additional treatment is placed on one of the projected parallel transversals in the $n_1 \times n_1$ part of $C_3(n_1 + n_2)$ and $C_4(n_1 + n_2)$, forming the $T:TT$ plan. This results in $n_1 + k$ elements or treatments in an $n_1 \times n_1$ square with the k additional treatments occurring n_1 times and the original n_1 treatments occurring $n_1 - k$ times. Each of the k additional treatments are in a randomized block arrangement among themselves and are in a supplemented balanced arrangement to the n_1 original treatments in the $T:TT$ plan. This class of plans for $n_1 + k$ treatments in an $n_1 \times n_1$ square is denoted as an $O:SS$ plan. (This plan could also be denoted as an *augmented* $T:TT$ plan to conform with notation used for other augmented plans presented later.)

To illustrate, let the $k = 3$ additional treatments be denoted as A, B, and C and use the $T:TT$ plan from $C_3(10)$ (Table 7.5) to obtain the plan in Table 7.13.

Table 7.13

0	A	B	3	C	5	6
A	B	4	C	6	0	1
B	5	C	0	1	2	A
6	C	1	2	3	A	B
C	2	3	4	A	B	0
3	4	5	A	B	1	C
5	6	A	B	2	C	4

A, B, and C each occur $n_1 = 7$ times and 0, 1, 2, 3, 4, 5, and 6 each occur $n_1 - k = 4$ times. If four appropriate transversals are projected—e.g., those remaining in $C_3(10)$—the resulting design is of the $T:TT$ type with $49-4(7) = 21$ entries; four additional treatments—A, B, C, and D—could be added to obtain a supplemented balanced plan with $7 + 4 = 11$ treatments. For example, in $C_1(10)$ (Table 7.3), project transversals corresponding to elements 0, 3, 5, and 6 of $L_1(7)$ (Table 7.4) and replace these transversals with elements A, B, C, and D respectively to obtain Table 7.14.

Table 7.14

A	1	2	B	4	C	D
2	3	B	5	C	D	A
4	B	6	C	D	A	3
B	0	C	D	A	4	5
1	C	D	A	5	6	B
C	D	A	6	0	B	2
D	A	0	1	B	3	C

In this plan the $T:TT$ plan with $n_1 = 7 = v, r = k = 3$, and $\lambda = 1$ was augmented with four additional treatments each occurring $n_1 = 7$ times. Alternatively, one could start with a $T:OO$ plan, as shown in Table 7.15, and augment this plan with treatments 0, 1, 2, 3, 4, 5, and 6 in such a manner that these augmented treatments form a balanced incomplete block design arrangement within rows and within columns. Table 7.15 utilizes the fact that the k projected transversals produce a balanced incomplete block design arrangement.

Table 7.15

A			B		C	D
		B		C	D	A
	B		C	D	A	
B		C	D	A		
	C	D	A			B
C	D	A			B	
D	A			B		C

If plans of the $P:PP$ type were augmented with additional treatments each occurring n_1 times, the resulting plan would be denoted as one with supplemented partial balance. Denoting this by S_p, the $O:S_pS_p$ plan for

$n_1 = 7$ treatments (each occurring five times) and two additional treatments A and B (each occurring $n_1 = 7$ times) may be obtained by removing two transversals, say transversals corresponding to elements 0 and 6 of $L_1(7)$ from $C_1(10)$, and replacing these with A and B as in Table 7.16. Also, one could start with a plan of the type in Table 7.17 and one could

Table 7.16

A	1	2	3	4	5	B
2	3	4	5	6	B	A
4	5	6	0	B	A	3
6	0	1	B	A	4	5
1	2	B	A	5	6	0
3	B	A	6	0	1	2
B	A	0	1	2	3	4

Table 7.17

A	B	C	D	E		
B	C	D	E			A
C	D	E			A	B
D	E			A	B	C
E			A	B	C	D
		A	B	C	D	E
	A	B	C	D	E	

augment this plan with treatments 0, 1, 2, 3, 4, 5, and 6 such that two of the three transversals required to obtain balance are selected. These treatments would form a partially balanced incomplete block design arrangement within rows and within columns; the resulting design (Table 7.18) is of the $O:S_pS_p$ type.

Table 7.18

A	B	C	D	E	5	6
B	C	D	E	6	0	A
C	D	E	0	1	A	B
D	E	1	2	A	B	C
E	2	3	A	B	C	D
3	4	A	B	C	D	E
5	A	B	C	D	E	4

8. Other Augmented Plans

Suppose that the basic design is the $n_1 \times n_1$ part of plans similar to $C_3(n_1 + n_2)$ and $C_4(n_1 + n_2)$ from which p transversals,

$$p = 1, 2, \ldots n_1 - 2,$$

have been projected and that the basic design is either of the $T:TT$ type or the $P:PP$ type. Denote the elements or treatments in the basic design as the original treatments. If the pn_1 empty cells in $C_3(n_1 + n_2)$ type plans are filled with pn_1 additional entries each occurring once and if the $p = k$ projected transversals form a balanced incomplete block design arrangement, the resulting plan is called an *augmented* $T:TT$ type of plan. An example of an augmented $T:TT$ plan with $pn_1 = 3(21)$ additional treatments is obtained from $C_3(10)$ (Table 7.5) as Table 7.19.

Table 7.19

0	a	b	3	c	5	6
d	e	4	f	6	0	1
g	5	h	0	1	2	i
6	j	1	2	3	k	l
m	2	3	4	n	o	0
3	4	5	p	q	1	r
5	6	s	t	2	u	4

The twenty-one augmented treatments are denoted by the letters $a, b, c, d, \ldots, t$, and u and each occur once. The treatments denoted by 0, 1, 2, 3, 4, 5, and 6 occur four times each.

Another form of augmentation for a $T:TT$ plan similar to the upper left-hand partitioning of $C_3(10)$ is to include $pn_1/r_a = 7$ treatments each occurring $r_a = 3$ times each. If the augmented treatments are included in such a manner as to form a $T:TT$ plan by themselves and these seven treatments are denoted by the letters a, b, c, d, e, f, and g, a plan like Table 7.20 results.

Table 7.20

0	b	c	3	e	5	6
c	d	4	f	6	0	1
e	5	g	0	1	2	d
6	a	1	2	3	e	f
b	2	3	4	f	g	0
3	4	5	g	a	1	c
5	6	a	b	2	d	4

Considering all fourteen treatments, the above 7×7 plan is of the $O:PP$ type and has a two-associate class partially balanced incomplete block design arrangement within rows and within columns.

From the 7×7 part of $C_1(10)$ (Table 7.3) several augmented plans may be obtained. Some of these are shown in Table 7.21.

Table 7.21

Transversals Projected	Augmented Treatment	Total Treatments	Replicates for Augmented Treatments	Replicates for Original Treatments
(no.)	(no.)	(no.)	(no.)	(no.)
1, 2, 3, 4, 5, 6	1	8	7, 14, 21, 28, 35, 42	6, 5, 4, 3, 2, 1
2, 4, 6	2	9	7, 14, 21	5, 3, 1
3, 6	3	10	7, 14	4, 1
4	4	11	7	3
5	5	12	7	2
6	6	13	7	1
1, 2, 3, 4, 5, 6	7	14	1, 2, 3, 4, 5, 6	6, 5, 4, 3, 2, 1
2, 4, 6	14	21	1, 2, 3	5, 3, 1
3, 6	21	28	1, 2	4, 1
4	28	35	1	3
5	35	42	1	2

In addition to the plans in Table 7.21 with two sets of replication, other augmented plans may be constructed for 3, 4, . . . sets of replication for the sets of treatments. For example, suppose that eight additional treatments are to be used. One treatment could be replicated $n_1 = 7$ times on one transversal; the other seven additional treatments could be replicated two times each and inserted on two transversals and in a partially balanced incomplete block design arrangement within rows and within columns; the original $n_1 = 7$ treatments could be included four times each in the $T:TT$ plan from $C_3(10)$. Thus there is supplemented balance between the original seven treatments and the treatment included seven times; there is supplemented partial balance between the additional treatment included seven times and the remaining augmented treatments which are included twice; the augmented treatment occurring $n_1 = 7$ times is in supplemented partial balance to the remaining fourteen treatments in the plan. This design is of the $O:S_pS_p$ type with three different numbers of replicates for the $1 + 7 + 7 = 15$ treatments.

The above illustrates the great diversity of augmented designs available for experimentation. Other augmented designs have been discussed by Federer (1956, 1960, 1961).

9. Plans for the Design of Successive Experiments

If two sets of treatments in two separate experiments are applied successively or simultaneously to the same set of experimental units, it is desirable to have plans in which the rows, columns, and each set of treatments are as nearly orthogonal as possible. Orthogonal experiments are more efficient than nonorthogonal ones in the sense that smaller variances of treatment differences are obtained. Plans with various properties among the two sets of treatments and the rows and columns have been constructed by various authors (see Hoblyn et al., 1954; Pearce, 1960, 1963; Clarke, 1963; Freeman, 1964; and Hedayat et al., 1970). Designs of the $O:OT:TOO$ and $O:OO:SSS$ types have been constructed, where the first three letters and the first colon have the meaning used previously and the three letters to the right of the second colon refer to the properties of the second set of treatments with respect to rows, columns, and the first set of treatments.

We may use plans of the form of $C_3(n_1 + n_2)$ and $C_4(n_1 + n_2)$ to construct 3×7, 4×7, 5×11, $6 \times 11, \ldots$, $k \times n_1$, and $(n_1 - k) \times n_1$ Youden square plans of the $O:OT:OTT$ type for successive experiments. For example, if the first seven rows and last three columns of $C_3(10)$ are superimposed on the corresponding rows and columns of $C_4(10)$ (Table 7.5), a 7-row $\times$ 3-column design of the $O:TO:TOT$ type results. Similarly superimposing the last three rows and the first seven columns of $C_3(10)$ on the corresponding ones of $C_4(10)$ produces a 3-row $\times$ 7-column design of the $O:OT:OTT$ type. The use of the remaining four transversals in $C_3(10)$ and $C_4(10)$ projected into rows and columns and superimposing one set on the other results in a 7-row $\times$ 4-column plan of the $O:TO:TOT$ type and a 3-row $\times$ 7-column plan of the $O:OT:OTT$ type.

Using the same procedure described above, we may construct 5-row $\times$ 11-column and 6-row $\times$ 11-column plans of the $O:OT:OTT$ type and 11-row $\times$ 5-column and 11-row $\times$ 6-column plans of the $O:TO:TOT$ type. In general, whenever a Youden square plan can be formed from the projection of k transversals into the last k rows of $C_3(n_1 + n_2)$ and $C_4(n_1 + n_2)$, superimposition of the corresponding parts produces a k-row $\times$ n_1-column $O:OT:OTT$ plan.

An alternative procedure for constructing $O:OT:OTT$ plans is to obtain a pair of orthogonal Latin squares such that one square is a cyclical permutation of the rows of the other. This is possible for all odd n_1 (see Hedayat and Federer, 1969). Then divide the n_1 rows into two sets of rows for both squares such that $n_1 - k$ rows and the remaining k rows form Youden squares. The same permutation of rows is used for both squares from the $O(n_1, 2)$ set. Using $L_1(7)$ (Table 7.4) and the Latin square of order 7 from $C_1(10)$ (Table 7.3) to form the $O(7, 2)$ set, the plan of Table 7.22 illustrates the procedure which produces plans of the $O:OT:OTT$ type.

Table 7.22

A1	B2	C3	D4	E5	F6	G7	
B3	C4	D5	E6	F7	G1	A2	Youden square plans for
E2	F3	G4	A5	B6	C7	D1	$v = 7 = n_1, k = 4 = r,$ and $\lambda = 2.$
G6	A7	B1	C2	D3	E4	F5	
C5	D6	E7	F1	G2	A3	B4	
D7	E1	F2	G3	A4	B5	C6	Youden square plans for $v = 7 = n_1, k = 3 = r,$
F4	G5	A6	B7	C1	D2	E3	and $\lambda = 1.$

If three sets of successive experiments are to be conducted with three sets of treatments, an $O(n_1, 3)$ set may be used to construct plans, e.g., of the $O:OT:OTT:OTTT$ type where the set of letters $OTTT$ refers to the relationship among the third set of treatments to the rows, the columns, the first set of treatments, and the second set of treatments. Likewise, $O(n_1, t)$ sets may be used to construct plans for t successive experiments on the same material.

Numerous additional plans of the augmented type may be constructed. The example in Table 7.23 illustrates one such type wherein an augmented $O:TT$ type plan is superimposed on a Latin square of order 7. The elements

Table 7.23

A0	Ba	Cb	D3	Ec	F5	G6
Bd	Ce	D4	Ef	F6	G0	A1
Cg	D5	Eh	F0	G1	A2	Bi
D6	Ej	F1	G2	A3	Bk	Cl
Em	F2	G3	A4	Bn	Co	D0
F3	G4	A5	Bp	Cq	D1	Er
G5	A6	Bs	Ct	D2	Eu	F4

A, B, C, D, E, F, and G replace elements 0, 1, 2, 3, 4, 5, and 6 in $L_1(7)$; the empty spaces in $C_3(10)$ are replaced by the letters a, b, c, ..., t, and u.

In addition to the above types of designs, the $n_1 \times n_1$ parts of plans like $L_1(n_1 + n_2)$ and $L_2(n_1 + n_2)$ formed by the sum composition method may be superimposed on each other to form an n_1-row $\times$ n_1-column design with $n_1 + n_2$ treatments in the first experiment and $n_1 + n_2$ treatments in the second experiment.

Other permutations of plans are possible and numerous, but the above suffices to demonstrate the usefulness of the sum composition method of constructing $O(n_1 + n_2, 2)$ sets and the common-parallel transversals in Latin squares for constructing several classes of experimental designs. These designs are useful for experimental material requiring control of heterogeneity from two sources of variation.

10. Summary

An experiment should be designed to satisfy the experimental considerations and the specified objectives. An experiment should not be designed to fit into known experimental designs *if* this results in a change in the desired findings. Consequently, new designs and concepts need to be developed to satisfy new experimental requirements which continue to arise as a result of new investigations and research interests.

Several new types of experimental designs and some new methods of constructing known designs are considered in this paper. The ideas of parallel transversals in a Latin square of order n and of common-parallel transversals in a pair of orthogonal Latin squares of order n proved useful in constructing new designs as well as some previously known. The ideas and the procedure involved in the sum composition method of constructing a pair of orthogonal Latin squares of order n proved very useful in the present work. The sum composition method is a new procedure for constructing a pair of orthogonal Latin squares and was discovered by A. Hedayat. It derives its name from the fact that use is made of pairs of orthogonal Latin squares of order n_1 and n_2, such that $n_1 + n_2 = n$ and $n_1 \geq 2n_2$, to produce the larger pair of orthogonal Latin squares of order n. The method makes use of the projection of parallel and common-parallel transversals into the last n_2 rows and columns of the square of order n.

It has been shown how to construct Youden square and balanced incomplete block designs using the sum composition method as well as plans of partially balanced Latin rectangle designs, $T : TT$ type designs, supplemented balanced and partially balanced designs, augmented designs, and designs for successive experiments conducted on the same experimental units.

References

Clarke, G. M. 1963. A second set of treatments in a Youden square design. *Biometrics* 19: 98–104.

Federer, W. T. 1956. Augmented (or hoonuiaku) designs. *Hawaiian Planters' Record* 55: 191–208.

———. 1960. Augmented designs for two-, three-, and higher-way elimination of heterogeneity. Paper presented and distributed at Annual Statistics Meetings, Palo Alto, Calif., Aug. 24, 1960.

———. 1961. Augmented designs with one-way elimination of heterogeneity. *Biometrics* 17: 447–73.

Federer, W. T., A. Hedayat, E. T. Parker, B. L. Raktoe, E. Seiden, and R. J. Turyn. 1969. Some techniques for constructing mutually orthogonal Latin squares. Math. Res. Center, U.S. Army and Univ. of Wisc., MRC Tech. Summ. Rept. 1030, and pp. 673–796 of Proc. Fifteenth Conference Design Expt. Army Res. Development Testing, ARO-D Rept. 70-2, 1970.

Freeman, G. H. 1964. The addition of further treatments to the Latin square designs. *Biometrics* 20: 713–29.

Hedayat, A., and W. T. Federer. 1969. An application of group theory to the existence of orthogonal Latin squares. *Biometrika* 56: 547–51.

———. 1970. An easy method of constructing partially replicated Latin square designs of order *n* for all *n* > 2. *Biometrics* 26: 327–30.

Hedayat, A., E. T. Parker, and W. T. Federer. 1970. The existence and construction of two families of designs for two successive experiments. *Biometrika* 57: 351–5.

Hedayat, A., and E. Seiden. 1970. On a method of sum composition of orthogonal Latin squares. (In press.)

Hoblyn, T. N., S. C. Pearce, and G. H. Freeman. 1954. Some considerations in the design of successive experiments in fruit plantations. *Biometrics* 10: 503–15.

Keller, K. J. 1969. A computer approach to the construction of orthogonal Latin squares. B.S. thesis, Biometrics Unit, Cornell Univ., Ithaca.

Pearce, S. C. 1960. Supplemented balance. *Biometrika* 47: 263–71.

———. 1963. The use and classification of non-orthogonal designs. *J. Roy. Statist. Soc.* Ser. A, 126: 353–78.

8

A Test of Fit for Continuous Distributions Based on Generalized Minimum Chi-Square

JOHN GURLAND and RAM C. DAHIYA

1. Introduction

A TEST OF FIT for continuous distributions is developed in this paper, based on generalized minimum χ^2 techniques. Although the Pearson χ^2 test of fit is widely used in the case of discrete distributions, there are difficulties in applying it in the case of continuous distributions. A discussion of these difficulties is included in the paper by Dahiya and Gurland (1970a). The results discussed in this paper were motivated by a desire to develop a test free of the complications associated with the Pearson χ^2 test. In particular, the question of how to form class intervals does not arise in the test of fit presented here. Furthermore, the asymptotic distribution is exactly that of a χ^2, in contradistinction to the asymptotic distribution of the statistic employed in the Pearson χ^2 test when the estimators of parameters are obtained from the ungrouped sample (cf. Chernoff and Lehmann, 1954).

The asymptotic nonnull distribution of the test statistic proposed here is developed for general alternatives. As a special case the asymptotic power is obtained for testing normality against several specific alternative families of distributions. The power of this test is compared with that of a

JOHN GURLAND is Professor, Department of Statistics, University of Wisconsin, Madison.
RAM C. DAHIYA is Assistant Professor, Department of Mathematics and Statistics, University of Massachusetts, Amherst.
This work was supported in part by the National Science Foundation, the Wisconsin Alumni Research Foundation, and the United States Army under Contract No. DA-31-124-ARO-D-462.

modified form of the Pearson χ^2 test based on random intervals presented by Dahiya and Gurland (1970a, 1970b).

Although the test of fit presented here is for continuous distributions, the method based on minimum χ^2 techniques is quite general and can in fact be adapted to discrete distributions. Hinz and Gurland (1970) have applied such techniques to develop a test of fit for the negative binomial and other contagious distributions.

2. Formulation of a Test Statistic Based on Sample Moments

First we consider the problem in a general context and show how to construct a statistic for testing the fit of a hypothesized distribution based on a set of sample moments. In a subsequent section the result obtained here will be applied to develop a test of normality.

Let $X_1, X_2, \ldots, X_n$ be a random sample from a certain distribution with probability density function (p.d.f.)

$$p_X(x|\theta), \tag{2.1}$$

where θ is a parameter vector of q components, i.e.,

$$\theta' = [\theta_1, \theta_2, \ldots, \theta_q]. \tag{2.2}$$

Denote the jth raw sample moment by

$$m'_j = \frac{1}{n} \sum_{i=1}^{n} X_i^j. \tag{2.3}$$

Let

$$m' = [m'_1, m'_2, \ldots, m'_s], \tag{2.4}$$

where s, $(s > q)$ is a fixed number that remains to be specified. (A low value of s is generally desirable due to the large sampling fluctuations of higher order moments.) Under the assumption that the $2s$th order moment of X exists, we can easily show, by making use of the central limit theorem, that the asymptotic distribution of $\sqrt{n}(m - \mu)$ is normal,

$$N(0; G) \tag{2.5}$$

where the vector μ is the population counterpart of m, given by

$$\mu' = [\mu'_1, \mu'_2, \ldots, \mu'_s] \tag{2.6}$$

and the $s \times s$ covariance matrix G is given by

$$G = (g_{ij}) = (\mu'_{i+j} - \mu'_i \mu'_j). \tag{2.7}$$

If $h_1, h_2, \ldots, h_s$ be s functions of m, i.e.,

$$h_i = h_i(m'_1, m'_2, \ldots, m'_s) \qquad i = 1, 2, \ldots, s \tag{2.8}$$

such that their population counterparts

$$\zeta_i = h_i(\mu'_1, \mu'_2, \ldots, \mu'_s) \qquad i = 1, 2, \ldots, s \tag{2.9}$$

are differentiable to the second order with respect to $\mu'_1, \mu'_2, \ldots, \mu'_s$, then the asymptotic distribution of

$$\sqrt{n}(h - \zeta) \tag{2.10}$$

is given by

$$N(0; \Sigma), \tag{2.11}$$

where

$$\begin{aligned}
h' &= [h_1, h_2, \ldots, h_s] \\
\zeta' &= [\zeta_1, \zeta_2, \ldots, \zeta_s] \\
\Sigma &= JGJ'
\end{aligned} \tag{2.12}$$

and J is the $s \times s$ Jacobian matrix $[\partial \zeta_i / \partial \mu'_j]$.

From this result it follows that the asymptotic distribution of

$$Q = n(h - \zeta)' \, \Sigma^{-1} \, (h - \zeta) \tag{2.13}$$

is that of χ_s^2. Furthermore, if $\hat{\Sigma}$ is a consistent estimator of Σ, which is obtained from Σ on replacing parameters by maximum likelihood or some other consistent estimators, then according to Gurland (1948) and Barankin and Gurland (1951) the asymptotic distribution of

$$Q^* = n(h - \zeta)' \, \hat{\Sigma}^{-1} \, (h - \zeta) \tag{2.14}$$

is the same as the asymptotic distribution of Q.

Now suppose we select functions h_i such that ζ_i are linear functions of the parameters $\theta_1, \theta_2, \ldots, \theta_q$, i.e.,

$$\zeta = W\theta, \tag{2.15}$$

where W is a $s \times q$ matrix of known constants. In such a case we can find an estimator for θ by minimizing the expression for Q^* in (2.14). This estimator, $\hat{\theta}$ say, is given by

$$\hat{\theta} = (W' \, \hat{\Sigma}^{-1} \, W)^{-1} \, W' \, \hat{\Sigma}^{-1} h. \tag{2.16}$$

Let

$$\hat{\zeta} = W\hat{\theta} \tag{2.17}$$

and

$$\hat{Q} = n(h - \hat{\zeta})' \, \hat{\Sigma}^{-1} \, (h - \hat{\zeta}). \tag{2.18}$$

Now let

$$\begin{aligned}
\hat{R} &= W(W' \, \hat{\Sigma}^{-1} W)^{-1} W' \, \hat{\Sigma}^{-1} \\
\hat{A} &= \hat{\Sigma}^{-1} \, (I - \hat{R}).
\end{aligned} \tag{2.19}$$

Then

$$\hat{Q} = n(h - \hat{R}h)' \hat{\Sigma}^{-1} (h - \hat{R}h) = nh'(I - \hat{R})' \hat{\Sigma}^{-1} (I - \hat{R})h$$
$$= nh'\hat{A}h. \tag{2.20}$$

From results of Gurland (1948) and Barankin and Gurland (1951) the asymptotic distribution of $nh'\hat{A}h$ is the same as the asymptotic distribution of $nh'Ah$, where A is obtained from $\hat{A}$ on replacing $\hat{\Sigma}$ by Σ. To find the distribution of $nh'Ah$, we make use of the following lemma.

LEMMA 2.1. If X is distributed as $N(\mu; \Sigma)$ and B is a matrix such that ΣB is idempotent, then the distribution of $X'BX$ is noncentral χ^2 with r degrees of freedom and noncentrality parameter λ denoted by $\chi^2_{r,\lambda}$, where r is the rank(B) and $\lambda = \mu'B\mu$.

PROOF. Let P be a nonsingular matrix such that

$$P\Sigma P' = I, \tag{2.21}$$

an identity matrix. On making use of the transformation

$$Y = PX, \tag{2.22}$$

it follows that Y is distributed as $N(P\mu; I)$ and

$$X'BX = Y'P'^{-1}BP^{-1}Y. \tag{2.23}$$

Now $P'^{-1}BP^{-1}$ is an idempotent matrix of rank r since $\Sigma B = P^{-1}P'^{-1}B$ is idempotent of rank r. Hence the distribution of $X'BX$ is $\chi^2_{r,\lambda}$, where

$$\lambda = (P\mu)'P'^{-1}BP^{-1}(P\mu) = \mu'B\mu. \tag{2.24}$$

This proves Lemma 2.1.

From the above lemma and also assuming W of full rank q, we see that the asymptotic null distribution of $nh'Ah$ is χ^2_{s-q} since ΣA is an idempotent matrix of rank $s - q$ and $\zeta'A\zeta = 0$, which can easily be verified. Thus it follows that the asymptotic distribution of $\hat{Q}$ is that of χ^2_{s-q}.

The statistic $\hat{Q}$ can be utilized for testing the fit of an assumed distribution. To ascertain how well such a test of fit behaves, its power against specific alternatives can be obtained from the nonnull distribution given in section 3.

3. Asymptotic Nonnull Distribution of $\hat{Q}$

The asymptotic nonnull distribution of $\hat{Q}$ turns out to be that of a weighted sum of independent noncentral χ^2 random variables each with one degree of freedom. A derivation of this result along with the precise weights and noncentralities is given in the following theorem.

THEOREM 3.1. Let the null and alternative hypotheses H_0, H_1 respectively be as follows:

$$H_0 : X \text{ has p.d.f. } p_X(x|\theta) \quad -\infty < x < \infty, \theta' = [\theta_1, \theta_2, \ldots, \theta_q] \qquad (3.1)$$

$$H_1 : X \text{ has p.d.f. } p_X^{(1)}(x|\gamma) \quad -\infty < x < \infty, \gamma' = [\gamma_1, \gamma_2, \ldots, \gamma_p], \qquad (3.2)$$

where p and q may differ. Here θ and γ are parameter vectors.

Then the asymptotic nonnull distribution of $\hat{Q}$ defined in (2.20) is of the form

$$\sum_1^{s-q} d_i \chi_{1, a_i}^2 \qquad (3.3)$$

The constants d_i are given by (3.7) and a_i by (3.8) and (3.10).

PROOF. Let us denote the matrix to which $\hat{\Sigma}$ converges under H_1 in probability by Σ^*, i.e.,

$$\hat{\Sigma} \xrightarrow{P} \Sigma^* \text{ under } H_1. \qquad (3.4)$$

Then Σ^* involves the parameter vector γ. Now the asymptotic nonnull distribution of $nh'\hat{A}h$ is the same as that of

$$Q^{(1)} = nh'A^*h, \qquad (3.5)$$

where A^* is obtained from $\hat{A}$ on replacing $\hat{\Sigma}$ by Σ^*. Let $\Sigma^{(1)}$ denote the asymptotic covariance matrix of $\sqrt{n}h$ under H_1 which can be found in the same way as Σ is found under H_0. Also if $\zeta^{(1)}$ denotes the population counterpart of h under H_1, the asymptotic nonnull distribution of $\sqrt{n}(h - \zeta^{(1)})$ is that of $N(0; \Sigma^{(1)})$. There exists a nonsingular matrix T and an orthogonal matrix P such that

$$T\Sigma^{(1)}T' = I \qquad (3.6)$$

and

$$PT'^{-1}A^*T^{-1}P' = D = \begin{bmatrix} d_1 & & & & & & & \\ & d_2 & & & & & O & \\ & & d_3 & & & & & \\ & & & \ddots & & & & \\ & & & & d_{s-q} & & & \\ & & & & & 0 & & \\ & & & & & & 0 & \\ & & & & & & & \ddots \\ & O & & & & & & & 0 \end{bmatrix} \qquad (3.7)$$

where I is the identity matrix and D is a diagonal such that the last q diagonal elements of D are zero. This is possible since rank $A^* = s - q$.

Let

$$u = PTh$$

and

$$\Psi = PT\zeta^{(1)}. \tag{3.8}$$

Then we have

$$nh'A*h = n(T^{-1}P'u)'A*(T^{-1}P'u) = nu'Du$$

$$= \sum_{1}^{s-q} d_i(\sqrt{n}u_i)^2. \tag{3.9}$$

Since the asymptotic distribution of $\sqrt{n}(h - \zeta^{(1)})$ is $N(0;\Sigma^{(1)})$, it follows that the asymptotic distribution of $\sqrt{n}(u - \Psi)$ is $N(0;I)$. Hence the asymptotic distribution of $nh'A*h$ is that of

$$\sum_{1}^{s-q} d_i\chi_1^2, a_i,$$

where

$$a_i = \sqrt{n}\Psi_i \qquad i = 1, \ldots, s - q \tag{3.10}$$

and Ψ_i is the ith element of Ψ. This proves the theorem.

4. Test of Fit for Normal Distribution Based on $\hat{Q}$

We shall now consider a test of fit based on $\hat{Q}$ when the null distribution is normal, i.e., X has p.d.f.

$$p_X(x|\theta) = \frac{1}{\sqrt{2\pi\theta_2}} \exp\left\{ -\frac{(x - \theta_1)^2}{2\theta_2} \right\}$$

$$-\infty < x < \infty, -\infty < \theta_1 < \infty, \theta_2 > 0. \tag{4.1}$$

Let $m_2, m_3,$ and m_4 be second, third, and fourth central sample moments respectively. The statistics b_1, b_2 given by

$$b_1 = m_3/m_2^{3/2}, \qquad b_2 = m_4/m_2^2 \tag{4.2}$$

are sometimes employed for testing normality by means of skewness and kurtosis. Instead of considering these two statistics separately, it appears more rational to formulate a single statistic involving the first four moments. This motivates our selection of functions h_i based on the first four sample moments. The mean, variance, third, and fourth central moments of X are respectively given by:

$$\mu_1' = \theta_1; \mu_2 = \theta_2; \mu_3 = 0; \mu_4 = 3\theta_2^2. \tag{4.3}$$

If we define

$$\theta_2^* = \log \theta_2$$

$$\zeta' = \left[\mu_1', \log \mu_2, \mu_3, \log\left(\frac{\mu_4}{3}\right) \right] \qquad (4.4)$$

then the elements of ζ are linear functions of the parameters θ_1 and θ_2^*. We can now write

$$\zeta = W\theta^* \qquad (4.5)$$

with

$$W = \begin{bmatrix} 1 & 0 \\ 0 & 1 \\ 0 & 0 \\ 0 & 2 \end{bmatrix}, \qquad \theta^* = \begin{bmatrix} \theta_1 \\ \\ \\ \theta_2^* \end{bmatrix}.$$

The corresponding h_i functions are given by

$$h_1 = m_1'; \, h_2 = \log m_2; \, h_3 = m_3; \, h_4 = \log\left(\frac{m_4}{3}\right) \qquad (4.6)$$

where m_1' is the sample mean and m_2, m_3, m_4 denote second, third, and fourth central sample moments respectively, as previously indicated.

The transformation from sample raw moments to functions h_i is achieved in two stages, i.e., from m_1', m_2', m_3', m_4' to m_1', m_2, m_3, m_4 and then finally to h_1, h_2, h_3, h_4. In the notations of section 3, $\sqrt{n}(h - \zeta)$ is asymptotic $N(0; \Sigma)$, where

$$\Sigma = J_2 J_1 G J_1' J_2' \qquad (4.7)$$

$$J_1 = \begin{bmatrix} 1 & 0 & 0 & 0 \\ 0 & 1 & 0 & 0 \\ -3\theta_2 & 0 & 1 & 0 \\ 0 & 0 & 0 & 1 \end{bmatrix}; \qquad J_2 = \begin{bmatrix} 1 & 0 & 0 & 0 \\ 0 & 1/\theta_2 & 0 & 0 \\ 0 & 0 & 1 & 0 \\ 0 & 0 & 0 & 1/(3\theta_2^2) \end{bmatrix} \qquad (4.8)$$

$$G = \begin{bmatrix} \theta_2 & 0 & 3\theta_2^2 & 0 \\ 0 & 2\theta_2^2 & 0 & 12\theta_2^3 \\ 3\theta_2^2 & 0 & 15\theta_2^3 & 0 \\ 0 & 12\theta_2^3 & 0 & 96\theta_2^4 \end{bmatrix}. \qquad (4.9)$$

After simplification we obtain:

$$\Sigma = \begin{bmatrix} \theta_2 & 0 & 0 & 0 \\ 0 & 2 & 0 & 4 \\ 0 & 0 & 6\theta_2^3 & 0 \\ 0 & 4 & 0 & 32/3 \end{bmatrix}. \qquad (4.10)$$

Now let

$$\hat{\Sigma} = [\Sigma]_{\theta_2 = m_2}, \tag{4.11}$$

where m_2 is the maximum likelihood estimator of θ_2. Then a statistic $\hat{Q}$ for testing normality is given by

$$\hat{Q} = nh'\hat{A}h, \tag{4.12}$$

where

$$\hat{A} = \hat{\Sigma}^{-1}(I - \hat{R})$$
$$\hat{R} = W(W'\hat{\Sigma}^{-1}W)^{-1}W'\hat{\Sigma}^{-1}. \tag{4.13}$$

After simplification we can show that

$$\hat{A} = \begin{bmatrix} 0 & 0 & 0 & 0 \\ 0 & 1.5 & 0 & -.75 \\ 0 & 0 & 1/(6m_2^3) & 0 \\ 0 & -.75 & 0 & .375 \end{bmatrix}. \tag{4.14}$$

Hence a simplified form of $\hat{Q}$ is given by

$$\hat{Q} = nu'\hat{B}u, \tag{4.15}$$

where

$$u' = [h_2, h_3, h_4] = \left[\log{(m_2)}, m_3, \log\left(\frac{m_4}{3}\right) \right] \tag{4.16}$$

and

$$\hat{B} = \begin{bmatrix} 1.5 & 0 & -.75 \\ 0 & 1/(6m_2^3) & 0 \\ -.75 & 0 & .375 \end{bmatrix}. \tag{4.17}$$

The statistic $\hat{Q}$ in (4.15) can easily be computed on a desk calculator.

The asymptotic distribution of $\hat{Q}$ is χ_2^2, since here $s = 4$ and $q = 2$ in the notations of section 2. Thus to carry out a test of fit for normality at a particular level of significance, one merely requires the corresponding critical point of the χ_2^2 distribution.

5. Power of the Test of Normality

Let $p_X^{(1)}(x|\gamma)$, where $\gamma' = [\gamma_1, \gamma_2, \ldots, \gamma_p]$ is a parameter vector, denote a general alternative to the null hypothesis of normality. If we denote its ith raw moment by $\mu_i^{(1)'}$ and the corresponding central moment by $\mu_i^{(1)}$, then the asymptotic nonnull distribution of $\sqrt{n}(h - \zeta^{(1)})$ is $N(0, \Sigma^{(1)})$,

where

$$\zeta^{(1)'} = \left[\mu_1^{(1)'}, \log \mu_2^{(1)}, \mu_3^{(1)}, \log \left(\frac{\mu_4^{(1)}}{3} \right) \right]$$

$$\Sigma^{(1)} = (J_2^{(1)} J_1^{(1)}) G^{(1)} (J_2^{(1)} J_1^{(1)})'$$

$$G^{(1)} = (\mu_{i+j}^{(1)} - \mu_i^{(1)} \mu_j^{(1)'}) \qquad i,j, = 1, 2, 3, 4 \qquad (5.1)$$

with

$$J^{(1)} = \begin{bmatrix} 1 & 0 & 0 & 0 \\ a_{2,1} & 1 & 0 & 0 \\ a_{3,1} & a_{3,2} & 1 & 0 \\ a_{4,1} & a_{4,2} & a_{4,3} & 1 \end{bmatrix} \qquad (5.2)$$

with

$$a_{2,1} = -2\mu_1^{(1)'}$$
$$a_{3,1} = 3(2\mu_1^{(1)'2} - \mu_2^{(1)'})$$
$$a_{3,2} = -3\mu_1^{(1)'}$$
$$a_{4,1} = 4(-3\mu_1^{(1)'3} + 3\mu_1^{(1)'}\mu_2^{(1)'} - \mu_3^{(1)'})$$
$$a_{4,2} = 6\mu_1^{(1)'2}; a_{4,3} = -4\mu_1^{(1)'}$$

$$J_2^{(1)} = \begin{bmatrix} 1 & 0 & 0 & 0 \\ 0 & 1/\mu_2^{(1)} & 0 & 0 \\ 0 & 0 & 1 & 0 \\ 0 & 0 & 0 & 1/\mu_4^{(1)} \end{bmatrix}. \qquad (5.3)$$

Σ and $\Sigma^{(1)}$ are the asymptotic covariance matrices of $\sqrt{n}h$ under H_0 and H_1 respectively, and it gives an insight about the test on examining how the structure of these two matrices differs depending on the distribution assumed. In fact the sensitivity of the test based on $\hat{Q}$ will depend on the difference in the structure of these two matrices.

Now we shall prove an attractive invariant property of the test based on $\hat{Q}$ defined by (4.15).

THEOREM 5.1. The power of the test based on $\hat{Q}$, defined by (4.15) is invariant with respect to the location and scale parameters of the alternative distribution.

PROOF. Since $m_2 \overset{P}{\to} \mu_2^{(1)}$ under H_1, the asymptotic nonnull distribution of $\hat{Q} = nu'\hat{B}u$ is the same as the asymptotic nonnull distribution of

$$Q^{(1)} = nu'B^{(1)}u \qquad (5.4)$$

where

$$
B^{(1)} = \begin{bmatrix} 1.50 & 0 & -.75 \\ 0 & 1/(6\mu_2^{(1)})^3 & 0 \\ -.75 & 0 & .375 \end{bmatrix}. \tag{5.5}
$$

The distribution of u is invariant with respect to the location parameter and $B^{(1)}$ does not involve this parameter, hence it follows that the asymptotic nonnull distribution of $Q^{(1)}$ does not involve the location parameter.

Now let β be the scale parameter in the alternative distribution of X. If we take

$$
Y = \frac{X}{\beta}, \tag{5.6}
$$

the distribution of Y does not involve β.

Let $V' = V_1, V_2, V_3$ be such that

$$
\begin{aligned}
V_1 &= u_1 - 2\log\beta = \log m_2(y) \\
V_2 &= u_2/\beta^3 \quad\;\; = m_3(y) \\
V_3 &= u_3 - 4\log\beta = \log m_4(y) - \log 3,
\end{aligned} \tag{5.7}
$$

where

$$
m_i(y) = m_i/\beta^i, \qquad i = 2, 3, 4. \tag{5.8}
$$

Then the distribution of V does not involve the parameter β since the distributions of $m_2(y)$, $m_3(y)$, and $m_4(y)$ do not involve this parameter. If $\mu_2(Y)$ be the variance of Y, then we have

$$
\mu_2^{(1)} = \mu_2(Y)\beta^2 \tag{5.9}
$$

and

$$
Q^{(1)} = u'B^{(1)}u = (V_1 + 2\log\beta,\ V_2\beta^3,\ V_3 + 4\log\beta)B^{(1)} \begin{bmatrix} V_1 + 2\log\beta \\ V_2\beta^3 \\ V_3 + 4\log\beta \end{bmatrix}
$$

$$
= V'B^*V, \tag{5.10}
$$

where

$$
B^* = \begin{bmatrix} 1.50 & 0 & -.75 \\ 0 & 1/[6\mu_2^3(Y)] & 0 \\ -.75 & 0 & .375 \end{bmatrix}.
$$

It is surprising that although the scale parameter β is involved in u and $B^{(1)}$, it cancels out in $u'B^{(1)}u$ as is evident in (5.10).

Since the distribution of V does not involve β and since B^* is also free of this parameter, the asymptotic distribution of $Q^{(1)}$ and hence that of $\hat{Q}$ does not involve the scale parameter β. This completes the proof of Theorem 5.1.

6. Calculation of Power for the Test of Normality Based on $\hat{Q}$

For studying the behavior of the test of fit for normality based on $\hat{Q}$, we have carried out power computations for several alternative families of distributions. The null hypothesis has been stated in (4.1), and the test statistic $\hat{Q}$ has been formulated in (4.15).

The following alternative distributions A_1, A_2, A_3, A_4, A_5 are considered.

A_1: exponential

$$p_X^{(1)}(x|\gamma) = \frac{1}{\gamma_2} \exp\left\{-\frac{x - \gamma_1}{\gamma_2}\right\} \qquad x \geq \gamma_1$$

$$-\infty < \gamma_1 < \infty, \gamma_2 > 0$$

A_2: double exponential

$$p_X^{(1)}(x|\gamma) = \frac{1}{2\gamma_2} \exp\left\{-\frac{|x - \gamma_1|}{\gamma_2}\right\} \qquad -\infty < x < \infty$$

$$-\infty < \gamma_1 < \infty, \gamma_2 > 0$$

A_3: logistic

$$p_X^{(1)}(x|\gamma) = \frac{\exp\left\{\dfrac{x - \gamma_1}{\gamma_2}\right\}}{\gamma_2\left[1 + \exp\left\{\dfrac{x - \gamma_1}{\gamma_2}\right\}\right]^2} \qquad -\infty < x < \infty$$

$$-\infty < \gamma_1 < \infty, \gamma_2 > 0$$

A_4: Pearson Type III

$$p_X^{(1)}(x|\gamma) = \frac{\exp\left\{-\dfrac{x - \gamma_1}{\gamma_2}\right\}}{\gamma_2\Gamma(\beta)}\left(\frac{x - \gamma_1}{\gamma_2}\right)^{\beta - 1} \qquad x > \gamma_1$$

$$-\infty < \gamma_1 < \infty, \gamma_2 > 0, \beta > 0.$$

A_5: "power distribution"

$$p_X^{(1)}(x|\gamma) = \frac{\exp\left\{-\frac{1}{2}\left|\frac{x - \gamma_1}{\gamma_2}\right|^{2/1+\beta}\right\}}{\gamma_2 \Gamma\left(1 + \frac{1 + \beta}{2}\right) 2^{[1 + (1 + \beta)/2]}} \quad -\infty < x < \infty$$

$$-\infty < \gamma_1 < \infty, \gamma_2 > 0, \beta > -1.$$

All the alternatives A_1–A_5 inclusive involve unknown parameters γ_1 and γ_2, which are location and scale respectively. Thus the power will be the same for all possible values of γ_1 and γ_2 according to the result proved in Theorem 8.2.

The asymptotic power is given by

$$P\left[\sum_1^2 d_i \chi_{1, a_i}^2 \geq \chi_2^2(\alpha)\right] \tag{6.1}$$

where the asymptotic nonnull distribution of $\hat{Q}$ is that of $\sum_1^2 d_i \chi_{1, a_i}^2$, as proved in Theorem 5.1, and $\chi_2^2(\alpha)$ is the $100(1 - \alpha)$ percent point of the χ_2^2 distribution.

A generalization of Gurland's (1955, 1956) Laguerre series expansion has been given by Kotz et al. (1967) for the distribution of quadratic forms in noncentral normal variates. We make use of this expansion to compute the power given by (6.1). These calculations have been carried out for sample sizes $n = 50, 75, 100$ and the two levels of significance $\alpha = .05$, .01. The results appear in Table 8.1 for all the alternatives A_1–A_5, with several different specified values of the parameter β in the case of A_4 and A_5 as indicated in the table.

A modified form of the Pearson χ^2 test has been considered by Dahiya and Gurland (1970b), where the test statistic is denoted by X_R^2. According to this modification the estimators obtained from the ungrouped sample are utilized in determining the class interval end points as well as in the test statistic X_R^2. For convenience in making some comparisons with the $\hat{Q}$ test, the values of power of the X_R^2 test against the alternatives listed in Table 8.1 are included for those cases corresponding to sample sizes $n = 50, 100$ which are available from Dahiya and Gurland (1970b). These values are enclosed in parentheses and are based in each case on the number of class intervals giving the maximum power in Tables 1 and 2 of Dahiya and Gurland (1970b). For example, in the case of alternative A_1 the power of the X_R^2 test attains a maximum value of 1.000 for sample sizes $n = 50, 100$ when the number of class intervals is 7, and in the case of alternative A_2 its power attains maximum values .547, .800 corresponding to sample sizes 50, 100 respectively based on three class intervals.

Table 8.1

Power of the $\hat{Q}$ Test for Normality

	Alternative	$\alpha = .05$			$\alpha = .01$		
		$n = 50$	75	100	50	75	100
A_1:	exponential	.927 (1.000)	.941	.953 (1.000)	.892	.913	.930
A_2:	double exponential	.833 (.547)	.858	.879 (.800)	.754	.789	.818
A_3:	logistic	.606 (.128)	.631	.654 (.180)	.465	.495	.523
A_4:	with $\beta = .5$	.966		.972	.976	.949	.957	.964
A_4:	with $\beta = 2.0$	.865		.898	.923	.804	.849	.883
A_4:	with $\beta = 2.5$	.839 (.502)	.879	.909 (.864)	.765	.820	.861
A_4:	with $\beta = 3.0$	.814 (.391)	.860	.895 (.716)	.732	.791	.837
A_4:	with $\beta = 3.5$	.790 (.318)	.841	.881 (.597)	.698	.762	.814
A_4:	with $\beta = 4.0$	.767 (.268)	.822	.865 (.506)	.667	.734	.789
A_4:	with $\beta = 5.0$	.722 (.205)	.784	.834 (.381)	.608	.680	.741
A_4:	with $\beta = 10.0$	.548		.617	.678	.407	.473	.534
A_5:	with $\beta = 3.0$	.996		.996	.997	.994	.995	.995
A_5:	with $\beta = 2.0$	.974		.976	.978	.960	.963	.966
A_5:	with $\beta = .95$	.815		.842	.866	.730	.767	.799
A_5:	with $\beta = .75$	.721 (.376)	.759	.792 (.603)	.604	.652	.695
A_5:	with $\beta = .50$	.527 (.211)	.571	.611 (.343)	.372	.418	.462
A_5:	with $\beta = .25$	.249		.271	.292	.117	.133	.149
A_5:	with $\beta = -.50$	.036 (.144)	.105	.216 (.262)	.002	.009	.029
A_5:	with $\beta = -.75$	.154		.483	.785	.008	.078	.280
A_5:	with $\beta = -.95$	.328 (.331)	.779	.969 (.583)	.027	.237	.621

n = sample size.
α = level of significance.
A_4 corresponds to the Pearson Type III distribution.
A_5 corresponds to the "power distribution."

It is evident on examining the values of power for the $\hat{Q}$ test in Table 8.1 that for most of the cases considered there its value is higher, and sometimes very much higher, than the value for the X_R^2 test.

As we examine Table 8.1 in detail, we note that for alternatives A_1 and A_2, namely the exponential and double exponential, the power is rather high. For the exponential the power is slightly lower than for that of the X_R^2 test, whereas the reverse is true for the double exponential. For a logistic alternative the difference in the power of the two tests is dramatic. For example, when $n = 100$, the power of the X_R^2 test with optimal number of classes $k = 3$ is .180 for $\alpha = .05$, whereas for the $\hat{Q}$ test the corresponding power is .654.

In regard to A_4, namely the Pearson Type III, it is evident from the table that the power is higher for low values of the parameter β and decreases slowly as β increases. The decrease in the power is explained by the fact that the alternative A_4 tends to normal as β becomes increasingly large. As evident from the few values of power of the X_R^2 test appearing for alternative A_4, it behaves similarly to the $\hat{Q}$ test, although its power is substantially less.

Alternative A_5 is considered in the table with values of β decreasing from 3.0 to $-.95$. Similar to the behavior of the X_R^2 test, the power increases as β increases for $\beta > 0$; it also increases as β decreases for $\beta < 0$, which behavior is explained by the fact that the normal distribution is a special case of the family A_5 with $\beta = 0$. For all the values of β considered here except $\beta = -.50$, the power of the $\hat{Q}$ test is obviously higher than that of the X_R^2 test.

7. Conclusion

The use of the statistic $\hat{Q}$ in testing for normality results in high values of power for many of the alternatives considered in Table 8.1. The form of $\hat{Q}$ for this test turns out to be relatively simple and could in fact be computed on a desk calculator. A modified form of the Pearson χ^2 statistic designated as X_R^2, which could also be used to test for normality as shown by Dahiya and Gurland (1970a, 1970b), has been compared with the $\hat{Q}$ test for several cases of the alternatives considered in Table 8.1 and found to have lower power for the most part.

References

Barankin, E. W., and J. Gurland. 1951. On asymptotically normal efficient estimators. *Univ. Calif. Publ. Stat.* 1: 89–129.

Chernoff, H., and E. L. Lehmann. 1954. The use of maximum likelihood estimates in χ^2 goodness of fit. *Ann. Math. Stat.* 25 : 579–86.

Dahiya, R. C., and J. Gurland. 1970a. Pearson chi-square test of fit with random intervals. I. Null case. MRC Tech. Summ. Rept. 1046.

———. 1970b. Pearson chi-square test of fit with random intervals. II. Non-null case. MRC Tech. Summ. Rep. 1051.

Gurland, J. 1948. Best asymptotically normal estimates. Unpubl. Ph.D. thesis, Univ. Calif., Berkeley.

———. 1955. Distribution of definite and of indefinite quadratic forms. *Ann. Math. Stat.* 26 : 122–27. (Correction. 1962. *Ann. Math. Stat.* 33 : 813.)

———. 1956. Quadratic forms in normally distributed random variables. *Sankhya* 17 : 37–50.

Hinz, P., and J. Gurland. 1970. A test of fit for the negative binomial and other contagious distributions. *J. Am. Statist. Assoc.* 65 : 887–903.

Kotz, S., N. L. Johnson, and D. W. Boyd. 1967. Series representations of distributions of quadratic forms in normal variables. II. Non-central case. *Ann. Math. Stat.* 38 : 838–48.

9

A Computer Program for the Mixed Analysis of Variance Model Based on Maximum Likelihood

H. O. HARTLEY and W. K. VAUGHN

1. Introduction

IN THIS PAPER we present a computer program for an analysis of variance of unbalanced data assumed to arise from a "mixed model." The analysis is based upon the principle of maximum likelihood estimation developed by Hartley and Rao (1967). To fix the ideas, it will be necessary to summarize the specification of the model and the estimation theory by maximum likelihood given by these authors. This is done in sections 1, 2, and 3. Section 4 then spells out in some detail the computational procedure developed. In section 5, we apply the numerical procedure to obtain point estimates of the components of variance involved in the mixed model. The examples chosen comprise both situations for balanced data (when comparison will be made with conventional analysis of variance estimates) as well as unbalanced data. The comparisons for balanced data show excellent agreement for all those situations in which maximum likelihood estimation agrees with the analysis of variance estimates on theoretical grounds. In the remaining situations good agreement is maintained. While we do not advocate the use of maximum likelihood for balanced data, the comparisons should inspire confidence for use with unbalanced data. Vaughn (1970) also contains details of the computer code as well as

H. O. HARTLEY is Director, Institute of Statistics, Texas A & M University, College Station.

WILLIAM K. VAUGHN is Assistant Professor of Biostatistics, Vanderbilt University, Nashville, Tennessee.

formulas for the asymptotic variances and covariances of the estimates of the ratios of the components of variance. These are of considerable importance in the estimation of measures of heritability and related studies.

2. The General Mixed Model

The specification of the general mixed analysis of variance model will be sufficiently general to cover most of the problems arising from unbalanced data. The linear model discussed herein is given by

$$y = X\alpha + U_1 b_1 + U_2 b_2 + \cdots + U_c b_c + e \tag{2.1}$$

where

X = an $n \times k$ matrix of known fixed numbers, $k \leq n$;
U_i = an $n \times m_i$ matrix of known fixed numbers, $m_i \leq n$;
α = a $k \times 1$ vector of unknown constants;
b_i = an $m_i \times 1$ vector of independent variables from $N(0, \sigma_i^2)$;
e = an $n \times 1$ vector of independent variables from $N(0, \sigma^2)$.

The random vectors $b_1, b_2, \ldots, b_c$, and e are mutually independent.

Further, it is assumed that the design matrices X and U_i, $i = 1, 2, \ldots, c$ are all of full rank. In the model given by (2.1) the fixed effects and random effects are separated so that α contains all levels of all fixed effects, and the c random factors are separated so that all elements of b_i have the same variance σ_i^2.

An additional important assumption about the design matrices is made which assures that the likelihood will tend to zero as any of the ratios $\gamma_i = \sigma_i^2/\sigma^2$ tend to infinity (see Hartley and Rao, 1967.) This is the following assumption of estimability. Denote by $m = \sum_{i=1}^{c} m_i$ the total number of levels in all c random components. Then the adjoined $n \times (k + m)$ matrix $M = [X|U_1|U_2|\ldots|U_c]$ is assumed to have as a base an $n \times r$ matrix W of the form $W = [X|U^*]$, where the $n \times (r - k)$ matrix U^* must contain at least one column from each U_1 so that $k + c \leq r \leq k + m$.

3. The Likelihood Equations

From the definition of y in (2.1) it is clear that y follows a multivariate normal distribution with mean $X\alpha$ and variance-covariance matrix

$$\sigma^2 H = [I + \gamma_1 U_1 U_1' + \gamma_2 U_2 U_2' + \cdots + \gamma_c U_c U_c'], \tag{3.1}$$

where

$$\gamma_i = \sigma_i^2/\sigma^2. \tag{3.2}$$

Then the likelihood of y is given by

$$L = (2\pi)^{-1/2n} \sigma^{-n} |H|^{-1/2} \exp\left[-\frac{1}{2\sigma^2} (y - X\alpha)' H^{-1}(y - X\alpha) \right]. \quad (3.3)$$

Writing $\lambda = \ln L$ we have

$$\lambda = -1/2n \ln(2\pi) - \frac{n}{2} \ln \sigma^2 - 1/2 \ln |H| - \frac{1}{2\sigma^2} (y - X\alpha)' H^{-1}(y - X\alpha),$$
$$(3.4)$$

and differentiating λ with respect to α, σ^2, and γ_i yields the equations

$$\frac{\partial \lambda}{\partial \alpha} = -\frac{1}{2\sigma^2} \left[-2X' H^{-1} y + 2(X' H^{-1} X)\alpha \right] = 0, \quad (3.5)$$

$$\frac{\partial \lambda}{\partial \sigma^2} = -\frac{n}{2\sigma^2} + \frac{1}{2\sigma^4} (y - X\alpha)' H^{-1}(y - X\alpha) = 0, \quad (3.6)$$

and

$$\frac{\partial \lambda}{\partial \gamma_i} = -1/2 \operatorname{tr}\left(H^{-1} \frac{\partial H}{\partial \gamma_i} \right) - \frac{1}{2\sigma^2} (y - X\alpha)' \frac{\partial H^{-1}}{\partial \gamma_i} (y - X\alpha)$$

$$= -1/2 \operatorname{tr}(H^{-1} U_i U_i') + \frac{1}{2\sigma^2} \left[(y - X\alpha)' H^{-1} U_i U_i' H^{-1}(y - X\alpha) \right]$$
$$= 0. \quad (3.7)$$

The maximum likelihood estimators for α and σ^2 in terms of the unknown γ_i are obtained from (3.5) and (3.6). They are

$$\tilde{\alpha}(\gamma_i) = (X' H^{-1} X)^{-1} X' H^{-1} y \quad (3.8)$$

and

$$n\tilde{\sigma}^2(\gamma_i) = y' H^{-1} y - y' H^{-1} X(X' H^{-1} X)^{-1} X' H^{-1} y. \quad (3.9)$$

However, the solution of (3.7), $\partial \lambda / \partial \gamma_i = 0$, cannot be found explicitly for $\gamma_1, \gamma_2, \ldots, \gamma_c$, thus making it necessary to employ numerical techniques.

4. Solution by Steepest Ascent

Substitution of (3.8) and (3.9) in (3.7) yields the c simultaneous nonlinear equations

$$\frac{\partial \lambda}{\partial \gamma_i} [\tilde{\alpha}(\gamma_i), \tilde{\sigma}^2(\gamma_i), \gamma_i] = -1/2 \operatorname{tr}(H^{-1} U_i U_i')$$

$$+ \frac{1}{2\tilde{\sigma}^2(\gamma_i)} [y - X\tilde{\alpha}(\gamma_i)]' H^{-1} U_i U_i' H^{-1}[y - X\tilde{\alpha}(\gamma_i)] = 0 \quad (4.1)$$

for the c values of the γ_i.

The solution to this system of equations can be obtained as the asymptotic limits of a system of c simultaneous differential equations, the equations of steepest ascent given by

$$\frac{d\gamma_i}{dt} = \frac{\partial\lambda}{\partial\gamma_i} [\tilde{\alpha}(\gamma_i), \tilde{\sigma}^2(\gamma_i), \gamma_i], \tag{4.2}$$

where the variable of integration t is auxiliary and the numerical integration commences at $t = 0$ with trial values $_0\gamma_i$ usually chosen as consistent estimators of the γ_i.

The solution $\gamma_i(t)$ converges to a solution point $\tilde{\gamma}_i$, which is a root of

$$\frac{d\gamma_i}{dt} = \frac{\partial\lambda}{\partial\gamma_i} = 0.$$

See Hartley and Rao (1967) for proof of convergence.

A modification of the steepest ascent will ensure that $\gamma_i \geq 0$ along the path of integration. Defining

$$\tau_i = \gamma_i^{1/2} \tag{4.3}$$

which is symmetrical at $\tau_i = 0$, we see that if τ_i is used as a parameter in place of γ_i, that

$$\frac{\partial\lambda}{\partial\tau_i} = \frac{\partial\lambda}{\partial\gamma_i} 2\tau_i. \tag{4.4}$$

Thus, the steepest ascent differential equations can be replaced by

$$\frac{d\tau_i}{dt} = \frac{\partial\lambda}{\partial\gamma_i} [\tilde{\alpha}(\gamma_i), \tilde{\sigma}^2(\gamma_i), \gamma_i] 2\tau_i. \tag{4.5}$$

Again the integration would commence at positive values $_0\gamma_i$; but should the path of integration reach a point where one or more of the $\tau_i = 0$, the integration would continue along the boundary $\tau_i = 0$ until the Runge-Kutta procedure would allow the τ_i to become positive again. This procedure ignores and avoids any possible solutions of the likelihood equations with $\tilde{\gamma}_i < 0$.

5. Application of the Runge-Kutta Procedure

Polynomial Approximation

The technique selected for the numerical integration of the system of c simultaneous differential equations given by (4.2) is a fifth-order Runge-Kutta procedure. Basically, any Runge-Kutta procedure provides an approximation to a truncated Taylor series expansion of the independent variables. For the fifth-order Runge-Kutta method the approximation is carried out in such a way that it agrees with the Taylor series expansion through terms involving h^5. To apply a fifth-order Runge-Kutta procedure to the system of steepest ascent equations, it is necessary to evaluate (4.1) six times for every iteration. Since (4.1) involves H^{-1} and a large number of iterations may be required for convergence, excessive amounts of computer time would be required to obtain a solution to the system of equations. For this reason a second-degree polynomial of the form

$$\frac{d\tau_i}{dt} = b_0^{(i)} + \sum_{j=1}^{c} b_j^{(i)}\delta_j + \sum_{\substack{j=1 \\ j \leq k}}^{c} \sum_{k=1}^{c} b_{jk}^{(i)}\delta_j\delta_k \qquad (5.1)$$

is used to approximate the right-hand sides of the equations of steepest ascent. Where $b_0^{(i)}$, $b_j^{(i)}$, and $b_{jk}^{(i)}$ are coefficients to be estimated and the $\delta_j, j = 1, 2, \ldots, c$ represent a coded point on a grid in the c-dimensional span of the τ_i. The criterion used to fit the polynomial to the equations of steepest ascent is least squares. The following steps are taken when fitting the polynomial approximation:

1. Since there are $(c + 1)(c + 2)/2$ coefficients to be estimated, the same number of points or more must be selected on the grid in the δ-space. In fact one more than the necessary number is used to obtain an estimate of the residuals. The set of points selected must be chosen so that the matrix Δ defined in (3) below has full-column rank. Consequently, the matrix $\Delta'\Delta$ can be inverted.

2. Defining $_0\tau_i = _0\gamma_i^{1/2}$ to be the initial trial value of the τ_i for the numerical integration and $\Delta_0\tau_i$ to be the grid increment in the τ-space, the points in the τ-space corresponding to the points in the δ-space are then found from the equation

$$\tau_i = \delta_i\Delta_0\tau_i + _0\tau_i \qquad (5.2)$$

3. Defining F_i to be the right-hand side of (4.5) evaluated at the grid points in the τ-space and Δ to be the matrix of squares and cross products of the Δ's whose ith row is

$$1, \delta_1\delta_2 \ldots \delta_c\delta_1^2\delta_2^2 \ldots \delta_c^2, \delta_1\delta_2, \delta_1\delta_3 \ldots \delta_{c-1}\delta_c$$

and $\hat{b}^{(i)}$ to be the vector of estimates of the coefficients for the ith equation (4.5), i.e.,

$$\hat{b}^{(i)\prime} = [\hat{b}_0^{(i)}, \hat{b}_1^{(i)}\hat{b}_2^{(i)} \ldots \hat{b}_c^{(i)}\hat{b}_{11}^{(i)}\hat{b}_{22}^{(i)} \ldots \hat{b}_{cc}^{(i)}\hat{b}_{12}^{(i)}\hat{b}_{13}^{(i)} \ldots \hat{b}_{c-1,c}^{(i)}],$$

the least squares solutions[1] are found from

$$\hat{b}^{(i)} = (\Delta'\Delta)^{-1}\Delta'F_i, \qquad i = 1, 2, \ldots, c. \qquad (5.3)$$

1. Least squares is used to obtain a mathematical approximation to a mathematical function. The justification of this procedure must be sought by monitoring the truncation of the approximation obtained.

Upon obtaining the least squares solutions of the coefficients of the polynomial approximation for each equation of the system, a Runge-Kutta procedure is now applied using these approximations as right-hand sides.[2] When this Runge-Kutta procedure yields a set of $\tilde{\tau}_i$ so that $|d\tau_i/dt| < \varepsilon_1$ for every i, the procedure terminates. Then if the estimates of τ_i obtained from the present cycle, say $\tilde{\tau}_i'$, are sufficiently close to the initial trial values for the cycle, say $\tilde{\tau}_i''$, i.e., if $|\tau_i' - \tau_i''| < \varepsilon_2$ for all values of i, convergence is established and the estimates of the variances and covariances can be computed. If convergence is not established, the current cycle of the Runge-Kutta procedure is terminated, another polynomial approximation is obtained, and with these right-hand sides a new Runge-Kutta cycle is started using the terminal values τ_i of the previous cycle as initial trial values for the new cycle.

Selection of Optimum Step Size for Runge-Kutta

The selection of the step size h, i.e., the increment in the variable of integration t, in a Runge-Kutta procedure is important since it governs the rate of convergence as well as the accuracy of the final solution. For example, if the step size is too small, convergence may be very slow. To choose a step size, the empirical principle of forcing the second-order term in the Taylor's series expansion to be one-tenth the first term was used. This gives for the first-order term

$$h\frac{d\tau_i}{dt} = h\frac{\partial\lambda}{\partial\tau_i}$$

and for the second-order term

$$\frac{h^2}{2}\frac{d^2\tau_i}{dt^2} = \frac{h^2}{2}\sum_j \frac{\partial^2\lambda}{\partial\tau_j\partial\tau_i}\frac{\partial\lambda}{\partial\tau_j}$$

and leads after some algebra to the rule

$$h = \frac{.2\sqrt{\sum_{i=1}^{c}[b_0^{(i)}]^2}}{\sqrt{\sum_{i=1}^{c}\left[\sum_{j=1}^{c}\frac{b_0^{(j)}b_j^{(i)}}{\Delta_0\tau_j}\right]^2}}.$$

A computer program has been implemented making use of the above derivations to solve the likelihood equations (3.5), (3.6), and (3.7). Documentation for the computer program is given in Vaughn (1970).

2. The cycle is monitored and as soon as $|\delta_i| \geqslant 1$ for at least one i it is restarted with a large $\Delta_0\tau_i$.

6. Examples for Point Estimation

Introduction

This section is concerned with applying the techniques derived in sections 2 to 5 to specific examples. The majority of the examples are small, balanced data examples to facilitate comparisons between maximum likelihood and analysis of variance estimators. While analysis of variance estimators are unbiased, such is not always the case with maximum likelihood; but agreement between analysis of variance and maximum likelihood is obtained in the balanced case when maximum likelihood yields unbiased estimators.

The Twofold Nested Model

Snedecor and Cochran (1967, p. 286) cite data on the calcium concentration in turnip greens. Four plants were taken at random, then three leaves were randomly selected from each plant. From each leaf two samples of 100 mg each were taken, and the calcium content was determined by microchemical methods giving rise to the data in Table 9.1.

The model used for this analysis is

$$y_{ijk} = \mu + a_i + b_{ij} + e_{ijk}, \tag{6.1}$$

Table 9.1
Calcium Concentration in Turnip Greens

Plant	Leaf	Determinations	
1	1	3.28	3.09
	2	3.52	3.48
	3	2.88	2.80
2	1	2.46	2.44
	2	1.87	1.92
	3	2.19	2.19
3	1	2.77	2.66
	2	3.74	3.44
	3	2.55	2.55
4	1	3.78	3.87
	2	4.07	4.12
	3	3.31	3.31

where

a_i represents the effect of the ith level of plants;

b_{ij} represents the effect of jth leaf from the ith plant; and

e_{ijk} is the effect of the kth determination from the jth leaf from the ith plant, and $i = 1, 2, 3, 4, j = 1, 2, 3,$ and $k = 1, 2$.

The following assumptions are made:

1. $a_i \sim N(0, \sigma_a^2)$,
2. $b_{ij} \sim N(0, \sigma_b^2)$,
3. $e_{ijk} \sim N(0, \sigma_e^2)$,
4. a_i, b_{ij}, and e_{ijk} are all mutually independent.

Table 9.2 gives the analysis of variance for the above data.

Table 9.2

Analysis of Variance for Turnip Green Data

Source of Variation	Degrees of Freedom	Mean Square	Expected Mean Square
Plants	3	2.520115267	$\sigma_e^2 + 2\sigma_b^2 + 6\sigma_a^2$
Leaves/plants	8	.328775	$\sigma_e^2 + 2\sigma_b^2$
Determinations/leaves	12	.0066541667	σ_e^2

The analysis of variance estimates can be obtained from Table 9.2 by equating the mean square column to the expected mean square column and solving for the unknown parameters. This gives

$\hat{\sigma}_a^2 = (2.520115267 - .328775)/6 = .3652233778$
$\hat{\sigma}_b^2 = (.328775 - .0066541667)/2 = .1610604167$
$\hat{\sigma}_e^2 = .0066541667$.

From these we obtain the estimates of γ_a and γ_b as $\hat{\gamma}_a = 54.8864$ and $\hat{\gamma}_b = 24.204$.

The twofold nested model given by (6.1) may be rewritten, using the notation of (2.1), as

$$y = X\mu + U_1 b_1 + U_2 b_2 + e, \tag{6.2}$$

where $n = 24$, $m_1 = 4$, $m_2 = 12$, $c = 2$, $k = 1$, and where X is the 24-element unitary column vector and U_1 and U_2 are the usual 24×4 and 24×12 design matrices of 1's and 0's representing "plants" and "leaves within plants" respectively. Finally, the vectors of effect variables b_1 and b_2 are defined by

$$b_1' = [a_1, a_2, a_3, a_4]$$
$$b_2' = [b_{11}, b_{12}, b_{13}, b_{21}, b_{22}, b_{23}, b_{31}, b_{32}, b_{33}, b_{41}, b_{42}, b_{43}].$$

To obtain the maximum likelihood estimates for the parameters of the mixed model, several complete Runge-Kutta cycles (i.e., refitting of the polynomial approximation) were necessary to achieve convergence for this example. The first complete cycle will be discussed in detail. The steps of this cycle are:

1. All necessary constants and design matrices as well as the δ-grid are input to the computer program. From the δ-grid the τ-grid is obtained, using the initial trial values $_0\tau_1 = (39)^{1/2} {}_0\tau_2 = (24)^{1/2}$ and the grid increments $\Delta {}_0\tau_1 = 5$, and $\Delta {}_0\tau_2 = 5$.
2. From $_0\tau_1$ and $_0\tau_2$ initial values of $\tilde{\alpha}(_0\tau_i)$ and $\tilde{\sigma}^2(_0\tau_i)$ are obtained. For this example they are $\tilde{\alpha}(_0\tau_i) = 3.0121$ and $\tilde{\sigma}^2(_0\tau_i) = .0066612$.
3. Using the estimates of α and σ^2 together with the grid obtained from the initial trial values of the τ_i and the grid increments, the polynomial approximations are obtained as described above. For the first cycle for this example they are:

$$\frac{d\tau_1}{dt} = .0044 - .0474\delta_1 + .022\delta_2 + .00089\delta_1^2 - .0098\delta_2^2 + .0084\delta_1\delta_2$$
(6.3)

$$\frac{d\tau_2}{dt} = .00699 + .0165\delta_1 - .214\delta_2 - .00057\delta_1^2 + .0526\delta_2^2 - .0106\delta_1\delta_2$$
(6.4)

4. The Runge-Kutta procedure is now applied to this system of differential equations, yielding at the end of the first cycle, $_1\tilde{y}_1 = 39.093$, $_1\tilde{y}_2 = 24.19$, $_1\tilde{\sigma}_e^2 = .00666$, and $_1\tilde{\mu} = 3.0121$.
5. Since convergence has not been established, another cycle is started with the initial trial values obtained from (4) above.

For this example a total of three complete cycles was necessary to establish convergence. The number of cycles required for convergence may vary and will always depend on the initial trial values for the first cycle. For this example the final values were $\tilde{y}_a = 39.095$, $\tilde{y}_b = 24.1999$, $\sigma_e^2 = .0066549$, and $\tilde{\mu} = 3.0121$.

It is well known that the analysis of variance procedure produces estimates that are unbiased. While in some cases maximum likelihood also provides unbiased estimates, there is no guarantee that this is the case. In this example it is obvious that $\tilde{y}_a$ is biased, while $\tilde{y}_b$ as well as $\tilde{\sigma}_e^2$ are unbiased. Hartley and Rao (1967) show that their procedure gives the following maximum likelihood estimates for the balanced twofold nested model of example one:

$$4(\tilde{\sigma}_e^2 + 2\tilde{\sigma}_b^2 + 6\tilde{\sigma}_a^2) = \sum_{ijk} (\bar{y}_i \ldots - \bar{y} \ldots)^2, \tag{6.5}$$

$$(\tilde{\sigma}_e^2 + 2\tilde{\sigma}_b^2) = \sum_{ijk} (\bar{y}_{ij\cdot} - \bar{y}_i \cdot\cdot)^2/8, \tag{6.6}$$

$$\tilde{\sigma}_e^2 = \sum_{ijk} (y_{ijk} - \bar{y}_{ij\cdot})^2/12. \tag{6.7}$$

Table 9.3

Comparison of Maximum Likelihood and Analysis of Variance Estimators for Turnip Green Data

Estimates	γ_a	γ_b	σ_e^2
Analysis of variance	54.8864	24.204	.0066542
Maximum likelihood estimators from computer program	39.095	24.1999	.0066549
Maximum likelihood estimators from Table (9.2)	39.106	24.204	.0066542

We see from Table 9.3 that the analysis of variance estimate of γ_a and the maximum likelihood estimate of γ_a are not the same. As indicated in (6.5), this happens because the maximum likelihood estimate is biased. However, if the sum of squares for plants in line one of Table 9.2 is divided by 4 instead of 3, thus making the analysis of variance estimate comparable to maximum likelihood, we see from lines two and three of Table 9.3 that there is very close agreement between $\tilde{\gamma}_a$ obtained from the two different methods. Since maximum likelihood gives unbiased estimates of γ_b and σ_e^2, there is no need for this adjustment for comparison. In all cases where the maximum likelihood estimate should agree with the analysis of variance estimate, the two agree to at least two decimal places and the estimates of the error mean square agree to five places. Indeed, if a more stringent criterion for convergence is imposed in the computer program, better agreement can be attained. Although this technique of maximum likelihood does not guarantee a global maximum of the likelihood function for this example, it was in fact obtained.

Unbalanced One-Way Classification

Ostle (1963, p. 287) cites data on the moisture content of pine boards. Five storage conditions were studied to determine the effect on the moisture content of white pine lumber. Table 9.4 gives the data arising from this example. The model used to analyze this data is

$$y_{ij} = \mu + a_i + e_{ij}, \tag{6.8}$$

Table 9.4

Moisture Content of Fourteen Pine Boards

Storage Conditions				
1	2	3	4	5
7.3	5.4	8.1	7.9	7.1
8.3	7.4	6.4	9.5	...
7.6	7.1	...	10.0	...
8.4	...	...	...	...
8.3	...	...	...	...

where

a_i represents the effect of the ith level of storage conditions;

e_{ij} represents the effect of the jth board subjected to the ith storage condition, $i = 1, \ldots, 5, j = 1, 2, \ldots, n_i$;

$a_i \sim N(0, \sigma_a^2)$;

$e_{ij} \sim N(0, \sigma_e^2)$; and

a_i and e_{ij} are all mutually independent.

The analysis of variance for this data is given in Table 9.5. The analysis of variance estimates for σ_a^2, σ_e^2, and γ_a are $\hat{\sigma}_a^2 = .708, \hat{\sigma}_e^2 = .80$, and $\hat{\gamma}_a = .885$.

Table 9.5

Analysis of Variance for Pine Board Data

Source of Variation	Degrees of Freedom	Mean Squares	Expected Mean Square
Storage conditions	4	2.67	$\sigma_e^2 + 2.64\sigma_a^2$
Experimental error	9	.80	σ_e^2

Writing the model (6.1) using the notation of (2.1), we have

$$y = X\mu + U_1 b_1 + e,$$

where $n = 14$, $m_1 = 5$, $c = 1$, $k = 1$, and X is the 14-element unitary vector and U_1 the 14×5 design matrix representing storage conditions while $b_1 = [a_1, a_2, a_3, a_4, a_5]$.

For the initial cycle for this example the trial value chosen was $_0\tau_1 = (_0\gamma_1)^{1/2} = (.9)^{1/2}$, and the grid increment was $\Delta \, _0\tau_1 = .85$. Table 9.6 gives a concise presentation of what happens during each cycle for this example.

Table 9.7 gives a comparison of the analysis of variance and maximum likelihood estimates for this example.

Table 9.6
Runge-Kutta Cycles for Pine Board Data

Cycle	$\tilde{\gamma}_a$	Polynomial Approximation*		
0	.9	$-.477$	$-1.53\delta_1$	$+.736\delta_1^2$
1	.666	.014	$-.988\delta_1$	$-.218\delta_1^2$
2	.675	$-.009$	$-1.15\delta_1$	$+.153\delta_1^2$
3	.669	.0042	$-1.06\delta_1$	$+.059\delta_1^2$
4	.672	$-.0021$	$-1.11\delta_1$	$+.106\delta_1^2$
5	.671	.00082	$-1.09\delta_1$	$+.085\delta_1^2$

*These approximations change from cycle to cycle because of the changes in origin and width of the grid in the τ-space resulting in different τ-δ relations.

Table 9.7
Comparison for One-Way Classification

Estimates	γ_a	σ_e^2
Analysis of variance	.885	.708
Maximum likelihood	.671	.773

Comparisons between maximum likelihood and analysis of variance are difficult for unbalanced data. Even for this simplest case of the one-way classification the likelihood equations cannot be solved explicitly for the estimates of σ_a^2 and σ_e^2 and hence γ_a. However, for this example maximum likelihood does not give answers too different from those from the customary analysis of variance.

Twofold Nested Model When One Variance Ratio Is Zero

Snedecor and Cochran (1967, p. 289) cite data on pig breeding. Five sires are to be evaluated in pig raising. Each sire is mated to a random group of dams, each mating producing a litter of pigs. Table 9.8 gives the average daily gain of two pigs from each litter.

In this example Snedecor regarded "sires" as a fixed effect. However, for purposes of illustrating the maximum likelihood technique when one variance ratio is zero, the same model and assumption are used as in (6.1). Table 9.9 gives the analysis of variance for this data.

Analysis of variance estimates of σ_a^2, σ_b^2, and σ_e^2 are obtained from Table 9.9. Since the mean square for dams within sires is larger than the

Table 9.8
Average Daily Gain of Two Pigs

Sire	Dam	Pig Gains	
1	1	2.77	2.38
	2	2.58	2.94
2	1	2.28	2.22
	2	3.01	2.61
3	1	2.36	2.71
	2	2.72	2.74
4	1	2.87	2.46
	2	2.31	2.24
5	1	2.74	2.56
	2	2.50	2.48

Table 9.9
Analysis of Variance for Pig Data

Source of Variation	Degrees of Freedom	Mean Square	Expected Mean Square
Sires	4	.0249325	$\sigma_e^2 + 2\sigma_b^2 + 4\sigma_a^2$
Dams/same sire	5	.11271	$\sigma_e^2 + 2\sigma_b^2$
Pairs/same dam	10	.0387	σ_e^2

mean square for sires, the analysis of variance method gives a negative estimate for σ_a^2. The estimate generally used when this happens is $\hat{\sigma}_a^2 = 0$. We can then obtain the estimates of σ_b^2 in two ways as follows:

1. The estimate, $\hat{\sigma}_b^2$, can be obtained from

$$_1\hat{\sigma}_b^2 = (.11271 - .0387)/2 = .037005$$

which is the usual analysis of variance estimate.

2. Since $\hat{\sigma}_a^2 = 0$ and assuming this implies $\sigma_a^2 = 0$, the mean square for sires has the same expectation as the mean square for dams/same sire. This suggests a pooling of the two sums of squares which gives

$$_2\hat{\sigma}_b^2 = (.073698 - .0387)/2 = .017499.$$

From these two estimates $_1\hat{\gamma}_b$ and $_2\hat{\gamma}_b$ are $_1\hat{\gamma}_b = .9562$ and $_2\hat{\gamma}_b = .45217$.

Turning now to the computer program for this example, the initial trial values were $_0\tau_1 = (_0\gamma_1)^{1/2} = (1)^{1/2}$, $_0\tau_2 = (_0\gamma_2)^{1/2} = 1$, $\Delta\,_0\tau_1 = 1$, and $\Delta\,_0\tau_2 = .9$. From these initial trial values four complete cycles were required to establish convergence. The final maximum likelihood estimates are $\tilde{\gamma}_a = 0$, $\tilde{\gamma}_b = .35696$, $\tilde{\sigma}_e^2 = .0387$, and $\tilde{\mu} = 2.5740$. The computer program solves (3.6) and (4.4); and since $\tilde{\gamma}_a = 0$, the computer program solves $\partial\lambda/\partial\sigma^2 = 0$, $\partial\lambda/\partial\gamma_b = 0$, and $\tilde{\gamma}_a = 0$. Hartley and Rao (1967, p. 100–101) spell out the above likelihood equations for the special case of a balanced twofold nested model. From these it can be verified that σ_e^2 should be estimated by the "pairs/same dam" mean square and $\sigma_e^2 + 2\sigma_b^2$ by the "pooled mean square" "sires" + "dams/same sire," using the "wrong" degrees of freedom $5 + 5$ as a divisor. This computation yields exactly $\tilde{\gamma}_b^2 = 0.35695$, confirming the computer program solution exactly. The comparisons are summarized in Table 9.10.

Table 9.10

Comparisons for Pig Data

Estimates	γ_a	γ_b	σ_e^2
AOV$_1$	0	.9562	.0387
AOV$_2$	0	.45217	.0387
AOV$_3$*	0	.35695	.0387
MLE	0	.35696	.0387

*Based on between-dams sum of squares/10.

Two-Way Classification with Interaction

Bowker and Lieberman (1963, p. 362) cite data on the variability among ovens used in life-testing various electronic components. Three ovens and two temperatures normally used for life-testing of electronic components are selected. A single type of component is selected and operated in an oven until it fails. Table 9.11 gives the data arising from this experiment.

Table 9.11

Electronic Component Data

Temperature	Oven		
	1	2	3
550°F	237	208	192
	254	178	186
	246	187	183
600°F	178	146	142
	179	145	125
	183	141	136

The model used for this analysis is $y_{ijk} = \mu + a_i + b_j + c_{ij} + e_{ijk}$, where $i = 1, \ldots, A, j = 1, 2, \ldots, B, k = 1, \ldots, N$,

$$a_i \sim \text{NID}(0, \sigma_a^2),$$
$$b_j \sim \text{NID}(0, \sigma_b^2),$$
$$c_{ij} \sim \text{NID}(0, \sigma_c^2),$$
$$e_{ijk} \sim \text{NID}(0, \sigma_e^2),$$

and a_i, b_j, c_{ij}, and e_{ijk} are all mutually independent. Table 9.12 gives the analysis of variance for this data.

Table 9.12

Analysis of Variance for Electronic Components

Source of Variation	Degrees of Freedom	Mean Square	Expected Mean Square
Oven	2	4823.17	$\sigma^2 + 3\sigma_c^2 + 6\sigma_a^2$
Temperature	1	13667.56	$\sigma^2 + 3\sigma_c^2 + 9\sigma_b^2$
Oven × temperature	2	137.39	$\sigma^2 + 3\sigma_c^2$
Error	12	69.78	σ^2

From Table 9.12 the analysis of variance estimates are

$$\hat{\gamma}_c = .323$$
$$\hat{\gamma}_b = 21.54$$
$$\hat{\gamma}_a = 11.19$$
$$\hat{\sigma}_e^2 = 69.78.$$

The initial trial values selected for this example are

$$_0\tau_1 = (_0\gamma_1)^{1/2} = (9)^{1/2}, \, _0\tau_2 = (_0\gamma_2)^{1/2} = (12)^{1/2}, \, _0\tau_3 = (_0\gamma_3)^{1/2} = (.5)^{1/2}$$
$$\Delta \, _0\tau_1 = 5, \Delta \, _0\tau_2 = 6, \Delta \, _0\tau_3 = .4.$$

The polynomial approximations for the first cycle are

$$\frac{d\tau_1}{dt} = .047 - .633\delta_1 + .0879\delta_2 + .026\delta_3 + .391\delta_1^2$$
$$- .04\delta_2^2 - .025\delta_3^2 - .069\delta_1\delta_2 + .0375\delta_1\delta_3 + .036\delta_2\delta_3,$$

$$\frac{d\tau_2}{dt} = .0398 + .082\delta_1 - .389\delta_2 + .022\delta_3 - .0349\delta_1^2$$
$$+ .23\delta_2^2 - .0189\delta_3^2 - .0557\delta_1\delta_2 + .035\delta_1\delta_3 + .0167\delta_2\delta_3,$$

$$\frac{d\tau_3}{dt} = -.31 + .024\delta_1 + .522\delta_2 - .629\delta_3 + .00757\delta_1^2$$
$$- .0267\delta_2^2 + .286\delta_3^2 + .082\delta_1\delta_2 - .082\delta_1\delta_3 - .0187\delta_2\delta_3.$$

At the end of the first cycle the revised estimates are

$$\tilde{\gamma}_a = 9.296, \tilde{\gamma}_b = 12.48, \tilde{\gamma}_c = .337.$$

At the end of the fifteenth cycle convergence is established and the maximum likelihood estimates are

$$\tilde{\gamma}_a = 9.54, \tilde{\gamma}_b = 12.82, \tilde{\gamma}_c = .326,$$
$$\tilde{\sigma}_e^2 = 69.75,$$
$$\tilde{\mu} = 180.33.$$

Comparing the maximum likelihood estimates with the analysis of variance estimates, we see that $\tilde{\gamma}_c$ and $\hat{\gamma}_c$ and $\tilde{\sigma}_e^2$ and $\hat{\sigma}_e^2$ agree quite well, while $\tilde{\gamma}_a$ and $\tilde{\gamma}_b$ do not agree with $\hat{\gamma}_a$ and $\hat{\gamma}_b$. This failure to agree occurs because $\tilde{\gamma}_a$ and $\tilde{\gamma}_b$ are biased.

References

Bowker, A. H., and G. J. Lieberman. 1963. *Engineering statistics.* Englewood Cliff, N.J.: Prentice-Hall.

Hartley, H. O., and J. N. K. Rao. 1967. Maximum likelihood estimation for the mixed analysis of variance model. *Biometrika* 54: 93–108.

Ostle, B. 1963. *Statistics in research.* Ames: Iowa State Univ. Press.

Snedecor, G. W., and W. G. Cochran. 1967. *Statistical methods,* 6th ed. Ames: Iowa State Univ. Press.

Vaughn, W. K. 1970. A technique for maximum likelihood estimation in mixed models. Ph.D. thesis, Texas A & M Univ., College Station.

10

On an Experiment in Forecasting
a Tree Crop by Counts and Measurements

RAYMOND J. JESSEN

1. Introduction

A SERIES OF SURVEYS were carried out in the 1953–54, 1954–55, and 1955–56 seasons for oranges in Florida, with the primary object of devising and testing new methods, more accurate than those currently available, to obtain early-season and within-season forecasts of ensuing crops. This study provided a means to estimate total trees and objective methods for estimating total weight of fruits on trees with a minimum of fruit picking. The methods of sampling and field procedures are described in this paper, and some comments are made on their usefulness.

Florida oranges have their main "bloom" in February and March. Some trees on some occasions have one or more "late blooms" which may occur at intervals of four to six weeks subsequent to the main bloom. After the bloom a number of the fruits, usually a substantial fraction, are dropped. By early July the surviving fruits are large enough to be identified, and their survival rates to harvest are high enough so that "fruit set" at this time is a useful indicator of the crop to be harvested. Three types are commonly distinguished: the early type are picked in October and November, the midseason type in December through February, and the late types (mostly Valencias) from February through June. Since all types are used

RAYMOND J. JESSEN is Professor of Business Statistics, University of California, Los Angeles.

The experimental program discussed in this paper was sponsored by the Birds-Eye Division of General Foods Corporation. Because of proprietary interests, key procedures and findings were held confidential for five years.

for frozen concentrate and most oranges in Florida end up as such, the aggregation of the three types is quite useful for many purposes.

The first official forecast for the new crop is released (announced) on or about October 10. This forecast, produced by the Florida Crop and Live-stock Reporting Service (FCLRS)—a joint agency of the departments of agriculture of the United States and of Florida—is based on a survey carried out during September. The data are obtained from questionnaires sent to growers. Each grower is asked to report his actual production for the past season and to estimate his production for the current season. Results of this survey are compared with those of previous years and adjusted for certain biases of judgment which may have appeared. The actual size of each crop in the past is fairly accurately determined, since processors and shippers are required to report quantities handled by their facilities. On November 10 and each following month through June a revised forecast is made, based on whatever new information becomes available during the season—such as growers' revised judgments on their grove's production, actual production from picked growers, industry data on total oranges processed or shipped by various dates during the season, etc.

Accurate information on the size of the new crop is very helpful to both growers and processors because the price is rather sensitive to supply (e.g., during the 1960's a 20% change in production resulted in a 30% change in the on-tree price). Changes in the month-to-month revised forecasts within a season can result in rather important changes in the value of oranges still on the tree. Early and accurate knowledge of the size of a crop can help greatly in bringing about orderly wholesaling and picking on the part of growers and orderly processing, storage, and retailing on the part of processors and other handlers.

An analysis of the accuracy record of the official October forecasts over the period 1939–40 to 1951–52 gave a mean square error of forecast of 8.8% (of the actual crop size). There was little evidence of bias. Since this error was regarded as rather large, it was hoped that a scheme could be developed to bring it down to about 3.5% or less, on a modest budget, and also to provide information useful to frozen fruit concentrators on the quality of fruits (such as juice yield, sweetness, acid, etc.).

2. The Program in General

After considering several alternatives such as (1) a projection based on a time series of past production, including one involving the separate projections of the number of bearing trees and yield per tree; (2) projections from a long series of "frame count" and "size" data; (3) projections based on weather data; and (4) use of "expert crop estimators" to estimate by eye the potential crop on a representative sample of groves, it was decided to try a scheme involving direct counts and measurements. The scheme follows in sketch form.

The forecaster in the first season is of the form,

$$Y = TB_H \tag{2.1}$$

where

T = the number of bearing trees in Florida.
B_H = the average number of boxes actually harvested per tree.
Y = the total boxes of fruit harvested.

Since in 1952 the number of bearing trees T was unknown; it was necessary to devise a scheme to estimate it as well as B_H. Operationally it was convenient to also componentize T and B_H. Thus,

$$T = XT_X \tag{2.2}$$

where

X = the total number of orange trees in the state according to the best information available at the time—in this case the FCLRS—which is presumed to be inaccurate.
T = the true number of orange trees in the state, which is unknown.
$T_X = T/X$.

The average harvested boxes of fruit per tree can be expressed as

$$B_H = (F_H = F_I - F_D - F_U)V_F \cdot W_V \cdot B_W \tag{2.3}$$

where

F_I = the number of fruits hanging on the average tree at a convenient beginning point in the maturing season.
F_D = the number of F_I fruits that drop to the ground (or become obviously unharvestable) prior to harvest.
F_U = the number of F_I fruits left on the tree after harvest.
V_F = the average volume (size) of harvested fruits.
W_V = the average weight per unit volume of harvested fruits.
B_W = the number of 90-pound boxes in an average unit volume of harvested fruit. (In Florida a box of oranges is 90 pounds of fruit, regardless of size.)

Each of these factors is to be predicted from counts and measurements taken at one or more times from July until harvest. The details are given in the next section. In the second and subsequent years forecasters other than (2.1) are feasible. These will also be discussed later.

The basic ideas for the program were worked out in September 1953, including a scheme for selecting sample branches on a tree (Jessen, 1955). During December a frame of areas was established to facilitate the selection of a sample for estimating the number of bearing trees and for providing a sample on which to count and measure fruits. The first field operation, carried out in January and February 1954, consisted of locating sample "lines" (tree rows), identifying the type of tree in each planting position within these lines, designating the particular trees belonging to the sample, selecting a sample of two branches on each sample tree and counting the fruit on them, and selecting a sample among these fruits and measuring their circumference for volume. Since 68% of the early and midseason (E & M) trees were already picked by this time, the initial survey dealt essentially with Valencias. A forecast of the 1953–54 Valencia crop was made from this survey.

The calendar and operations schedule for the three crop seasons involved were as follows:

1953–54: A forecast for the Valencia crop only. Field work on fruit counts took place through January to February 7, 1954. Size and drop counts were made at this time only.

1954–55: Forecasts were made on both E & M's and Valencias. Field work on fruit counts took place in July 1954. Drop and gleaning counts (unharvested fruits) and size and specific gravity measurements were made in October, December, February, and April. Measurements on fruit quality (juice, acid, and Brix) were taken in the October to April surveys.

1955–56: Forecasts were made on both E & M's and Valencias. Field work on fruit counts took place in September 1955. Drop and gleaning counts and size and specific gravity measurements were made in September, October, December, February, and April. Quality measurements were also taken in the following months.

3. The Samples

The data for the forecasts were obtained from a series of samples tied together in a single "bundle." Since it was felt that the number of bearing orange trees was not known accurately enough for the purposes at hand and it appeared highly desirable to have a sample of trees with known selection probabilities in order to have a self-contained means for evaluating accuracy, it was decided to construct an area sampling frame to deal with both matters.

General

Considering the limitations of resources and time and the general nature of the problem at hand, it appeared that a sample of 2,000 trees located in the vicinities of about 500 locations in the citrus area of Florida would serve as a useful scaffolding on which to plan the detailed structure of the samples. With an estimated 20 million trees in the state, this represented an overall sampling rate of 1 in 10,000. Within this framework the next step was to design a set of procedures which would produce a set of 2,000 trees selected from all trees with equal probabilities and consisting of clusters of as near 4 as possible at each of the 500 locations.

Strata (Counties)

Since about 90% of the approximately 20 million orange trees in Florida were situated in 16 of the state's 67 counties, it was decided to confine the survey to this relatively compact geographic area.[1] The county became a logical stratum, and the distribution of bearing orange trees by county published by the FCLRS was used for allocating the finally determined locations among the county strata.

Primaries (Area Sections)

A frame was constructed for each of the counties using the U.S. Public Land Survey section (an area approximately one mile square), which is fairly easily identifiable and listable in the Florida citrus area. It was desirable to obtain some information—even if fairly crude—on the distribution of the county's orange trees among the sections, which for all counties is quite lumpy. To do this, two schemes were followed. In Pasco and Polk counties information supplied by the Florida State Plant Board (based on its disease inspection records) was used. For our purposes this data had to be taken off maps and records and assigned to appropriate sections. In the remaining 14 counties information was obtained by using aerial photographs on file at each of the county agricultural offices. One of our workers viewed each of the photographs (covering about 9 square miles) and after identifying the appropriate area on a county map showing the section boundaries, estimated by direct judgment by eye the approximate area of each section covered by groves. With a small amount of

1. The counties were: Brevard, De Soto, Hardee, Highlands, Hillsborough, Indian River, Lake, Manatee, Marion, Orange, Pasco, Pinellas, Polk, St. Lucie, Seminole, Volusia.

practice this can be done quite rapidly and accurately. A listing of all sections within each county together with a measure of the land devoted to groves (or in the case of Polk and Pasco counties, the number of orange trees according to inspection records) was prepared for each of the 16 counties. Using the information as a measure of size, sections or "locations" were selected with probability proportional to size by a systematic draw from each county, hence achieving some geographic stratification within counties. An aerial photograph was obtained for each of the 493 selected sections. These are relatively low in cost and at a scale of 3 inches to the mile provide an accurate (at the time they are taken), detailed map of the sample area and its vicinity.

Secondaries (Sectors or "Groves")

At the second stage of sampling a grove or some convenient sector of tree plantings was used. These units were identified on the aerial photo and outlined on a translucent paper overlay prepared for the purpose. All land visibly planted to grove (practically all groves in this area are citrus, and all but the most recently set trees are visible on a photo) was put into this frame of sectors. Large solid plantings are usually divisible into 40-acre tracts and were so sectorized. An "eyeball" estimate made of the area in each sector was used as a measure of size for selecting sample sectors.

The number of sectors chosen for the sample was made to depend on the fraction of all citrus trees that are oranges (information with suitable accuracy being available only on a county basis). The scheme for doing this and some supporting logic is given in the next section. The number varied from 1 to 5. Selection was systematic with probability proportional to size from eyeball estimates.

Tertiaries (Rows)

The third-stage unit was a row of trees appearing on the aerial photograph. The "standard" row was 1/4 mile long (across a 40-acre tract) and contained about 50 trees. In the sample case where the sector was a solid square of trees, a convenient orientation was chosen to permit easy access from visible roads, the rows were counted under a low-powered magnifying glass, and a random selection was made. Unambiguous directions for locating the selected row were written as well as sketched on the translucent overlay as say "row 22 from northwest corner." For irregular-shaped sectors, fractions of rows were pieced together to obtain a composite row of about 50 tree positions.

To keep the number of trees in the ultimate sample as near 4 as possible, two measures were adopted. The first was to select an orientation such that rows ran across areas in the photos which had a heterogeneous appearance such as strips of dark- and light-colored trees. These heterogeneous strips may indicate a mixture of citrus, such as tangerines and grapefruit (a not unknown practice in many parts of the citrus area), age differences, etc. This measure should prevent the selection of a row of solid tangerines or grapefruit in these mixed planting cases. Second, the number of rows (and hence essentially the number of sectors) assigned to a particular location depended on the fraction of trees believed to be orange in that county. More lines were assigned in cases where the fraction was low. The assignments varied from five in Indian River (where about 44% of the trees were orange) to two in Marion County (where 91% were orange).

Quartiaries (Planting Positions)

The field worker was instructed to locate each sample row and indicate on a record for that purpose the occupancy status of each planting position. He would walk down the row identifying each position as to whether it was "empty," "grapefruit," "Valencia," "palmetto," "young citrus–type unknown," etc. This operation was essentially a complete listing of all positions of the sample rows with their relevant characteristics.

From this listing a sample of positions was selected by taking a systematic sample of 1 in k by following specified instructions like "take #5, 25, etc." The sampling rate was computed to give an overall rate of approximately 4 in 40,000 in selecting specific trees. The objective was to obtain an expected 4 orange trees at each location. All other trees and blanks falling in the sample were ignored.

Quintiaries (Tree Limbs)

At the site of an orange tree falling into the sample, the next step was to obtain a sample of limbs on which to make fruit counts. An average-size orange tree in an average year has about 1,000 fruits in September. It seemed appropriate to estimate the correct number by sampling.

Within each sample tree two branches were selected with calculable probabilities, using the following procedure. Each branch at the first forking from the trunk was numbered by a lumber crayon, starting from the most northerly position and proceeding clockwise around the tree. The circumference of each limb was measured in inches, squared, and recorded. The cumulative totals of the branch measurements (which were $\pi^2 d^2$ or 4π times the cross-sectional areas) were determined and put on a relative basis by dividing by the total. A random number lying between 1 and 100 predesignated for each tree forking was examined to determine which branch fell into the sample. That branch was then followed to its forking and the previous procedure was repeated to determine the sample branch. This was continued until a point was reached where the fruits on the remaining branches could be conveniently counted. Usually this was done when about 20 fruits remained. The whole procedure was repeated independently of the first selection, except of course where the same branching was followed, in which case the same measurements could be used.

Sextiaries (Fruits)

All fruits attached to the sample limb constituted a frame for obtaining a sample for size measurement. A tag or pin (such as a small plastic clothes pin) was attached to the stem of each fruit on the limb. The pins were numbered serially $1, 2, 3, \ldots, N$; hence a systematic sample of 1/4 the fruits could be designated. The circumference of each of these fruits was measured by metal tape and recorded. An average of 5 per limb, and hence 10 per tree, was measured.

Double Sampling for Measuring Accuracy of Fruit Counts

Although accuracy in counting fruits on sample limbs (without damage to the limb and without picking or damaging fruit) was stressed with the field workers, inaccuracies in counts were nevertheless expected. Rather than requiring full recounts and perhaps other schemes and devices to ensure accurate counts, it was decided to use a double sampling scheme of measuring and adjusting for such errors, using specially trained persons. The subsample consisted of a 1/10 sample of locations with all sample limbs being recounted.

Doubling Sampling for a Sample of Picked Fruits

Using the fruits measured for volume as a frame, a small subsample was taken to obtain fruits for the determination of specific gravity. Measures of fruit quality such as Brix, acid, and juice yield were desired at various times during the maturing season. To obtain these characteristics, fruits were required for laboratory determinations. The size of this sample was varied during the course of the experimental program. In October 1955 it consisted of a subsample of 1/4 the locations, all trees in those locations, 1/2 the sample limbs, 1/16 the "calipered" (measured for volume) fruits, or a subsampling rate of 1/128. About 500 fruits (or two boxes) were picked.

Within-Season Paneling for Measuring Changes

The initial survey of the new crop was carried out in July, August, or September. This survey provided information on the number and average size of fruits hanging on the trees prior to the harvest season, which for some varieties of oranges (such as Valencias) was some 6–9 months away. To follow the growth and maturation of this crop, surveys were carried out during the season to measure such changes as number of fruits still hanging on the tree, changes in volume per fruit, specific gravity per unit volume, and various qualities of interest to the industry. These surveys were based on a subsample of the initial sample—usually 49 of the 493 locations, the same trees, the same sample limbs, and the same sample fruits. This paneling of as much of the sample as feasible was done with great care to minimize possible effects on the behavior of the sample. This required the elaborate use of marking, tagging, and record keeping and constant monitoring of the operation to ensure that everything was under control. Several checks were instituted to detect possible faults and to assess their effects.

Between-Season Paneling for Measuring Changes

In the second and third seasons a number of decisions had to be made on the appropriate sample design to measure changes through time. In general the advantages of holding to a panel of trees were given great weight in arriving at final decisions. The basic plan was to stay with a panel of sections indefinitely, adding new sections to accommodate the new groves coming into production in those sections of the sampling frame previously listed as having "no area in citrus," i.e., in the "zero section." New groves in the "nonzero sections" could be appropriately handled within the sample sections. They could be detected by on-the-ground inspection surveys or through newer photographs. Very large new plantings might require other measures. However, due to the shortness of this experiment the problem of a continuous scheme was not fully faced.

Within the three seasons the following schemes were instituted. Since the 1953–54 season was incomplete on E & M oranges, it provided only a partial base for the 1954–55 season. In the second season a full match of trees and sample branches was made where possible. In the third season, 1955–56, one of the two sample limbs was dropped and a new one selected. Data were kept separate by sample limb (whether old or new) to test for possible bias as well as to permit use of more efficient estimators. Some comments on this experience will be given later.

4. The Data

Field operations during January 1954 consisted of visiting each location and, with the aid of the aerial photographs and the attached overlays, locating the sample rows; marking them at roadside; walking down a sample row and by visual examination of each tree classifying it; locating and marking the sample trees; selecting and marking the sample limbs— leaving crayon marks on the tree, showing the numbering of each branch at each forking in the probability trail up to sample limb; tagging the individual fruits on the sample limb; calipering the sample fruits; counting the fruits and picking the samples, if any, for the laboratory weighing; counting the number of fruits on the ground (ground counts); and if the tree had been picked, counting the number of fruits left unharvested. In addition on 1/10 of the subsamples of locations, these steps were repeated by a small group of checkers to ascertain the quality of work done and included the careful recounting of fruits.

In July the first survey was carried out on the new crop. In October a program was initiated of picking a sample of fruits (about 600 fruits each run) to determine such quality characteristics as Brix, acid, and juice content. This survey was repeated roughly every two months during the maturing season for the two crop years.

The number of fruits hanging on the trees changes continuously through the season. After June no new ones appear, but a continual dropping of fruits occurs due to such known reasons as disease and to various unknown reasons. Since the droppage may vary from season to season, it is a factor in forecasting the total crop. During the experimental program the droppage was measured indirectly by counting the fruits on the ground under the sample trees at each visit. Where the grove was obviously diseased or otherwise worked since the last visit, this was noted in the field report. It was hoped these ground counts would supply useful data for forecasting, especially after experience over several seasons was obtained.

The program of data collection for the 2 1/2 seasons is shown in Table 10.1. The volume of data obtained is presented in Table 10.2.

Although the original plan called for a sample of 1 in 10,000 of the state's trees with 4 trees at each of 500 locations, the sample turned out to be 493 locations with an average of 3.35 trees each. This attrition is due in part to revised orange tree estimates used in allocating locations to counties, rounding losses in accepting only simple integral sampling rates within sample rows, failure to keep row lengths up to full measure of size requirements, and in a few cases denial of access to groves by the growers. The loss does not appear to be a source of significant bias, however.

Table 10.1
Schedule of Surveys Made and Operations Performed

Season	July	Aug.	Sept.	Oct.	Nov.	Dec.	Jan.	Feb.	Mar.	Apr.	May
1953–54	...	...	...	...	...	...	FC	...	...	...	...
	...	...	...	...	...	...	FW	...	...	...	...
	...	...	...	...	...	...	DP	...	...	...	...
	...	...	...	...	...	...	NP	...	...	...	...
1954–55	FC	...	...	FW	...	FW	...	FW	...	FW	...
	FW	...	...	DP	...	DP	...	DP	...	DP	...
	...	...	...	Q	...	Q	...	Q	...	Q	...
	...	...	...	...	...	NP	...	NP	...	NP	...
1955–56	...	...	FC	FW	...	FW	...	FW	...	FW	...
	...	...	FW	DP	...	DP	...	DP	...	DP	...
	...	...	DP	Q	...	Q	...	Q	...	Q	...
	...	...	Q	...	...	NP	...	NP	...	NP	...

FC = fruit count on limbs.
FW = fruits picked for weight determination.
DP = count of fruits dropped to ground.
NP = count of unharvested fruits in picked and spot-picked trees.
 Q = fruits picked for weight determination also used for quality determination: Brix, acid, and juice.

Table 10.2
Composition of the Samples

		Number in the Samples		
Unit	Total in the 19 Counties	Early and mid-season oranges	Valencia oranges	All oranges
Locations (sections)	14,032	338	307	493
Rows (groves)	...	...	...	1,375
Positions in rows	...	...	...	66,533
Orange trees in positions	25×10^6*	...	...	45,737
Trees in fruit count	...	862	780	1,642
Sample limbs	...	1,724	1,560	3,284
Fruits on sample limbs	20×10^9*	27,000*	31,000*	58,000
Fruits "calipered"	...	6,900	7,600	14,500
Fruits picked	...	345†	380†	725†

*Approximate.
†Each run.

5. Predictors and Predictions

In the first season the predictor of the harvested crop was of the form

$$\hat{Y} = \hat{T}\hat{B}_T \tag{5.1}$$

where

$\hat{T}$ = the estimated number of bearing trees at the beginning of the season,

$\hat{B}_T$ = the forecasted mean yield of fruit harvested from the $\hat{T}$ trees,

$\hat{Y}$ = the predicted harvested production.

In the second and subsequent seasons, ratio and regression type predictors are also available. For example, we can use as a forecaster of Y''' the production of year two, the ratio type:

$$\hat{Y}''_{\text{Ratio}} = Y' \frac{\hat{B}''_H}{\hat{B}'_H} + \hat{Y}^* \tag{5.2}$$

or the regression type:

$$\hat{Y}''_{\text{Regr.}} = T'[\hat{B}''_H + b(B'_H - \hat{B}'_H)] + \hat{Y}^* \tag{5.3}$$

where

Y' = the actual harvested production in the first year,

$\hat{B}'_H$ = may be the mean weight of fruit in boxes per tree in the sample at say September last year,

$\hat{B}''_H$ = same as $\hat{B}'_H$ except for year two,

B'_H = the actual B_H last year = $\dfrac{Y'}{T'}$,

T' = actual number of trees in the universe in year one,

$\hat{Y}^*$ = the estimated production from groves just entering production in year two,

b = the coefficient of regression of the weight of fruit this year on that of last year by location.

In this section, some detail will be given on the manner in which these predictors were formulated and estimates made. Also, some indications will be given of their sampling variability.

Estimator of T

The problem of predicting the size of a crop sometime before it fully materializes can in this case be split into two parts: (1) estimating the number of trees in existence at the beginning of the season and (2) forecasting the yield of fruits per tree that will be harvested from those trees by the end of the season. To estimate the number of trees, we have the estimator,

$$\hat{T} = X \frac{1}{\sum\limits_{h=1}^{L} N_h} \sum_{h=1}^{L} \sum_{j=1}^{N_h} r_{hi} = X \hat{T}_X \tag{5.4}$$

where

$$r_{hi} = \frac{X_h'}{X_h} \cdot \frac{X_{hi}''}{X_{hi}'} \sum_{j=1}^{m_{hi}} \frac{X_{hij}'''}{X_{hij}''} \sum_{k=1}^{1_{hij}} \frac{X_{hijk}^{iv}}{X_{hijk}'''} \sum_{j=1}^{m_{hi}} 1_{hij}$$

and

$$\hat{T}_X = \frac{1}{\sum\limits_{h=1}^{L} N_h} \sum_{h=1}^{L} \sum_{i=1}^{N_h} r_{hi} = \bar{r}_{hi}$$

and where

$X_h'/X_h =$ for county h the ratio of eyeball citrus acres to official estimated number of orange trees (from FCLRS).

$X_{hi}''/X_{hi}' =$ for the ith location in county h the ratio of eyeball sector acres to eyeball citrus acres.

$X_{hij}'''/X_{hij}'' =$ for the jth sector ("grove") in hith location, the ratio of positions (from photo counts) to eyeball sector acres.

$X_{hijk}^{iv}/X_{hijk}''' =$ for the kth row in the hijth sector, the ratio of the number of observed orange trees to positions.

$1_{hij} =$ the number of rows in the hijth sector (usually 1).

$m_{hi} =$ the number of sectors in the hith location.

$L =$ the number of strata (counties).

$X =$ the total trees in the 19-county region as officially estimated by the FCLRS for the 1952–53 season.

$\hat{T}_X =$ the estimated ratio of observed trees to official trees as indicated by the survey.

Since we have two types of X_{hi}', that in Pasco and Polk counties being the number of trees based on inspection records and in the remaining counties eyeball citrus acres, we may distinguish the two subuniverses. Thus

$$X^{(16)} = X^{(14)} + X^{(2)} \tag{5.7}$$

then

$$T^{(16)} = X^{(14)} T_X^{(14)} + X^{(2)} T_X^{(2)}$$
$$= (13,964,000)(0.988) + (5,638,000)(0.958)$$
$$= 13,733,000 + 5,431,000$$
$$= 19,164,000. \tag{5.8}$$

Correcting for the excluded counties, we have $\hat{T}^{(U)} = 21,170,000 \pm 546,000$ (or 2.6%) trees as the estimated number of bearing[2] trees for December 1954. For Valencias alone, the estimate was $10,150,000 \pm 330,000$ trees.

Forecaster of B_H

The forecasted number of boxes of fruit that will be picked from the average tree can be obtained by replacing the parameters in (2.3) by their estimates or forecasts. Thus

$$\hat{B}_H = (\hat{F}_I - \hat{F}_D - \hat{F}_U)\hat{V}_F \cdot \hat{W}_V \cdot B_W. \tag{5.9}$$

To determine $\hat{F}_I$, the estimated average number of fruit per tree at the beginning of the forecast season (say July–September), we need first to estimate the number of fruit in each sample tree. Since sample limbs were selected with varying but calculable probabilities and if F_t is the number of fruits counted on a limb obtained in trial t which had a probability of selection P_t, then an unbiased estimate of F, the actual fruits on the tree, is given by

$$F = \frac{F_t}{P_t} \tag{5.10}$$

and for g trials,

$$\hat{F} = \frac{1}{g} \sum_{t=1}^{g} \frac{F_t}{P_t}. \tag{5.11}$$

Since the sample of orange trees was one assuring essentially equal probability for all trees, a simple mean of $\hat{F}$ over the sample will provide an unbiased estimate of F_I. Hence,

$$\hat{F}_I = \text{sample mean of } \hat{F}. \tag{5.12}$$

An estimate of $H\%$, the fraction of the September fruits which survive to maturity and are harvested, is obtained by estimating its complement.

---

2. The figure given includes adjustments for Hillsborough, Marion, Orange, and Seminole counties to include trees now bearing that would not appear in aerial photos available at the time, since they were taken prior to May 1949. This adjustment added about 500,000 trees.

The complement of this factor, i.e., the fraction of fruits lost, can be split into two parts, droppage and harvest loss. Droppage was measured by taking counts of fruits falling to the ground at various times between September and harvest, whereas harvest loss was measured by counting the number of fruits left on a tree after picking.

The next factor V_F, the average volume of fruits, was obtained by taking the circumference of a sample of fruits around their equator and converting this measure into volume on the assumption the fruit is a sphere. For $\hat{V}_F$ a simple mean over the self-weighting sample was taken.

For $\hat{W}_V$, average weight of fruit per unit volume, an estimate was obtained by the simple ratio of the total weight of fruits picked by the sum of the volumes. The sample was self-weighting.

Since a box of oranges is defined as 90 pounds of fruit, the factor B_W is required to weight units used (grams) into 90-pound boxes.

In the first season field operations were confined to January, hence a number of assumptions were made to obtain a forecast for that season. To do this, $H\%$ for Valencias was assumed to be about the same as for E & M's, the season for which was about 2/3 completed. $\hat{V}_F$ was based on average size of harvested Valencias over past seasons based on industry data (Grower's Administrative Committee), and $\hat{W}_V$ was based on E & M data. As a result the following forecast was calculated:

$$\hat{B}_T = (\hat{F}_I - \hat{F}_D - \hat{F}_U)\hat{V}_F \cdot \hat{W}_V \cdot B_W$$
$$(\hat{B}_T)_{\text{Val}} = (727.7 - 14.5)(13.33)(16.246)(1/40,824)$$
$$= 3.835 \pm 0.151 \quad (\text{or } 3.9\%).$$

And for the forecasted production to be harvested,

$$(\hat{Y})_{\text{Val}} = (\hat{T})_{\text{Val}} \cdot (\hat{B}_T)_{\text{Val}}$$
$$= (10,150,000)(3.835)$$
$$= 38,928,000 \pm 1,985,000 \quad (\text{or } 5.1\%).$$

Extension to the Second and Third Seasons

Because of a late start, the work in the 1953–54 season was limited to getting an estimate of the number of trees in the state and to obtaining a trial run on the late oranges (Valencias). Since about 1/3 of the E & M's were still unpicked, limb counts of these were also obtained as well as harvest losses on the "picked" 2/3. Work on the 1954–55 season began with fruit counts; size and specific gravity measurements in July on the full sample, followed by surveys on size; and specific gravity and fruit quality measurements in October, December, February, and April. During this season the sample of sectors and rows remained unchanged. However, rows falling in groves which in the 1953–54 season were not yet of bearing age were reexamined to include those then bearing and to designate the trees then included in the sample for fruit counts. Except in those counties where aerial photos were too old (i.e., were taken more than four years previously), the sample of sectors and rows provided the proper inclusion of new bearing trees. (In the cases where photos were too old somewhat less desirable methods had to be used to take new groves into account.)

Aside from the new trees coming into the sample as they matured and some trees dropping out because of abandonment or disease, the sample of trees during the three-season period remained unchanged. The limb sample consisted of two limbs within each tree, labeled replicates 1 and 2. These were kept in the sample years one and two but in year three the replicate limbs labeled 1 were dropped and replaced with replicates labeled 3, so that in year three one-half the sample constituted the same limbs and all the trees of years one and two. This feature permitted some opportunities for more complex and, it was hoped, more precise estimators.

The ratio estimator of (5.2) can be written in terms of the components. Thus,

$$\hat{Y}''_{\text{Ratio}} = Y' \frac{\hat{F}''_I \cdot \hat{V}''_F \cdot \hat{W}''_V}{\hat{F}'_I \cdot \hat{V}'_F \cdot \hat{W}'_V} + \hat{Y}* \tag{5.13}$$

$$= Y' \frac{(\hat{W}'')_t}{(\hat{W}')_t} + \hat{Y}*. \tag{5.14}$$

It will be noted that (5.13) can be regarded as the previous year's production multiplied by the ratios of fruit count change, size change, and change in specific gravity. We can express this quantity as simply the change from the previous year of the estimated weight of total fruit hanging on the tree at some similar season time t. The term $\hat{Y}*$ was added to account for estimated production due to trees coming into bearing in the "old photo" counties. If aerial photos were taken say at least every three years, it would be possible to include all bearing trees in the sample of locations. By properly revising the selection of sections with new measures of size X'_{hi}, with a minimum loss of those locations currently in the sample, it would be possible to keep the panel continuously representative with a minimum disturbance from the adding and subtracting of locations.

In the current program a regression estimator was preferred over the ratio type. It had the form of (5.3).

There are rather strong reasons to panelize—i.e., to match units over time—as much of the sample as possible for statistical efficiency as well as for operating quality and economy. The correlations between seasons for fruit counts on the same limbs were observed to be 0.39 for E & M's and

0.53 for Valencias. When these counts were aggregated over all trees in each location the correlation coefficients had the values of 0.63 and 0.68 respectively (based on the 1954–55 versus 1955–56 seasons).

If the experimental program had been continued, it would have been possible to use yet another type of regression estimator. With three or more seasons an estimator can be based on the relationship between current measurements or forecasts against the final production on some common basis such as trees. Thus, for example, the estimated total production for year T, the historical regression estimator, is of the form:

$$\hat{Y}_{\text{Hist}} = \hat{T} \cdot \hat{B}_{\text{Hist}} \tag{5.15}$$

where $\hat{B}_{\text{Hist}}$ is the value of $\hat{B}_H$ estimated from a line fitting for a series of experiences, the actual turnout B_T on the indicators such as $(\hat{B}_T)_t$, $(\hat{W}_T)_t$, or $(F_I - F_D)_t$, etc., for survey time t. This estimator should remove any biases in procedure or concept that persist over time.

The Forecasts Compared

An important objective of the program was to develop early-season forecasts with reasonable precision. Some measure of success is indicated by comparing the survey's forecasts with those of the FCLRS and the final harvest figures, as shown in Table 10.3. It will be noted the FCLRS was hard to beat in the 1955–56 season and the survey was fairly lucky considering the sampling errors. It was encouraging to note sampling error on the October forecast was decreasing as time went on.

Table 10.3

October Forecasts of Orange Production; FCLRS and Survey Compared (in millions of boxes)

Season	Type of Oranges Forecasted	Survey Forecast, SE and RSE*	FCLRS Forecast, SE and RSE*	FCLRS Final harvest
1953–54	Valencias only	38.9† ± 2.0 (5.1%)	34.0 ± 4.2 (12.4%)	41.1
1954–55	All-around	85.6 ± 3.3 (3.8%)	93.6 ± 7.8 (8.8%)	85.9
1955–56	All-around	84.0 ± 2.5 (3.0%)	88.2 ± 8.0 (8.8%)	88.2

*Standard error and relative standard error respectively. Errors for FCLRS calculated from historical data over past eleven years.
†Made in February 1954, not October 1953.

The Tree Estimates Compared

A census of citrus groves was taken in the middle 1950s to determine among other things the total number of trees by class of citrus, by variety, and by age group. In the case of four counties the tree census provides figures for the 1954–55 season, a period for which the survey had its own estimates. Also available for comparison are figures provided by the FCLRS. Table 10.4 gives all figures.

Table 10.4

Tree Numbers Compared, December 1954
(in thousands)

	Number of Bearing Trees								
	FCLRS			Survey			Tree census		
County	E & M	Val	Total	E & M	Val	Total	E & M	Val	Total
Highlands	358	546	904	207	584	791	244	652	896
Lake	1,732	1,039	2,771	2,074	1,555	3,629	1,864	1,724	3,588
Pinellas	260	280	540	223	212	435	164	241	405
Polk	2,371	2,683	5,054	1,670	2,921	4,591	1,866	2,736	4,602
4-county total	4,721	4,548	9,269	4,174	5,272	9,446	4,138	5,353	9,491

6. Some Analyses and Remarks

An analysis of sources of variation and their contribution to the reliability of estimates suggests a number of modifications in the design of the surveys and in the procedures used.

Stratification

Locations are an important source of variation for most of the items dealt with. Geographic stratification further than county, and what is offered by a rough listing structure within county, is suggested. The delineation of known homogeneous districts would help. In addition, age of tree has an important effect on yield. Either post- or prestratification by age group, using many strata, should produce a significant increase in precision. Useful information on number and ages of trees is now available with reasonable frequency and precision of updating.

Optimization in Sample Allocation

An analysis of the data indicates that the variance of estimated numbers of fruits on a tree is proportional to tree size (cross-sectional area of the trunk). This being the case, we may examine the policy of taking two sample limbs per tree to estimate total fruits.

If we regard a sample tree structure as one wherein at each forking there are two branches and where this goes on through k stages, then the number of end limbs is given by the relationship:

$$N = 2^k. \tag{6.1}$$

Let us also regard the last-stage limbs as the only ones that bear fruit, and each branch at a given stage is of the same size. Then using the limb selection technique of this survey, the probability of selecting a branch is $1/N$ and the estimated total fruits from a single replicate is

$$\hat{Y} = Ny. \tag{6.2}$$

If we have two trees, one of k_1 stages and the other of k_2 stages, we wish to estimate the total fruits for the two trees, obtained by

$$\hat{Y} = \frac{1}{n_1} \Sigma \hat{Y}_1 + \frac{1}{n_2} \Sigma \hat{Y}_2 \tag{6.3}$$

where n_1 and n_2 are the sample sizes (with replacement) used in each tree respectively. Minimizing the variance of $\hat{Y}$ with the constraint that $n = n_1 + n_2$, that c_1 and c_2 are the costs of selecting a sample limb and counting fruits on it for tree 1 and 2, respectively, and that $c = n_1 c_1 + n_2 c_2$ is held fixed, we find $n_h \alpha (N_h \sigma_h)/\sqrt{c_h}$, the same result as for standard stratification. The usual relationships of k and N with A, the cross-sectional area of the trunk are:

$$N \alpha A \tag{6.4}$$

and

$$2^k \alpha A. \tag{6.5}$$

In our surveys we observed that $\text{Var}(\hat{Y}) \alpha A$, which means that $N^2 \sigma^2 \alpha A$ and hence the best allocation occurs when

$$n_h \alpha \frac{\sqrt{A_h}}{\sqrt{c_h}}. \tag{6.6}$$

Cost consists of two main components, that required to select the limb and that required to count fruits. The former is dependent on the number of forkings that are met along the randomized pathway to the sample limb

Table 10.5

Relationship of Trunk Size to Tree Age

	Tree Age (years)			
Trunk Size	8	16	23	30
Trunk area, A (sq in)	23.9	47.8	71.7	95.6
$\sqrt{A}$	4.89	6.90	8.47	9.78
Relative $\sqrt{A}$	1.00	1.41	1.73	2.00

and therefore increases with size of tree; the latter is essentially independent of tree size. The cost for each replicate is given by $c = c_c + kc_S$. Since $k\alpha \log A$, it increases rather rapidly with increases of A for small trees but slacks off considerably for the larger trees. Moreover, the component kc_S is likely to be much smaller than c_c. Hence c does not increase much for larger trees. It appears that c will be somewhere between a constant and proportional to A, hence

$$n_h \alpha \begin{cases} \text{constant or} \\ \sqrt{A_h} \end{cases}.$$

From Table 10.5 we may note that the size (cross-sectional area of the trunk) of a 16-year old tree is about two times that of an 8-year old tree; a 23-year old tree is three times the size, and a 30-year old tree is four times as large. The results suggest that within the size range including most trees, perhaps the older trees should be sampled about twice as heavily as the younger ones. However, this recommendation is fairly speculative and greatly dependent on procedures used in the selection of limbs and making fruit counts.

The variance between replications of limb counts was an important source of variation in the estimation of the total crop. The data in Table 10.6 is from the 1953–54 survey. Two replicates per tree were used, each

Table 10.6

Between-Limb Variability of Estimated Fruits on a Tree

Type of Orange	Number of Trees in Sample	Variance between Replicates within Trees	Mean	Coefficient of Variation
E & M	278	511,845	659.3	1.18
Valencia	753	462,951	747.0	0.83

providing an estimate of total fruits. Hence the coefficient of variation is about 100%. In the full sample there were about 3,500 replications; consequently, this source of error was brought down to about 2.5%—still a large error for precise estimation. Adding one or more replicates to the larger trees could be of help.

Reference

Jessen, R. J. 1955. Determining the fruit count on a tree by randomized branch sampling. *Biometrics* 2:99–109.

11

Theories of Inference and Data Analysis

OSCAR KEMPTHORNE

1. Introduction

To MAKE BROAD STATEMENTS on the nature of all human activity is most hazardous; but to place the present essay in context, this will be attempted. It is certain that one of the major problems of humanity is to make reasonable accommodation with itself and with the nonhuman world which determines to an important extent the nature of human life. The only way this can be done is to develop models of the world that can be used to develop understanding, predictive capacity, and suggestions for human actions that will produce desired human ends. The role of mathematics in this endeavor is obviously tremendous. It has been said, perhaps facetiously, that "God was a mathematician" (perhaps "was" should be replaced by "is"). The hope of great thinkers for centuries or even millenia was that the sole key to understanding of the world was mathematics. The world was, it was thought, governed by numbers; one needs only to note the tremendous force that certain numbers possess for some minds, e.g., the numbers 2, 3, 4, 5, 7. In particular, the prime numbers seemed to have a unique role. Even the poet Yeats was fascinated with the numbers game or the partition game, though one may surmise that he would have objected most strenuously to being described in these terms.

OSCAR KEMPTHORNE is Professor of Statistics and Distinguished Professor of Sciences and Humanities, Iowa State University, Ames.

Journal Paper J-6810 of the Iowa Agriculture and Home Economics Experiment Station, Ames, Project 890.

Certainly, mathematics is the queen of science because the only way we can use a model precisely is to formulate it mathematically. Also, but quite irrelevant to the present context, mathematics is a beautiful art form.

The development of mathematics has proceeded in a more or less steady fashion from, say, 3,000 years ago. There were, to be sure, centuries in which little development took place, but the overall picture is one of continuous development.

This paper is concerned with another development of human intellectual processes of much more recent origin. The relation of mathematical systems to the real world is not at all necessary and obvious, as we can see vividly in the mathematical treatment of the infinite. A big question is how one establishes some sort of correspondence between the "world of mathematics" and the "real world."

Among the earliest deep applications, and from some viewpoints the dominant one, was that of the concepts of calculus to Newtonian mechanics. Observations of motion showed that distance was related to time t by $s = ct^2$ "to a reasonable extent." The data never in fact fitted the mathematical relationship perfectly, but the lack of fit was "swept under the rug," justifiably as the whole future development has shown, by an appeal to errors of technique and measurement.

This process of dismissing lack of fit of a mathematical model by mere hand-waving was highly successful as long as workers were dealing with "small" systems. But this particular process had to come to an end, and we can bring this into sharp focus by noting that the field of physics is now quite deeply involved in recognition of "errors." It is relevant to note also that the recognition of errors and the development of a way of living with errors was made relatively early in astronomy by great mathematicians, including Laplace and Gauss. It is interesting that the main impetus for the early development of modes of human adaptation to "error" (or "wandering") or variability came from an activity of no deep relevance to human affairs, but which was relevant merely to individuals who wished to play games of chance. There is a lesson to all of us in noting this because we can ask who would have thought that the theory of playing dice would be relevant to the whole knowledge process.

The big impetus for development of ways of learning in the presence of variability was the obvious need to develop models of organic life. It is easy to imagine purifying a chemical mixture by the various chemical processes and then to speculate on what would happen with a pure substance, though in fact such a program is not viable when pushed to the limit. But any such program does not get off the ground when one is dealing with biological organisms. One cannot "purify" a mouse, one cannot develop two mice that are "identical," and one has to deal with mice as entities exhibiting variability. The beginnings of a direct recognition of the existence of variability that could not be removed were made, it appears, by Quetelet early in the nineteenth century. Progress was slow, and toward the end of that century the ideas were pursued by individuals one could name on the fingers of one hand, especially Galton, Weldon, and Pearson. George Snedecor has written a very informative essay on the beginnings of biometry (Snedecor, 1954). But the really big events in this whole line of development were two, the publication in 1925 by R. A. Fisher of *Statistical Methods for Research Workers* and the publication in 1937 by G. W. Snedecor of *Statistical Methods*, the latter being prompted highly by the former.

I am indebted to Dr. Malin of the Institute for Scientific Information for the following information. This institute prepares the very valuable *Science Citation Index*. Snedecor's *Statistical Methods* was the most cited book in 1961, and from 1964 to 1967 (the last year for which the index information is available) was the second most cited. The first most cited book in these years was *Methods of Enzymology* edited by S. P. Colowick and N. O. Kaplan. The numbers of citations of Snedecor's book, given to me by Dr. Malin are also interesting:

1961	361	1965	684	1967	880
1964	621	1966	806		

So the use of Snedecor also grew, though not necessarily at the same rate as the scientific literature. This book, now with W. G. Cochran as co-author, has gone through six editions since 1937 and has become to a remarkable extent the pharmacopoeia of collectors of data in the sciences.

For the past fifteen years or so the field of statistics has undergone controversies of the deepest kind. There are workers in the "foundations of statistics" (a subject Irwin Bross says should be banned from professional statistical meetings) who appear to take the view that "statistical methods" is utter nonsense and that the people who follow these methods are stupid or have been misled by their proponents. These people are, it is implied, caught up in mores of scientific procedure that have no real logical basis, just as in past centuries peoples were totally involved in the worship of Homeric gods that determined modes of thought and accommodation to the real world.

G. E. P. Box stated in his discussion of the paper of Birnbaum (1962): "I believe for instance, that it would be very difficult to persuade an intelligent physicist that current statistical practice was sensible, but that there would be much less difficulty with an approach via likelihood and Bayes' theorem." It may interest readers to study the paper of the physicists

Orear and Cassel (1970) presented at the Waterloo Symposium on Foundations of Statistical Inference. My impression is that these physicists were bothered by exactly the same type of question that has concerned biologists and is treated in Snedecor's *Statistical Methods*. These men seemed to me to be asking questions of significance of the sort addressed in simpler situations by the chi-square goodness of fit test.

It seems essential then to examine the situation. Is Snedecor's *Statistical Methods*, a snare and a delusion? It may be surmised by those who know the author of this essay that the conclusion to be reached will be in the negative. It is surely the case that there are many deep underlying questions that have not been resolved. It is also the opinion of this writer that there are aspects common to most presentations of statistical methods that may reasonably be questioned.

In his 1954 essay Snedecor said, "Some of you, perhaps many of you, are young enough to be members of the third generation of biometricians. What will be its theme and who will be its leader? After the Pearsonian era of correlation and the Fisherian era of the designed small sample, what prophet will inspire your loyalty and what topic will command your imagination? Will it be something growing out of the theory of games or out of sequential analysis or decision functions? Or is the voice of one crying in the wilderness yet to be heard?"

The purpose of this paper is to discuss the history of ideas and to comment on present-day controversies.

2. The Origin of the Controversies

A broad overview of the history of statistics shows that while the world of scientific investigation uninterruptedly has used the general ideas of statistical methods as exemplified by Snedecor, two major and one minor alternative movements have been present:

1. From the middle thirties the decision theory viewpoint starting off with errors of Type I and Type II and leading up to a prescription of rules of inductive behavior based on costs, risks, utilities, and so on.
2. From the fifties (though Jeffreys was proposing a somewhat similar viewpoint from the thirties), the Bayesian viewpoint.
3. A nearly stillborn but not actually dead viewpoint based on a principle called the likelihood principle.

There are to be sure other approaches, one of which is the fiducial argument of R. A. Fisher. But this and others have received little support.

Underlying the whole problem is the question of the meaning of probability. To determine the history of the term probability and its use is an immense task and one may hope that a historian of science will undertake it. This history can be sketched inadequately but hopefully in a not misleading way as follows: ·

1. The use of probability before say A.D. 1000 or 1200 as a verbal indication of unsureness.
2. The use of probability as relative frequency with gambling devices extending from perhaps the 1400s to the present time.
3. The use of some sort of Bayes postulate to yield results like Laplace's Rule of Succession, which states a probability without, however, any real discussion of how one should use it except with the implication that the probability obtained has the same status as a probability with a tested die, for example.
4. The rejection of this by Boole and Venn, along with rejection by Venn, at least, of probability as a measurement of belief. For example, Venn says (1866, p. 138), "The conception then of the science of Probability as a science of the laws of belief seems to break down at every point." It is remarkable to this author how close the ideas of Venn are to the ideas in *Statistical Methods*. According to Venn, probability was frequency in a large number of objects or a *series* of them, and, later in his book, he asks how we are to obtain an appropriate series for a question we ask. There seems no doubt that this is a question of data analysis of just the sort that *Statistical Methods* tackles. The last chapters of Venn's book discuss elementary statistical methods at a level of sophistication not unreasonable for 1866 and with an outlook not at all dissonant with modern practice.
5. The idea of Keynes (1921) that there are logical probabilities. It is most interesting, however, that the latter half of Keynes's book is a discussion appropriate perhaps to 1900 or 1910 of statistical methods.
6. The theory of belief of Ramsey (1926) later followed by de Finetti and Savage (1954). But Ramsey in his last pages refers in a commendatory way to Fisherian ideas.
7. The ideas of von Mises which appear to be a sort of axiomatization of the idea of relative frequency. The same sort of idea was put forward by Reichenbach (1949 and earlier).
8. The resurgence of rational belief ideas as in Jeffreys (1961) and the more recent exposition of Lindley (1965).

It is interesting to review all these different ideas in depth, because those who have written from one point of view have by and large managed to give very telling arguments against all other points of view.

Here are a few examples. Jeffreys (1961, p. 11) is quite caustic in his book about the idea of probability as the limit of relative frequency. Reichenbach (1949, p. 68) on the other hand says, "In order to develop the frequency interpretation, we define probability as the limit of frequency within an infinite sequence. The definition follows a path that was pointed out by S. D. Poisson in 1837. In 1854, it was used by George Boole, and in recent times it was brought to the fore by Richard von Mises, who defended it successfully against critical objections." Reichenbach also says (p. 367); "The criterion for the justification of an interpretation (of probability) lies in its adequacy for the purpose of prediction." We may contrast this with the statement of Savage (1954, p. 59), "That theory (the theory of personal probability, normatively interpreted) is a code of consistency for the person applying it, not a system of predictions about the world around him."

With regard to the case of observing m successes in $(m + n)$ trials and the model of independent Bernoulli trials with probability q, Keynes (1921, p. 387) said, "I see no justification for the assumption that all possible values of the ratio q are equally likely." We may contrast this to the suggested use of exactly this prior by Jeffreys (1961) and others.

It will be instructive to study the criticisms of Ramsey (1926, pp. 160–65) of the probability theory of Keynes, that "between any two propositions taken as premiss and conclusion, there holds one and only one relation of a certain sort called probability relations."

With regard to the ideas of subjective probability, it will be instructive likewise to read Chapter 6 of Venn (1866). He says (p. 119), "There is a large body of writers, including some of the most eminent authorities upon this subject who state or imply that we are distinctly conscious of such variation of the amount of belief, and that this state of our minds can be measured and determined with almost the same accuracy as the external events to which they refer." He gives (p. 124) as a position, against which strong objections can be brought, "That our belief of every proposition is a thing which we can, strictly speaking, be said to measure; that there must be a certain amount of it in every case, which we can realize in consciousness and refer to some standard so as to pronounce upon its value."

The one idea that everyone seems to accept in the last resort is the idea of probability as frequency with a tested chance device. In regard to one of his postulates Savage, one of the leaders of personal probability, says (p. 33), "On the other hand, such a postulate could be made relatively acceptable by observing that it will obtain if there is a coin that the person is firmly convinced is fair, i.e., a coin such that any finite sequence of heads and tails is for him no more probable than any other sequence of the same length." In the view of this writer, this is veering as close to a frequency probability as can be done without actually mentioning the idea. How does a person become convinced that a coin is fair? Furthermore, Savage and others refer to randomization (p. 66), and one may reasonably speculate whether Savage means what most people mean, which is surely a frequency-in-repetitions idea.

It has been said to the author that when a student asked Professor Hotelling his views about prior probability arguments, Professor Hotelling replied that they resurface about every fifteen years. It seems also a statistical fact that they appeal to youth and lose their force with age and experience.

The origin of the controversies seems to lie, after all is said and done, in the fact that some wish to make of probability a definite objective fact of the real world, whereas others go to the other extreme and make of it a purely personal reaction to the personal history of an individual. The truth seems to be very much in the middle. The simplest way to generate controversy is to set up a straw man, and the above two views are of this type. Probability is a matter of judgment and, as such, subjective, but the facts on which it should be based are partly objective. It would aid the resolution of the controversy to ban the words "subjective" and "objective."

3. Theories of Inference

The controversies in statistics of the past two decades have dealt with the following formal problem: Given a set of data D known to be random with a known probability law $P(D, \theta)$, with θ belonging to some space Θ, to draw conclusions about the unknown value of θ. I shall take the view later that this usually is a highly artificial problem. It is necessary, however, to discuss the approaches because for many of the protagonists this formal problem is the basic or fundamental problem of statistics.

Bayesian Methods

To apply this class of methods, one must by hook or by crook specify a prior distribution for the parameter θ. If, for example, the parameter space Θ is the real line or segments of the real line, we must write down a probability measure over the space Θ which we can represent by the density $f(\theta)d\theta$. Then given datum D with probability under the model equal to $P(D, \theta)$, we are to compute the posterior distribution of θ which will have density

$$f_{post}(\theta)d\theta = \frac{P(D, \theta)f(\theta)d\theta}{\int_{\Theta} P(D, \theta)f(\theta)d\theta}.$$

This prescription is beautifully easy. Indeed, the ease of the computation has, in my opinion, led many to view it as the panacea.

There are two types of Bayesian developments. One type, of which the leader is Jeffreys, takes the view that some sort of logical analysis of the situation will lead to an appropriate prior distribution for θ. However, this approach must, it is felt, be highly embarrassing to its proponents because the mathematical principles applied to yield a prior seem to be highly ad hoc and because the prior that one is led to use depends strongly on the question one is asking. If the aim of the exercise is to obtain a probability distribution for θ after knowledge of the data, one uses one prior distribution. If, however, one has been led by some path, devious or not, to think that θ might be θ_0 and one wishes to question this, then it is suggested by Jeffreys that one use the prior:

$$P(\theta = \theta_0) = 1/2$$

$$P(\theta \neq \theta_0) = 1/2$$

and $P(k \leq \theta \leq k + \Delta k | \theta \neq \theta_0) = f(k)\Delta k$, where $f(k)$ is chosen by some rule. Cornfield (1969) merely chooses an $f(k)$ which is mathematically convenient and familiar. Furthermore, if θ happens to be a vector, (θ_1, θ_2), the Jeffreys prescription for a prior on $\theta = (\theta_1, \theta_2)$ depends on one's aims.

The Jeffreys procedures would carry some appeal as leading to a *logical* probability if there were a set of principles having great appeal that one would use for all aspects of the model-data situation. Part of the Jeffreys procedures is to use an improper prior distribution, e.g., in the $N(\mu, \sigma^2)$ case, to use as a prior density for μ,

$$f(\mu)\, d\mu = d\mu, \quad -\infty < \mu < \infty.$$

The same improper prior distribution is advocated by Lindley. The student is referred to pages 18 and 19 of volume II of Lindley (1965) for a statement claiming to justify this, which he may or may not (like this author) find convincing. It has been questioned (e.g., by Hacking, 1965) how one can convolute a concept that *obviously* is not a probability concept with a probability concept and reach a probability concept. The difficulty does not, however, cause these authors concern, and *perhaps* they are correct.

But even if one accepts the Jeffreys approach, the following essential questions *must* be asked. Suppose I tell you that the probability derived in this way that a parameter θ is in the interval (3, 5) is 0.9; of what relevance is this to my further actions? If indeed this probability is 0.99, am I entitled to feel rather sure that θ lies in this interval? May I predict that θ is in the interval (3, 5) with 99% confidence (in some sense)? I shall take the view that my probability that a well-tested coin will not give seven heads in a row is 127/128. I shall be prepared to wager a small sum on this with a friend repeatedly when spending a few hours with nothing better to do. But what is the content of a Jeffreys probability of 127/128? Jeffreys has been highly critical of other views of probability but appears to lack the ability or is very reluctant to examine his own views similarly.

I am of the opinion that this type of *formal* data manipulation has no unique place in the logic of the sciences. It is interesting in the present context to note that Snedecor's *Statistical Methods* takes *no* cognizance of this approach, even though it has been in the literature in one form or another since the early thirties at least.

An alternative Bayesian approach is that of the "subjectivists" or the "personalistic decision" approach. But even in this there is variability. One version of this approach is that one should introspect and obtain a distribution of belief, prior to seeing the data, as to the value of the parameters which are unknown. So one might have the situation of a sample from $N(\mu, \sigma^2)$ with μ and σ^2 being unknown—or rather, uncertain. One is then to develop, by examining the history in mind, a belief on the two parameters. Following this prescription, I might find I think that μ has a triangular distribution with center 80 and a range of 20. I might also have the belief that μ/σ is uniformly distributed between .06 and .10. From this input I can generate a prior distribution for μ and σ^2, which can then be used to get a posterior distribution. Given this, I can do all the operations associated with probability distributions, such as obtaining the marginal distributions of μ and of σ, the posterior distribution of μ/σ, and whatever. This prescription is certainly performable.

A basic defect of this mode of approach for most problems of data analysis is that the probability model for the data is not known. Books on statistical methods give us suggestions on how to choose a probability model for data (e.g., by performing moment calculations or by exploring a class of transformations of the data as in the analysis of variance). The Bayesian prescription seems to be that one can incorporate all these aspects in the prior distribution. It seems clear that this leads to a horrendous mess in situations that are of very common occurrence. For example, with a two-way classification having say a common effect μ, r block effects, b_i, and s treatment effects, t_j, and the class of transformations $(y + c)^p$ where y is the actual observation, one has to specify a prior distribution on μ, the b_i effects, the t_j effects, c, and p. I can imagine no more sterile exercise of mental energy. If, on the other hand, one permits the data to determine c and p, I believe it is impossible to write down a prior distribution of the μ, b_i, and t_j. Furthermore, such a procedure would appear to destroy any logical basis for the data analysis and would merely be what is termed an ad hoc data massaging process.

Quite apart from this sort of criticism the prescription has the following defect. If the experiment gives ideas that are strongly in conflict with the experimenter's based on all the knowledge and opinions he has which are external to the experiment, the experimenter will tend to distrust the evidence in one way or another, *or* he may take the view that the prior he wrote down and used was poor in the sense that in formulating it he did not include ideas he should have included. This difficulty in the approach under discussion forces to our attention the idea that after we have performed the experiment, we have in our possession two "pieces" of information—what we thought before the experiment and what the experiment itself indicates. In my opinion the existence of both of these is critical.

The "personalistic decision" approach advocated strongly by Savage is developed from a very small set of axioms and is regarded by some as *the* proper approach to all problems. The strength of any such system of thought must be based on two modes of examination. One must examine the postulates and determine if they seem appropriate, and one must examine the consequences and determine if they are reasonable or plausible. Reliance only on the former is highly questionable because a decision to accept the axioms is a decision to accept all consequences of them. For this reason the development of axioms is very difficult. A mathematical theory is only a battery of statements that are mutually consistent in terms of Aristotelian logic.

Discussion of the Savage approach is critical because loosely speaking, it is often said, "this theory proves that a rational person *must* be a Bayesian." What are the ingredients of the Savage approach? To write a detailed commentary would take us far afield. The primary ingredient is a table of acts $\{A_i\}$ of states of nature $\{s_j\}$ (such as the possible values of a parameter), and consequences:

Unknown States of Nature

	s_1	s_2	$\cdots$	s_M
Act 1	d_{11}	d_{12}	$\cdots$	d_{1M}
2	d_{21}	d_{22}	$\cdots$	d_{2M}
$\vdots$	$\vdots$	$\vdots$	$\cdots$	$\vdots$
A	d_{A1}	d_{A2}	$\cdots$	d_{AM}

Savage assumes that you, the decision maker, can reach an ordering of the acts.

Here d_{ij} is the consequence of act A_i if the true state of nature is s_j. It might be, for instance, that you have a quarrel with your wife, while another d_{ij} might be that you go to the movies. At this stage this author already has colossal difficulty with any but the *most* trivial problems. Savage is aware of the difficulties but decides to give them no weight. The basic problem is that every act has consequences that extend into the indefinite future. The consequences of any act are uncertain. But implementation of the Savage axiom requires that one be able to specify all the consequences. If, indeed, we think about the consequences of a simple act such as having a haircut or buying an automobile, we find that we have to engage in a mass of speculation. The difficulties seem fantastically great to the point where one wonders how anyone of us ever chooses an act. But the fact is that we do. So it seems much more reasonable to make some psychological study of how individuals make choices *in the real situation for which consequences are not known perfectly.* Indeed, it is my view that the assumption in this theory that consequences for given states of nature can be specified totally and exactly, renders the theory essentially valueless.

In modes of development like that of Savage it is inevitable that the concept of utility enters. In some theories it enters essentially at the beginning and in others, like Savage's, later. An extensive very informative description and discussion of the whole idea is given by Luce and Raiffa (1957). They assume (pp. 23–31) that the decision maker has objective probabilities (i.e., probabilities of the status of those we obtain from a well-tested die) that there are prizes $A_1, A_2, \ldots, A_r$ with objective probabilities $p_1, p_2, \ldots, p_r$ that the decision maker can order the prizes, with say A_1 as the most preferred and A_r as the least preferred, and that for any other prize A_i the decision maker will be indifferent between A_i for certain or A_1 with probability u_i and A_r with probability $1 - u_i$. On the basis of such ideas, Luce and Raiffa conclude that every lottery has a utility equal to $\sum_{i=1}^r p_i u_i$ and that the decision maker will choose the lottery with maximum utility. This is a very elegant theory. It is essential, however, to note that preferences determine utilities and not vice versa. Luce and Raiffa say also that the problem of interpersonal comparisons of utility have *not* been solved, and on the experimental determination of utility for a particular person they say that observed ordering of prizes by an individual will often exhibit intransitivities. They say (p. 35), "One cannot expect the data to fit the model perfectly, but how does one determine which model they fit more closely and how does one measure how good the agreement is? Such problems pose the following intriguing and important statistical problems . . . ." This seems to indicate clearly that utility theory requires statistical ideas for its application to the real world. This suggests or even implies quite definitely that to base the fundamental ideas of statistical methods on utility theory places one in a vicious circle. I see no alternative to this. Luce and Raiffa also say (p. 36), "There can be no question that it is extremely difficult to determine a person's utility function even under the most ideal and idealized experimental conditions: one can almost say that it has yet to be done." It seems quite impossible to avoid the conclusion that utility theory cannot be part of the *basis* of statistical methods.

It is interesting to note remarks by Ramsey (1926, pp. 204–5) that are near the end of his paper—one regarded by Savage and others of the personalistic school as a strong precursor of their ideas:

> The science of statistics is concerned with abbreviating facts about numerous individuals which are interpreted as a random selection from an infinite "population". If the qualities concerned are discrete, this means simply that we consider the proportions of the observed individuals which have the qualities, and ascribe these proportions to the hypothetical population. If the qualities are continuous, we take the population to be of a convenient simple form containing various parameters which are then chosen to give the highest probability to the instances observed. In either case the probable error is calculated for such a sample from such a population. (For all this see Fisher.)

So it appears that Ramsey envisaged a real role for Fisherian ideas, though he rejected the idea of an infinite population.

The Prefiducial Ideas of Fisher

To review the totality of the ideas and suggestions put forward by R. A. Fisher before he became convinced of the fiducial argument would be to review essentially the whole of Snedecor's *Statistical Methods*. There is hardly a page of the book not bearing in one way or another the imprint of Fisher. It is relevant to note that the fiducial argument has not received real support from anyone other than Fisher. Some Fisherians have written on the ideas, but none has made the matter at all clear. In many instances it appears that the immense creativity of Fisher in many other directions causes us to be very wary of rejecting the fiducial idea, and this is reasonable. But we should be aware of the repeated occurrence of creative individuals becoming enamored of very curious ideas.

One aspect of Fisherian ideas is the notion of a "test of significance." Fisher never gave a clear unambiguous statement of what he meant by this. One might reasonably expect that Fisher's (1956) latest book *Statistical Methods and Scientific Inference*, would contain such a statement, but it does not. Later in this essay an attempt will be made to summarize intelligibly some of the ideas. Another aspect of Fisherian thought was a concept of "exhausting the information in the data," which led to the idea of maximum likelihood "estimation." The word "estimation" is placed in quotation marks here *because it is not at all clear what Fisher meant by estimation*. My best guess is that the Fisherian idea of estimation of a parameter θ is to obtain a *statistic* θ^* and a population of repetitions of the data, so that θ^* "contains all the information" on θ in a technical sense of the word. The problem here is that the Fisherian concept of information is applicable *only* to a certain class of situations which can be specified briefly and sufficiently precisely for most purposes as the class of multinomial situations in which the number of classes with nonzero probability for any θ does not depend on θ. For example, the case of an observation which is uniform from 0 to θ does not fall in the class.

Part of the obscurity of Fisher with regard to tests of significance arises in connection with the meaning of a significance level, which is in some cases a frequency in a population of repetitions, but is not in other cases. A second obscurity arises in the choice of population of repetitions, a problem which will be discussed later.

The Neyman-Pearson Theory of Testing Hypotheses

It seems clear that Fisher did not give any definitive theory of a test of significance and that the initial aim of Neyman and Pearson was to develop one. They accomplished this by making a restriction on the overall outlook, a restriction that appeared to many at that time not to involve any modification of the basic idea of significance testing. Their procedure was to convert significance testing into an accept-reject rule. So with a sample space $\mathcal{D}$, say, one is to pick out a part of the space $\mathcal{D}$, say, $\mathcal{R}$, which would be used as a rejection region. If the data point falls into $\mathcal{R}$, the hypothesis under test is to be rejected and otherwise to be accepted. It was natural with this to talk about errors of the two types and to require that $\mathcal{R}$ be chosen so that the frequency of Type I error would be less than or equal to α (i.e., $\mathcal{R}$ must have size α) and the frequency β of Type II error would be minimized.

The force of this idea has been tremendous to the point of dominating statistical research apart from Fisher for decades. Even those who have great affinity for Fisher and Fisherian ideas have accepted this in part.

The question that apparently was never addressed by Neyman and Pearson was how to choose α or, in later days, how to choose an α, β combination. If this were solved, the procedure obviously would be appropriate for acceptance sampling in which there are obviously just two decisions, to accept the lot or to reject it. From the viewpoint of the Neyman-Pearson theory of testing hypotheses—or as this author prefers, the Neyman-Pearson theory of accept-reject rules—an inspector is not permitted the following thought process. Suppose two particular data points D_1 and D_2 fall in the rejection region of size $\alpha = 0.5$. Suppose also that D_1 falls in the region of size $\alpha = 0.01$ and D_2 does not. Then it is very natural to take the view that D_1 disagrees with the null hypothesis more than D_2. But to use phraseology that is becoming current, this would be an *evidential* conclusion. It appears that no such conclusions are permitted in the Neyman-Pearson *theory*. Indeed, it can happen that a sample point is in the rejection region of size 0.01 and is not in the rejection region of size 0.05. It may be true that those who use the Neyman-Pearson theory will reach the evidential conclusion above, and indeed many of the ideas of the theory have been taken over and used in an evidential way. But nothing in the Neyman-Pearson theory permits this activity.

Another aspect of the Neyman-Pearson theory bringing to the fore its nonevidential nature is that given a situation, the theory says that one is to use one test and only one test. There is no basis for using one test to examine the data from one viewpoint and another test to examine the data from a different one. There seems to be nothing in the theory that will enable one to make a statistical statement analogous to the following type of biological statement: This leaf is like leaves of species A with regard to overall shape, like species B with regard to venation, like species C with regard to texture of the leaf skin, etc.

A problem of interpretation arises because the very great preponderance of writings by supporters of the Neyman-Pearson theory are strictly mathematical. It is exceedingly rare to find the theory addressed to any real problem other than acceptance sampling. Additionally, it seems that recent writings of Dr. Neyman (e.g., on modifying rainfall) do not reflect the Neyman-Pearson theory of testing hypotheses, but instead seem aimed at the evidential content of weather data and at the broad evidential use of statistical tests.

A great theoretical contribution of Neyman was the theory of confidence intervals (C.I.'s). The early history is interesting in that there was deep obscurity as to whether confidence intervals were the same as fiducial intervals. This was resolved in the negative after a few years. The theory of confidence intervals has become widely accepted. Books like Snedecor's *Statistical Methods* use the idea very frequently. The intended aim of the method is to give a theory of estimation based on ideas of classical probability, in fact to answer the Bayesian question of formation of opinion about possible values for the parameter without the unpleasant introduction of a prior.

The confidence interval solution to specifying parameter values that are reasonable to a specified extent in the light of the data has had varying support. This author has found himself oscillating on the matter. It is standard to teach that the method supplies a rule for associating a region say $R(D; \alpha)$ in the parameter space with each set of data D and a *particular prechosen* confidence coefficient α which has the property that the frequency with which $R(D; \alpha)$ contains the true parameter value θ in repetitions of D is α. It is concluded that the particular $R(D; \alpha)$ arising from a set

D contains θ with degree of confidence α. If we consider the proposition "$R(D; \alpha)$ contains θ" to be a proposition like "Result of a toss of a die is 6," we may say that the proposition has a truth value of α. But the procedure has a certain defect. The C.I. theory says that the relative frequency in repetitions with which $R(D; \alpha)$ contains θ is α. But it does *not* state that there may be aspects of the data that permit one to make a different assessment. This writer finds the example of Buehler and Feddersen (1963) compelling. It is seen that with a sample x_1, x_2 from $N(\mu, \sigma^2)$ the ordinary 50% Student's t interval is $x_{min} \leq \mu \leq x_{max}$. But Buehler and Feddersen show that there is an identifiable subset of data situations— i.e., a region say C in the x_1, x_2 plane, such that the frequency in repetitions that lie in C of the above interval containing θ is greater than or equal to 0.5181. So one can define a subclass of repetitions in which the confidence coefficient is not realized.

Other very simple examples in which C.I.'s are clearly of questionable value were given (Kempthorne, discussion on paper by J. Cornfield, 1969). The details will not be given here. However, we can hardly expect a scientist to attach a confidence coefficient of say 77.7% to the value of θ_0 for θ when just a little simple thought tells him to be 100% confident that $\theta = \theta_0$. Nor will the practicing scientist be comfortable if he is told that the interval 1.9 to 2.4 is a 95% interval, a 90% interval, an 80% interval, etc.

It is the existence of these features, which should not be regarded as abnormalities, that raises seriously the question of whether confidence intervals have the evidential content that people wish to ascribe to them.

The Theory of Decision

The Neyman-Pearson theory of testing hypotheses is regarded by Neyman (1970 and earlier) as the proper theory of inductive behavior. This theory of inductive behavior is a reflection of the fact that the theory is one of accept-reject rules. Neyman (1970) regards this theory as the beginning of the general theory of inductive behavior encompassed by Wald's decision theory.

The idea behind this whole development is that there is a space of decisions from which one is to be chosen on the basis of the data. It seems totally critical in this development that there is one correct decision in the decision space. The simple example is that of point estimation—e.g., with a sample $x_1, x_2, \ldots, x_n$ from $N(\mu, \sigma^2)$—to nominate a value for μ which one may denote by $\hat{\mu}$. It is assumed there is a loss function—e.g., $(\hat{\mu} - \mu)^2$—and a risk function which is equal to the expectation of the loss function plus costs of obtaining observations.

Application of this theory to any *real* problem of research and development seems to be incredibly difficult. There are several reasons that seem incontrovertible. In the first place, it is only in the simplest cases, such as examining a lot by looking at a random sample of its members, that one knows the probability model. This is not a defect of this approach alone, and it pervades the whole panoply of approaches. In the second place, the possible outcomes of some piece of research or development usually cannot be specified before data are obtained. So there usually is *no* definite space of decisions from which one decision is selected. In the third place, even if one were to force onto the situation a simple dichotomous decision space (such as to submit to the journals a claim that a treatment does or does not affect this), the losses are difficult if not impossible to evaluate. In the fourth place, one is surely involved in ideas of utility, and we have seen above how questionable this concept is. In the fifth place, it is not at all clear that there is a perfect decision in that, as the whole history of science shows, one can be totally confident that whatever model is discovered and used will prove erroneous.

I hazard the opinion that Snedecor's *Statistical Methods* has had some appeal to scientists and has not been modified in basic outlook by the development of decision theory, because decision theory deals with problems that are so simple (e.g., how to approach the problem of making scrambled eggs) and so simplified as to have no essential relevance to the problems of research and development. I am taking what may appear to be a "hard" line but am unable to see any merit in doing otherwise. If statistical theory is just an intellectual exercise for those working on it, well and good. These workers can take any system of axioms they deem appropriate. But it is totally unjustified that they denigrate statistical methods on the basis of their theories. It would be more appropriate and honest, I believe, if they would say that their theories have essentially no relevance to the problems addressed by statistical methods.

A general difficulty with evaluating decision theory is that it is so general that, on the one hand, application to real problems except the simplest is incredibly difficult depending on a specification that is rather tight of consequences; and on the other hand, it is possible that any human activity can be contained within it. It is possible to envisage a space of "tenability function," any one of which gives a measure of the extent to which any value θ of a parameter is "supported" by the data, and to then incorporate some ideas of loss and risk. However, such a program appears to be extremely difficult.

It may seem unreasonable to the reader to associate the above criticisms with the Neyman work. To justify the association, it is useful to quote Neyman (1970), ". . . the general problem of behavioristic theory of statistics: with reference to any given situation S_n, and to all the data pertinent thereto, to define an optimal decision function $d_n(\cdot)$ and to determine it, if it exists: if the originally defined optimal decision function does not exist (for example, uniformly most powerful tests exist but very rarely), define a compromise best statistical decision function, etc."

The Likelihood Principle

The remaining strongly advocated process of inference is the use of the likelihood principle. This was suggested in a very informal way by Fisher (1956), though his earlier writings perhaps contain indications that he was leaning in this direction. The principle has been advocated strongly by Barnard (1949) and Barnard et al. (1962). Exact and definitive statement of the principle has not been given. The clearest statement is perhaps that of Birnbaum (1962), which says that the evidential meaning of the outcome of any experiment is characterized completely by the likelihood function and is otherwise independent of the structure of the experiment. If, then, the data are denoted by D with probability $P(D, \theta)$, the likelihood principle states that the total evidential content of the data is given by the function $cP(D, \theta)$, where $1/c$ can be taken to be the supremum of $P(D, \theta)$ with regard to θ with D fixed. In this description the concept "evidential meaning" due to Birnbaum is purposely not defined so as to leave open what an explicit idea of evidence might be. The prescription of the advocates of this principle seems to be to look at the likelihood function and to judge the relative support for different parameter values by the values of the likelihood function.

It seems curious that this principle has been advocated when it fails totally for the very simple inferential situation where one has observed a finite sample of size n chosen at random from a finite population of size N say. In this case the parameters of the situation are say $x_1, x_2, \ldots, x_N$, the attributes of the N population members. What is random in this case is the occurrence of individuals in the sample, and the probability of this has no dependence on the parameters. But it is clear that observing a random sample tells one something about the population. This example is related, perhaps, to the example given by Birnbaum (1969) which apparently caused him to reach the conclusion that the likelihood principle is not acceptable. In Birnbaum's example there are two parameters μ and σ, and the likelihood function with a single observation gives huge relative likelihood to one value of σ regardless of what value is correct for the population being sampled.

Proofs have been given that the likelihood principle is implied by other seemingly compelling principles, e.g., by Birnbaum (1962, 1969). Furthermore, the proofs are said to be elementary and were regarded, it appears, as acceptable by several (see the discussion on Birnbaum's 1962 paper). To this author, however, the situation is highly obscure, and the proofs seem to have a very obscure semantic flavor. In particular it seems very curious that two principles that describe properties of a hypothesized entity called "evidential meaning," involving both the experiment structure and the data, can imply a principle asserting that the experiment structure is irrelevant except insofar as it determines the likelihood function.

4. The Nature of Statistical Methods

To persons familiar with books on statistical methods, the above greatly abbreviated review of the nature of theories of inference must seem highly tangential. To pinpoint why this is the case seems critical, but to do so to a definitive extent requires a very long essay. However, it is possible to abstract the essential features of statistical methods in a few summary statements.

The primitive problem of statistical methods, and indeed of inference, is that we have to form an opinion in a rational way about the nature of a finite population of N units, with each of which is associated a number say x by picking a subset of the units, examining these, and proceeding to form an opinion in a rational way. It is interesting and highly relevant that a uniformly accepted logical basis for doing this has so far not been attained. It is important to note that some of the neo-Bayesians and supporters of personalistic decision theory take the view that random sampling is an illogical process. So, for instance, one should conclude that the vast governmental processes that follow this general paradigm are a waste of time and money. It is really unfortunate that proponents of this view do not tell us how they would form opinions in the stated conditions. How would they obtain a subset of the population? (One Bayesian has said that he would do this "at haphazard.") What would they do with the results? At times it seems as though they take the view that the problem is unanswerable. It may be that the problem is unanswerable by their criteria, but in that case they should, it seems, change their criteria because it is totally obvious that the cumulative distribution function of a random sample from the population is an "estimate" of the population cumulative distribution function. The word estimate is placed in quotation marks for obvious reasons. It seems quite clear that the process of random sampling "works" because this is an empirically established fact. We do not really understand totally why it works, but we should reject out-of-hand any theory that says it cannot "work."

The second primitive aspect of statistical methods is that we rarely have random samples. We have a set of data. We make a judgment of an appropriate population from the data. So for example, given a set of observations on the number of squirrels seen by an observer in successive days, we make a judgment that the achieved data, which contain *no* random aspect per se, are *like* a random sample from a certain Poisson distribution say. What does it mean to say "a set of data is like a random sample from a Poisson distribution?" It seems quite clear that we envisage a population of random samples from a Poisson distribution we can realize partly by sampling with a "proven" random number generator. We then form the opinion that our set of data looks like a typical member of such a population of random samples. Our set of data has, for example, a mean that is not far from the mean of the population of random samples, and similarly for the variance and so on.

There seems to be no doubt that in doing this activity we are comparing our actual set of data to sets we might have obtained if we were sampling a Poisson population. It seems evident that if we denote the actual data by D_O and the population of data sets which would arise by random sampling from the population by $\mathscr{D}$, we have to form a distance function $d(D_O; \mathscr{D})$, the distance of D_O from $\mathscr{D}$. If our actual data consist of numbers $x_0, x_1, x_2, \ldots$, where x_i is the number of days in which i squirrels were observed, our data are represented by an infinite vector x_O with elements x_{Oi}, $i = 0, 1, 2, \ldots$. Our possible samples are represented by an infinity, say $\mathscr{D}$, of infinite vectors $\{x\}$. So our problem is to obtain a distance between one infinite vector x_O and a population of infinite vectors $\mathscr{D}$. We can form a distance in an infinity of ways. We could, for instance, use the average of $(x_{O5} - x_5)^2$, where the average is taken over the population $\mathscr{D}$, though this would not be a good way.

One natural way of accomplishing this is as follows. The actual data have to occur in the infinite population of data possible under the model. So one maps every possible data set onto the real line so that for every data set $\mathscr{D}$ there is a number $g(D)$. We then can take as the distance the average of $[g(D_O) - g(D)]^2$. Denoting this by $d(D_O; \mathscr{D})$, we shall obtain a real number in the case of a completely specified model. It might be, for instance, 5.7. We now have to decide if this is a large distance. To do this we can consider $d(D; \mathscr{D})$ for every possible D; we then have to form an opinion as to whether $d(D_O; \mathscr{D})$ is large relative to the assemblage of values $[d(D; \mathscr{D})]$. It now seems natural to associate with D_O the proportion of values from the assemblage $[d(D; \mathscr{D})]$ which are at least as large as $d(D_O; \mathscr{D})$. (The reader may object to the use of the word "natural" here. It is important therefore to state categorically that there is *no* way of avoiding words like "natural," "reasonable," "plausible," etc. Whatever we do, we are forced in the last resort to depend on an unanalyzable notion of "reasonable" or whatever. An axiomatic system as a representation of real-world phenomena is judged favorably if the axioms are in some sense "natural" or "reasonable.") If the proportion of such values is small, the investigator is entitled to be reluctant to accept the appropriateness of saying that D_O is like the possible realizations in $\mathscr{D}$.

It may be informative to illustrate this procedure in the simple case of r successes, n trials, with the model of independent Bernoulli trials. Let $g(r) = r$. Then $d(r;\mathscr{D}) = \mathrm{Av}(r - R)^2$, where R is $\mathrm{Bi}(n,p)$, which equals $r^2 - 2rE(R) + E(R^2)$ or $r^2 - 2rnp + n^2p^2 + np(1 - p)$. Because we envisage comparing this with the population of distances, we can drop the constant term $np(1 - p)$, and we shall say that the comparative distance is $(r - np)^2$. So our end result is to associate with the datum r the proportion of times $(R - np)^2$ equals or exceeds $(r - np)^2$, or to put matters in a more familiar mode, the proportion of times $(R - np)^2/np(1 - p)$ exceeds $(r - np)^2/np(1 - p)$. This will be given approximately by the χ^2 table. The reason for this somewhat tortuous development is that probability does not really enter the argument at all except in defining the population of repetitions. The number achieved here would commonly be called the significance level of the null hypothesis that the Bernoulli parameter is p as determined by the χ^2 test.

In somewhat more general terms we replace the data D_O by a number $g(D_O;\mathscr{D})$ and then find the frequency in the population of repetitions envisaged by the model that $g(D_O;\mathscr{D})$ is equaled or exceeded. The resulting number may be characterized as a degree of support for the hypothesized model on a scale such that zero means "no support" and one means "complete support." It has the primitive property that the probability beforehand that the support is less than or equal to α is equal to α if α is an achievable value.

Another set of words describing the process is as follows. Let D_O be the observed data. Let D_i denote a generic member of the population of data sets under the hypothesized model. Then we map the D_i onto the real line, as $y(D_i)$ and we associate with D_O the number α which is the proportion of D_i such that $y(D_i) \geq y(D_O)$.

As we have said, there are an infinity of ways of mapping data sets onto the real line. In the case of data $x_1, x_2, \ldots, x_n$ considered as being like a random sample from $N(\mu, \sigma^2)$ with μ and σ^2 specified, we can make mappings that are sensitive to a wide variety of aspects—e.g., inappropriate μ, inappropriate σ^2, inappropriate μ, σ^2 combination, inappropriate distribution (e.g., skewness kurtosis, outliers, etc.).

The overall output of almost all the methods of Snedecor's *Statistical Methods* consists of statements of classes of models (e.g., indexed by μ

or indexed by σ, etc.) which are *consonant* with the data to particular degrees as measured by the above type of process. The language used in the book for describing the process is not that given above. The distances are called *P*-values, and are abbreviated for the most part by the use of *, **, ***, denoting significant, highly significant, and very highly significant. It seems a fact that experimenters use *P*-values as indicating distances of data from models. It is extremely convenient to represent distances by the probability scale, though any monotone transformation of that scale would be as useful. Workers are now used to using the probability scale and find *P*-values informative. The use of *P*-values has been criticized, it appears, because users may tend to interpret them as probabilities of hypotheses; but this seems to be very far-fetched. A *P*-value is a statistic, in no way different from the way in which a sample mean or a sample variance is a statistic. The derivation is more complex to be sure because the statistic involves hypothesized values for one or more parameters and also involves the probability structure.

The use of confidence intervals has become widespread, but not, I think, because of the logical nature of the confidence interval argument at all. Rather, a useful 95% confidence interval on a parameter θ, contains every value θ^* say such that the "distance" of the data from the sets of data which would arise with θ equal to θ^* is equal to or less than the 5% "distance." It is unfortunate that a smaller *P*-value means a larger distance. This sometimes causes description to be clumsy.

If we adopt an idea such as the above, the modification of a probability distance by adding to it the value of a random variable given by an auxiliary device such as a die cannot improve matters and merely introduces noise. It is also apparent that the use of this distance-type of idea will not lead to the absurdities of the confidence interval argument of the type that a data set D_O is "significant at" the 5% level and not at the 10% level.

It is obvious that there should be no sharp distinction made between cases having a *P*-value of say 4.9% and those having a *P*-value of 5.1%—a distinction forced by the language of confidence interval theory.

There are many ways one can measure distance of data D_O from a model. It seems foolish to restrict oneself to only one measure of distance. If there is a single scalar parameter and the probability structure has a monotone likelihood ratio, a single way of measuring distance from θ_O for all values of θ on one side of θ_O can be shown to be good, but even in the scalar parameter case there is no single way of calculating distance which will be best for all purposes. It seems, furthermore, that the whole idea of unbiased tests, which has consumed very many years of highly mathematical intellectual endeavor, is a useless one except for the development of mathematical abilities and the production of doctoral theses and journal papers.

There is one major unresolved difficulty in the whole idea of measuring consonance of data with a model, and this has been passed over in the foregoing. Rather than to get the distance of D_O from $\mathscr{D}$, the collection of all possible data sets under the model, it seems plausible in some cases to obtain the distance of D_O from a subset $\mathscr{D}_O$ of the collection of all possible data sets, this subset being chosen to consist of those members of $\mathscr{D}$ which are like D_O in respects not involving unknowns. The idea of using this restricted subpopulation for comparative purposes is called the principle of ancillarity. It has strong appeal in some cases but cannot be applied unambiguously in other very simple situations. This author's view is that the choice of a family of repetitions with which to assess the observed data D_O is a matter for the experimenter and cannot be specified on the basis of mathematical principles. The work of Basu (1959, 1964) is highly relevant to this question.

The probability content of these outputs is in one sense extremely small. From another viewpoint it is large and dominates because without probability ideas, which are frequency ideas induced by the model or class of models, one would not be able to perform the process.

This way of viewing the matter sheds light on what aspects of mathematical statistics are really important, and it is an empirical fact that users of statistical methods do find these aspects relevant. The aspects are contained under the general rubric of "sampling distribution theory" which is as unambiguous an aspect of the whole business as one can imagine.

In this way of thinking, it is clear that consideration of what might have happened under the model is intrinsic. It is clear that a model consists of a family of repetitions of sampling from the hypothetical population. This is mentioned to contrast it to the view that it is only what one has observed that really matters, which in the case of a model with an unspecified parameter θ say declares that it is the achieved likelihood function and *only* the achieved likelihood function that has evidential value.

5. Comparison of Theories of Inference

It is informative, perhaps, to close this essay with a comparison of theories of inference. From one point of view, the differences among all the theories are quite insubstantial. To justify this view, consider the following. In all the theories of inference discussed above we have data D_O say, with a class of models $M(\theta)$ with θ in a set Θ and with a probability structure having probabilities $P(D; \theta)$. All the theories of inference lead to the calculation of statistics. With the Bayesian process each set of data D is mapped or transformed into a posterior distribution which if there were k possible values for θ, would be a statistic with k components and for θ belonging to an interval is a statistic with a nonenumerable infinity of components. One prior distribution gives one statistic, another prior distribution gives another statistic, and so on. With the likelihood process each set of data is mapped into a statistic of the same type, which is the likelihood function. With the Neyman-Pearson theory of accept-reject rules, each set of data is mapped into the space of two points, say $1 = $ Accept and $0 = $ Reject. With a significance test or the probability distance given above, each set of data is mapped onto the $(0, 1]$ real line segment. With maximum likelihood estimation each set of data is mapped into the parameter space and represented by $\hat{\theta}$, the maximum likelihood statistic. With the decision theory approach, with say k possible decisions, each set of data is mapped into one of k points; and if the points are represented by $i = 1, 2, \ldots, k$, each set of data is replaced by one of these values, again by a statistic.

It is apparent then that all the theories or processes of inference lead to the calculation of statistics of greater or less complexity.

What then are the real issues? The first seems rather definitely to be that of determining what statistics to calculate. We can calculate the Bayesian statistic with as many priors as we care to use, but there seems to be no compelling logic for the choice of any particular ones. The likelihood statistic has the virtue of being unique but does not seem to have any other compelling virtues. One can use many different Neyman-Pearson accept-reject rules, so that a set of data gives results such as "Accept according to Test 1," "Reject according to Test 2," "Accept according to Test 3," and so on. One can use many different significance tests, so that one has "P-value equals 0.71 by Test 1," "P-value equals 0.01 by Test 2," and so on. So we have this bewildering array of possible statistics. In all cases, the prescription for use of the statistic amounts in the last resort to "Look at it." This is what the advocates of the likelihood principle have told us to do. It is all that the Bayesians can tell us to do because their statistic has no predictive value. It is all that the significance tester can tell us to do, though some of us have gained appreciation of the process through repeated practice.

The arithmetical logic of any of the processes is, except in Fisher's fiducial hands, quite transparent and easy to understand. All we can do is to attempt an understanding of this for each process and, as in calculating statistics such as the mean or sample variance, to ask: What is the precision of the statistic you have calculated? It appears to be a dominant fact of the Bayesian and likelihood processes and of the decision theory process in its most common form that this sort of question is not admitted as having any relevance at all. Savage (1954, p. 257) says, "The doctrine is often expressed that a point estimate is of little or no value unless accompanied by an estimate of its own accuracy." He then goes on to say that he sees no point in the doctrine. I am firmly of the view that any statistic is of highly questionable value unless accompanied by an estimate of its accuracy. It follows from this that the question of what value a statistic might have taken under a model or class of models is always relevant, and this is true whether the statistic is a posterior distribution calculated with any particular prior distribution, is a likelihood function, or is a point in a decision space.

It seems that the divisions in statistics result almost completely from differences in attitude to the question of whether operating characteristics of data analysis procedures are important or not. The essence of science and of data collection appears to be that operating characteristics of procedures are important, and there seems to be no strong reason this should not also be true of data interpretation.

6. The Future of Statistical Methods and Data Analysis

Is "statistical methods" a snare and a delusion? I believe strongly the answer must be in the negative. Investigators obviously find the arithmetical processes described in the common texts highly informative; the processes enable the investigator to interact with the data. The processes enable the investigator to monitor his thinking and interpretation and to present his results in a form understandable to the audience he desires. The main defect of the usual presentations is, I think, that the process of interaction with data is not described and illustrated. Too often it is said that "the appropriate model for the data is such-and-such," with little or no reference to the fact that models have been determined from the data and from experience with data situations "like" the one under consideration. The usual texts do not make clear that probabilities are always conditional on a model that cannot be justified completely, making all probabilities partly "subjective." The Neyman-Pearson doctrine of errors of the two kinds has been allowed to occupy too dominant a role, but on the other hand the basic idea of operating characteristics of data analysis methods in hypothetical situations must be retained. The future of statistical methods lies in the appreciation of the investigator-data interaction process and the implementation of this process by means of the modern computer.

George Snedecor knew this many decades ago and was instrumental in setting up statistical courses and a university computing system, albeit of a very simple type compared to present-day versions. If this paper aids a little in explaining his life and work, it will have repaid a huge debt.

References

Barnard, G. W. 1949. Statistical inference. *J. Roy. Statist. Soc.* Ser. B, 11: 116–49.

Barnard, G. A., B. M. Jenkins, and C. B. Winsten. 1962. Likelihood inference and time series (with discussion). *J. Roy. Statist. Soc.* Ser. A, 125: 321–72.

Basu, D. 1959. The family of ancillary statistics. *Sankhya* 21: 247–56.

———. 1964. Recovery of ancillary information. *Sankhya* 26: 3–16.

Birnbaum, A. 1962. On the foundations of statistical inference (with discussion). *J. Am. Statist. Assoc.* 57: 269–306.

———. 1969. Concepts of statistical evidence. In *Philosophy, science, and method: Essays in honor of E. Nagel*, ed. S. Morganbesser et al. New York: St. Martins.

Buehler, R. J., and A. P. Feddersen. 1963. Note on a conditional property of Student's *t*. *Ann. Math. Stat.* 34: 1098–1100.

Cornfield, J. 1969. The Bayesian outlook and its application (with discussion). *Biometrics* 25: 617–57.

Fisher, R. A. 1956. *Statistical methods and scientific inference.* Edinburgh: Oliver & Boyd.

Hacking, I. 1965. *Logic of statistical inference.* New York: Cambridge Univ. Press.

Jeffreys, H. 1961. *The theory of probability,* 3rd ed. Clarendon, Tex.: Clarendon Press.

Keynes, J. M. 1921. *A treatise on probability.* Reprint. New York: Harper.

Lindley, D. V. 1965. *Introduction to probability and statistics from a Bayesian viewpoint.* New York: Cambridge Univ. Press.

Luce, R. D., and H. Raiffa. 1957. *Games and decisions.* New York: Wiley.

Neyman, J. 1970. Foundations of the behavioristic statistics. Paper presented at the Waterloo Symposium on Foundations of Statistical Inference.

Orear, J., and D. Cassel. 1970. Applications of statistical inference to physics. Paper presented at the Waterloo Symposium on Foundations of Statistical Inference.

Ramsey, F. P. 1926. Truth and probability. In *The foundations of mathematics and other logical essays.* London: Kegan Paul, Trench, Trubner, and Company.

Reichenbach, H. 1949. *The theory of probability.* Berkeley: Univ. Calif. Press.

Savage, L. J. 1954. *The foundations of statistics.* New York: Wiley.

Snedecor, G. W. 1954. Biometry, its makers and its concepts. In *Statistics and mathematics in biology,* ed. O. Kempthorne, et al., ch. 1. New York: Hafner.

Venn, J. A. 1866. *The logic of chance.* Reprint. Bronx, N.Y.: Chelsea.

12

The History and Future
of Statistics

M. G. KENDALL

1. Introduction

ONE OF O. HENRY'S CHARACTERS once defined statistics as the lowest grade of information known to exist; and I daresay that most people still think of it in those terms, as a colorless collection of numerical facts which can be twisted in all sorts of ways to support an argument of doubtful validity. In reality the science of statistics, as distinct from the raw material with which it works, is an all-embracing branch of scientific method, and in its broadest interpretation is almost scientific method itself. We find statisticians at work in every field where quantitative evidence is available. The very breadth of the subject makes an account of its history difficult to present in any ordered way. Statistics is not a subject which began at any identifiable point of time or has pursued a traceable line of development. On the contrary, it now embraces under a single discipline a range of subjects which pursued independent courses for centuries. For example, demographic statistics took their origin in the bills of mortality which came into being at the time of the Great Plague of the seventeenth century. Actuarial statistics developed from the assurance offices which grew up in the Protestant countries about the same time. A century earlier, gamblers managed to interest mathematicians in the laws of chance and founded the

M. G. KENDALL is Chairman, Scientific Control Systems Limited, London.

This paper is a condensed version of two lectures given in Seattle at the University of Washington while the author was visiting as Walker-Ames professor.

theory of probability. Astronomers discovered that their observations were subject to errors which had an identifiable pattern and set up the theory of errors of observation. In the middle of the nineteenth century Francis Galton and Karl Pearson discovered that biological material presented similar stable patterns of behavior. It is hard to think of any subject which has not made some kind of contribution to statistical theory—agriculture, astronomy, biology, chemistry, and so on through the alphabet. The remarkable thing, perhaps, is that these lines of development remained relatively independent for so long and only in the present century have been seen to have a common conceptual content.

Something can be done, however, to knit the threads together if we consider the evolution of the scientific approach, say from the time of the Renaissance which, notwithstanding our intellectual debt to Greece and Islam, really marks the end of the time when the western world no longer was content to rely on dogma and introspection but developed an experimental approach to natural phenomena. The early successes in the physical sciences by way of the discovery of laws—those of stretched strings, of reflection and refraction of light, of the expansion of gases, and so forth—led to a belief that the physical universe was deterministic and its rules of behavior could be expressed in fairly concise mathematical terms. The work of Isaac Newton on gravitation appeared for some time as a supreme example of this fact. And as further patterns of behavior came to light, in the laws of electricity and magnetism of chemistry and of heat flow, for example, the belief cemented itself into human thought to such an extent that at the end of the eighteenth century Laplace remarked that if we only knew the position and velocity of every particle in the universe at any one time, we could predict the whole of future history. This was perhaps felt to be an exaggeration, but the further successes in the physical sciences achieved during the nineteenth century did nothing to disturb the belief that the physical world was deterministic. It was generally felt that any observed anomalies or unexplained effects would ultimately yield to the deterministic approach.

The extraordinary successes in the physical sciences were not long in attracting the attention of men who were interested in the behavioral sciences. Around 1700 Queen Anne's physician, a Doctor John Arbuthnot, called attention to the remarkable constancy of the sex ratio of births. He did not enunciate his conclusions as a law, but attributed the phenomenon to Divine Providence, which I suppose comes to much the same thing to a deist. But his idea was pursued through the eighteenth century by a number of men who were impressed by the fact that although the fate of an individual was unpredictable, the behavior of an aggregate could be formulated in almost as precise terms as the law of physics, provided that the aggregate was large enough. And so, almost imperceptibly, mankind arrived at its second major conceptual advance—the realization that there were laws of aggregates even when laws pertaining to individuals could not be discerned.

Up to the beginning of the present century the physical and behavioral scientists had very few points of contact, but they then began to move rather closer together. It had already been realized that some apparently deterministic laws—or rather laws which were deterministic in the large, like Boyle's law relating pressure and volume in a gas—were only the macroemanation of the behavior of a swarm of particles and were, in fact, statistical laws. The discovery of radioactive substance indicated that in the subatomic field there were types of behavior which were either not deterministic at all or were observationally indistinguishable from random occurrence. It seemed that the primary particles had the irresponsibility of human beings in deciding whether to change state. Determinism in molecular physics received a shock from which it has not yet recovered.

Both lines of thought, then, in the physical as well as the behavioral sciences began to converge onto the theory of aggregate; at this point the mathematical probabilist came into his own. For some time he had been mainly concerned with the laws of chance as exemplified in gaming or errors of observation, with a few exceptions such as Maxwell's work on gases and Boltzmann's on heat. But now all phenomena came within his range. Moreover, he had to develop theories to deal, not with the static situations of classical probability, but with dynamic situations in which a system moved probabilistically through time. To this new subject was given the name "stochastic processes." Links were established with disciplines which had hitherto been regarded more as branches of philosophy than of empirical sciences—economics, psychology, and sociology. Chance had come to stay as an integral part of many of the systems which the scientist had to study.

As a natural consequence of this development, scientific inference and the validation of scientific hypotheses had to be reviewed. The statistician usually formulated his conclusions in terms of probability, not of rigorous deduction from accepted premises. He has, so to speak, to think probabilistically. The remarkable thing about chance is that it can be controlled, even though, on the face of it, to speak of the laws of chance sounds almost like a contradiction in terms. Indeed, by a kind of scientific judo trick we can use chance to increase certainty. For example, when we design an experiment to test for the reality of some effect of interest, we usually replicate it in such a way that the outcome is extremely improbable if the effect does not exist; the more improbable the outcome the more we reject the nonexistence of the effect. By this means we can quantify the degree of confidence we have in the reality of the effect. Although we still have some way to go, the physical, biological, and behavioral sciences now recognize, I think, that a great part of their methodology has common ground in statistical theory. We live in a world which is a mixture of deterministic and stochastic effects.

2. History of Statistical Method

Having sketched in the background I now want to go back and fill in the outlines by dealing with some aspects of the subject in greater detail. But before I do so, there is one point to be clarified. It concerns the changed nature of the word statistics itself. The ancients, who had a very rudimentary system of writing down numbers, must have kept accounts in some form or other, but their knowledge of what went on within the borders of their own countries was very imperfect; and on the rare occasions when anybody thought it worthwhile to take an inventory of his possessions—as Augustus, Charlemagne, and William the Conqueror did—the result was an uncoordinated and quantitative list. The first recorded use of the word "statistics" occurs in Latin in the middle of the sixteenth century; and it related, not to numerical information at all, but to matters concerning the state. It continued in use in this sense for two hundred years, and the so-called statistical accounts of the seventeenth and eighteenth centuries are collections of facts which read more like a Baedeker's guide than the statistical digests of today. Inevitably, however, those facts become more and more numerical in character and finally developed into summary tables. These were statistics, but they still related to matters of state concern.

The consequence is that the history of statistics as so defined has practically nothing to do with the history of statistics as we now understand it; or, at any rate, it has nothing to do with the history of statistical method. The only methodological content discernible before the middle of the seventeenth century is the tabular presentation. Even graphical presentation was surprisingly late in arriving—the pie chart dates from the middle of the eighteenth century, and there was no respectable graph of a time series before the end of that century.

I therefore choose to begin the history of statistical method in the reign of Charles II. If we have to divide the period of three hundred years from 1660 to the present day into sections, I would deal with the first two hundred years en bloc. During that time demographic statistics grew to a fair degree of advancement. Captain John Graunt (basing himself on the bills of mortality from the Great Plague), followed by a number of able amateurs like Gregory King and Sir William Petty, broke new ground, not by merely collecting numerical information, but by reasoning from it—for example, estimating the total population, the volume of exports and imports, the wealth of the country, and so forth.

Life insurance began to develop into a science about 1670 and we still possess a letter from Ludwig Huyghens to his brother Christian concluding "Live well. According to my calculations you will live to about $56\frac{1}{2}$." A little later, in 1693, the astronomer Edmund Halley (of Halley's comet) constructed the first comprehensive life table, basing himself on the experience of Breslau in Germany. Life tables, of course, are continually under revision, and new combinations of insurance and assurance are being developed; but basically Halley's table and the work done on it by actuaries and probabilists remains as the starting point of a long line of continuous development.

In the economic sphere, as information accumulated, there grew up rather slowly some of the standard techniques of modern descriptive statistics. We can, for example, trace the theory of index numbers back to about 1750, when Bishop Fleetwood examined the question whether a rule requiring a Fellow of an Oxford college to surrender his stipend if in receipt of more than £5 per annum from private means would still be in effect after a lapse of two centuries. The Bishop, incidentally, came to the conclusion that over that period money had lost five-sixths of its value, so the corresponding current limit should be £30 per annum. (It would be interesting to know what the present-day figure would be.) This is one of the first examples of a change of attitude in the seats of authority. In the middle ages and later any statistical information which was produced appeared either accidentally, as the by-product of some administrative necessity or in response to a specific enquiry. Governments now began to collect information on a more systematic basis, both from internal sources, and as international trade developed, from external sources. In particular, censuses of population became a regular feature of the developed countries. The first recorded census seems to have been that of Iceland in 1790. Britain took her first population census in 1801 and has done so regularly at ten-year intervals.

Theory of Probability

Concurrently with these developments in human affairs the mathematical theory of probability was actively developed. The subject has a most unusual time scale. Although man had been gambling with astragali and dice for several thousand years, the idea that unbiased dice had very definite patterns of behavior in the long run was very late in emerging; it seems that the honor of realizing that there were laws of chance in dice and card-playing lies with the Italian Girolamo Cardano, a remarkable combination of physician, mathematician, and gambler. This delay was probably due, among other things, to the fact that gambling was frowned on by the Church—medieval history is full of attempts to prohibit gaming, with results that might have been expected—and it was not until the middle of the seventeenth century that the subject attained respectability. From that point it developed very rapidly and culminated in Laplace's famous treatise on probability of 1812. The mathematicians who contributed to the subject were mostly interested in it for its own sake, but there were a few links with the practical world. For example, Abraham de Moivre, who discovered the normal distribution, earned his living in part by advising on annuities; Daniel Bernoulli wrote on mortality; Laplace himself applied probability to consider the likelihood that the planes in which the planets move were randomly oriented. Several writers, notably Siméon Poisson whose name is attached to one of the standard distributions of statistics, considered probability in the sense of a measure of the correctness of judgments. But on the whole the doctrine of chance was not yet ready for a complete merger with statistics.

Up to the middle of the nineteenth century we may discern two other almost independent lines of development. Karl Gauss founded the theory of errors of observation, and it is remarkable how much of his work in this field still lives. In the behavioral area a number of inquiring minds discovered patterns of stability in all kinds of recorded human activity. For example the Belgian astronomer Adolphe Quetelet, one of the founders of the Royal Statistical Society and a tutor of Queen Victoria's husband Prince Albert, was struck by the consistency with which events of small probability occur and, in his own words, appalled by the regularity with which suicides reproduce themselves.

At this point we are poised for developments of extreme importance and far-reaching consequences. Hitherto, statistical studies had been concerned with either the physical world or human affairs. Now the whole of life was brought under statistical notice. It was observed that measurements on the individuals of a species, although varying from one individual to another, had an identifiable frequency distribution. Measurements on leaves, oysters, flowers, bees, and the individual organs of the body all showed the same effect. It appeared that there could be developed a mathematics of life, not the traditional type of mathematics involving deterministic variables and functions, but a mathematics in which the basic element was the frequency distribution and the basic relationships were correlations. A new kind of dependence was observed, midway between functional dependence and independence, so that, for example, it was possible to say that the height of a son was neither fully dependent nor fully independent of the height of the father, but that given the latter, statements could be made in probability about the range of the former.

3. Emergence of Modern Statistical Development

At long last the various lines of development began to converge toward the subject of statistics as we know it today. The realization of the necessity of a science of aggregates—whether of human beings, physical objects, errors of measurement, throws of dice, or the weak endings of Shakespeare's verse—was the unifying theme. The further realization that reasoning about these aggregates was to be conducted in terms of probabilistic arguments brought in the doctrine of chances. Statistical societies grew up throughout the world. The American Statistical Association was founded in 1839. In 1861 Prince Albert, obviously well schooled by Quetelet, presided over the founding of the International Statistical Institute. The appellation of statistician remained for most people a term of reproach rather than of admiration, but the persons concerned recognized themselves as the exponents of a new discipline which had come to stay.

One of the difficulties of a historical review is that for expository reasons it is necessary to segment the subject and hence to convey the impression of discontinuity, when the developments are in fact continuous. If I were writing a history of modern developments I should have to begin rather arbitrarily about the year 1890. At this point a young man of thirty-five called Karl Pearson, who had made something of a name for himself as an applied mathematician, became interested in statistics and embarked on the long career which among other things founded the first group, at University College London, for the study of statistical method as such, mainly in a biological context. He founded *Biometrika* which is still one of the major statistical journals. Also at this point a forty-five-year-old language scholar, F. Y. Edgeworth, became interested in human values and hence developed into an economist and then, being equally interested in quantification, became a statistician. A young student of twenty named George Udny Yule joined Pearson and was to become the father of the modern theory of time series and the author of perhaps the best-known introductory text on theoretical statistics. And in 1890 also R. A. Fisher was born, and his impact on statistics throughout the world was to surpass them all.

We begin then in 1890, and although, as I have shown, it is possible to trace many of the ideas much further back, we may genuinely regard this as the starting point of modern developments. This was not merely a scientific accident. The two decades from 1890 to 1910 were ones of intense sociological and scientific ferment. Science was perceived, not as the occupation of the dilettante, but as an instrument by which man could shape his own future. It was not a coincidence that Pearson, the founder of modern statistics, was also the author of a well-known and influential book titled *The Grammar of Science* or that Francis Galton founded eugenics. Statistics became, in a sense, part of the science of life, and life itself was seen to be an evolutionary process whose laws could be expressed in statistical terms. We cannot now, perhaps, recapture the zest with which these new ideas were seized upon and developed. Even Karl Marx became a statistician at the close of his life, and so did Florence Nightingale.

Concept of Sampling

I have already referred to the emergence of the concept of statistical relationship, which still offers us a number of unsolved problems. The second major line of development of the past fifty years concerns the concept of sampling. In the majority of cases the information we have is only a subset of all possible information, either because we can examine only a part of the total which exists, as in a public opinion survey, or because we cannot generate all possible happenings, as when we throw dice. Nevertheless our inferences have to go from the sample to the population of origin. They are essentially probabilistic. And so we arrive at four of the main categories of development: (1) the theory of sampling itself—How do we get a good sample for a given outlay? (2) the theory of sampling distributions—How are the quantities of interest to us distributed over all possible samples? (3) the theory of estimation—How do we estimate parental population quantities from the limited information provided by a sample? and (4) the theory of testing hypotheses—When are data subject to error or natural variation consistent with a scientific model?

Everybody knows that a sample can be biased, either accidentally or deliberately, and one of the main problems in sampling is to ascertain the probabilities of different kinds of samples in particular cases. For this reason the statistician much prefers random samples; but randomness, under examination, is a very subtle concept, and truly random samples are far from easy to obtain. I shall refer later to the current and future developments in this field; an enormous amount of intellectual effort has been expended. The physicist often avoids the issue by assuming that his errors of observation behave in a random way; but as sampling spread into the behavioral sciences (and in particular all kinds of government and business), decisions were based on the examination of a sample of human beings, the techniques of drawing unbiased samples developed almost into a subject of its own, and there grew up a profession of sample designers as identifiable, in its way, as that of the house or boat designer.

The formalization of the sampling process and the derivation of sampling distributions are, however, only means to an end: estimation, the determination of errors of estimates and inference generally. Up to 1925 or thereabouts ideas on estimation, though sound in the main, were intuitive and uncoordinated. From that point on, the subject became systematic and acquired the characteristic of an exact science. We now know, for example, how to define and estimate bias. We know that in general there are many possible estimates of a particular quantity but that some are better than others in the sense of having smaller errors. We have various ways of finding the more efficient estimates and can sometimes even find a most efficient one. At the same time we can put error bands or confidence intervals around our estimates to be able to assert with preassigned probability how reliable they are. Much remains to be done in the more complicated situations of real life, but conceptually the problem of estimation and the control of sampling error advanced more in the twenty-five years from 1935 to 1950 than in the previous twenty-five centuries.

Theory of Testing Hypotheses

At the beginning of the century, ideas were very intuitive. One had a hypothesis which said that a certain experiment should give a certain result. The experiment was performed and gave a different result. If the difference was small, the hypothesis was not suspected, but if it was large, the hypothesis was rejected. But how small was small, and how large was large? And however large the difference might be, if no other hypothesis made it any smaller, what was the inference to be?

The history of scientific inference is a subject to which one could devote a whole book. For present purposes the subject may be regarded as starting with a famous paper by The Reverend Thomas Bayes, published in 1763. Bayes attempted to invert the probabilistic relationship between hypothesis and outcome. On a given hypothesis certain outcomes had determinate probabilities. In some sense, Bayes thought it should be possible to invert such statements and to state the probabilities of the various hypotheses on a given outcome. This idea was accepted, in the main, by the probabilists of the nineteenth century, though a few had misgivings. It was explicitly rejected by Sir Ronald Fisher who spent a lot of time, rather unsuccessfully, trying to put something in its place. Probably under Fisher's influence, which was very widespread, Bayesian inference fell into discredit until quite recently, when it was revived and now numbers some distinguished names among its adherents.

The basic work on hypothesis testing sprang from a similar rejection of the Bayesian approach and was almost entirely the work of two men, Egon Pearson (the son of Karl) and Jerzy Neyman. Like many scientific ideas, the basic concepts are simple enough when you see them. In essence, Neyman and Pearson pointed out that you never test a hypothesis all by itself, but only in comparison with some other hypothesis or set of hypotheses. Consequently you expose yourself to two kinds of risk, of rejecting the hypothesis under test when it is true or of accepting it when some alternative is true. We can control errors of the first kind by a probability distribution of the test statistic, and we seek to minimize errors of the second kind. Thus we set up a theory of hypothesis testing, looking for those tests which are better than others in a discriminatory sense; sometimes we are even able to find optimal tests.

All these developments extended the theory of the subject to an enormous extent and made some very exacting demands on the mathematicians. They were, however, accompanied by equally rewarding results in application. In particular, the statistician began to take a hand in experimental design; one might almost say he took over experimental design. Not content with leaving others to collect information and then being invited to make the best of material for which he could accept no responsibility, he made the point very effectively that one got the best value out of experiments if they were designed so that rigorous analysis could be applied to the results. Here again, the basic ideas are simple enough when you see them. If we are conducting a complex experiment to test a number of factors simultaneously (and if the factors interact, that is the sort of experiment we ought to be conducting), we do not wish to emerge with results from which the separate effects of the various factors cannot be disentangled. So one designs the experiment in such a way that disentanglement is possible. It is characteristic of our whole subject that this simple idea leads to the most recondite combination of mathematics involving group theory, finite geometries, and graph theory.

In nonexperimental situations, notably in economics, we cannot arrange for these neat experimental layouts, and the problem of disentangling causes and effects in a complex system becomes much more severe. The general problem of dealing with such systems, so far as statistics is concerned, goes under the name of "multivariate analysis." Typically we have a number of individuals on each of which is observed the values of a number of variables; the fundamental problem, as a rule, is to see the forest, not the trees. Thus, in market research we may have a hundred or more variables observed on several thousand human beings; or in economics we may have thirty or forty measurements on the economy over time. Our variables are, in general, correlated among themselves, and there may be —there certainly are in economics—some fairly complicated feedback effects which make the disentanglement of cause and effect particularly difficult to achieve. Many of the major problems arising in this area came to notice between 1930 and 1960, but for the most part their solution had to await the arrival of the electronic computer.

Theory of Time Series

There is, however, one further important development of the past to discuss before we contemplate the present state of statistics and look forward to the future. I refer to the theory of time series. Most of what I have said already relates, especially where probabilistic elements are involved, to static situations. Early attempts in the nineteenth century to analyze time series were conducted entirely in deterministic terms. A trend was represented by a polynomial in time, an oscillatory movement by a sum of harmonic terms. Attempts to represent such effects as the so-called "trade cycles" in terms of Fourier-type series were failures. There were oscillations, but they were neither regular in period nor constant in amplitude.

About 1926, Udny Yule provided a possible explanation of these pseudo-cyclical effects by putting a stochastic term into the equations of motion of a system. This was the origin of the modern methods of dealing with time series, in which values from one point to the next are not independent but are not purely deterministic either. Some Russian writers, notably Markoff and Slutzky, had considered stochastic elements in time series analysis and laid the foundations of a dynamics of probability, the so-called stochastic processes. Yule's independent work introduced the subject to Western Europe and the rest of the world. World War II caused an interruption in development in the sense that statisticians were usually too busy on other things and in any case published very little. But, as happened in other areas, the war actually gave a fillip to study of oscillatory time series of the stochastic type because a number of technical problems arising during the war (such as tracking aircraft by radar) required the study of precisely the type of series that Yule had studied in astronomy and economics. Since the war the subject has grown as rapidly as any. There are crazes and fashions in statistics, just as in any other field of human endeavor; but as stochastic processes were applied to an ever-increasing range of problems—queuing theory, inventory control, manpower studies, etc.—there could be no doubt that this branch of the subject had come to stay.

4. Recent Developments

This brings us to the last twenty years or so, from the end of World War II to 1969. We can look on that period in two ways: as the flowering of the seminal work of the previous fifty years and as the formative period of a new wave of advance.

For fifty years statistical calculations were based on the desk calculating machine. There were improvements, of course. Electric motors speeded up the pace, and the simpler arithmetical processes such as multiplication and division were automated. But basically the machine remained the same, an assembly of pinwheels which could remember only three rows of digits at a time and required the full-time attention of an operator. Practical statistics was conditioned by what such a machine—or in a few favored cases, a battery of such machines—could accomplish. In consequence theoretical advance was held back, not so much by the shortage of ideas or even of capable men to explore them as by the technological impossibility of performing the necessary calculations. The Golden Age of theoretical statistics was also the age of the desk computer. Perhaps this was not a net disadvantage. It generated, like all situations of scarcity, some very resourceful shortcuts, economies, and what are known unkindly and unfairly as quick and dirty methods. But it was undoubtedly still a barrier.

The Computer

The floodgate began to open about 1950 with the invention of the electronic computer. This instrument is remarkable, not so much for its fantastic speed, as for the fact that it has a capacious memory; and such a memory not only can store data but can hold a set of instructions so that the machine can be left to monitor its own work. I doubt whether any other technological advance ever has increased human power by such a large factor as has the computer. In 1945 I did some calculations on a set of time series that took me three weeks on a desk machine. A short while ago I had a similar set done and the work took three seconds, and before long we shall regard this as rather slow.

We therefore have to view the future of our subject against the power of the computer. This perhaps may sound like putting the cart before the horse. The computer after all does only what we tell it to do. But the point is that bright ideas do not fructify unless we can bring them to bear on numerical material, and for many of our outstanding problems, as we shall see, the computer is necessary.

Statisticians have been rather slow to take advantage of the computer, partly due to the fact that many have been unable to get access to the necessary machines, at least in Europe. But it has also been due to conservatism in habits of thought—the classical mathematician of the previous century, forgetting that the greatest mathematicians of the past such as Newton and Gauss were also expert numerical analysts, tried to elevate mathematics into a branch of abstract reasoning and affected to look down on numerical analysis as a very menial occupation. I hope that this attitude is dying out of its own accord. If not we shall have to kill it.

But there is more to be said about the relationship between mathematics and statistics. The early statisticians of the present century were competent at mathematics, but they were not great creative mathematicians. Karl Pearson was trained in mathematics, but Edgeworth was a classical scholar and Yule an engineer by training. Fisher, who *was* a creative mathematician, criticized his predecessors for the clumsiness of their style; but even he wrote in the tradition of English mathematics, which does not care much about extreme generalization or extreme rigor as long as it gets the right answer to its problems. The consequence was that, with a few exceptions, theoretical statistics in the forties could be understood by anybody with moderate mathematical attainment, say at the first-year undergraduate level.

I deeply regret to say that the situation has changed so much for the worse that the journals devoted to mathematical statistics are now completely unreadable. Most statisticians deplore the fact, but there is not very much they can do about it. I do not propose to make this an occasion to protest the way mathematicians are spoiling our subject or are in danger of repelling a large number of young men who could make useful contributions but are led to believe it is too difficult. My purpose in dwelling on the matter is twofold: to establish that theoretical statistics is *not* a part of pure mathematics and that our future lies not so much in the development of new mathematical ideas as in the exploitation of what we already have.

Let us first of all consider the impact of the computer on what we now understand by the solution of a problem. In the old days (stated in general but perhaps oversimplified terms) solving an equation meant writing down an expression in which the quantity we sought was on the left of an equation and the quantities in terms of which it was to be expressed were displayed explicitly on the right. Very often in more advanced cases those quantities on the right were in the form of rather intractable expressions such as integrals which had not been tabulated, but at least they were explicitly displayed.

Today however, this is not necessary and sometimes not even possible.

For example, over the past twenty years or so we have set up a mathematics of inequalities in what was called, until recently, linear programming. Problems in this area are solved by iterative methods, and the "solution" of any specified structured problem consists of a machine program based on some algorithm or other which will converge to an answer, given the numerical input. Again, the days when we used to print expensive tables of functions may well be past. The "table" of the future may be just a few feet of magnetic tape embodying a program which will work out the function to any required accuracy for any specified argument. The "solution" of a problem may then consist of a machine program which carries the investigator from a numerical input to a numerical output. Conceptually of course this is the same thing as a formula, but in practice it is rather different. Manipulative mathematics in this sense has to be replaced by the stringing together of a number of machine subroutines.

Let nobody infer that I am stating that the older mathematics is becoming obsolete. Some of the expedients to which it was driven may be so, but the mathematical statistician requires a full mathematical armory to bring his solving process to the point where the machine can take over if required. There is in fact hardly any branch of mathematics which has not been called into play by the statistician, and many of them have been stimulated to new extensions by the posing of statistical problems.

Simulation

In particular we are increasingly having to arrive at partial or approximate solutions of a problem by sampling methods. For example, if we want to know the sampling distribution of some statistic in circumstances where the formal mathematics are too difficult, we may generate a lot of random samples and construct the distribution empirically. We may not have the resources to construct all possible samples, and it may be theoretically impossible to do so; but we may construct enough to give us a sufficiently accurate picture of what the sampling distribution would be like if we could. Or again, we may write down a complicated econometric model and be unable to solve it in any explicit way; but we can try it under a variety of circumstances, sampling, as it were, from all its possible behaviors. This is known as simulation. Clearly one of the main frontiers of advance is going to be the study of complex systems by simulation processes.

Multivariate Analysis

I turn to a rather different growth area, that of multivariate analysis. The problem, or rather the set of problems for there are many in this field, is relatively new. In the past the statistician has often complained about the lack of information at his disposal. It is still true, especially in economics, that we may be woefully short of experience of the phenomena which we are studying over time. But in other areas the statistician is in danger of having too much information, or rather too many figures. The fatal facility with which unthinking minds can churn out stacks of offprint from the computer is a worry to anyone who is trying to see the forest, not the trees. This is not merely a question of being snowed under by observations on too many individuals or of being snowed under by analyses of data in scores of possible ways. It is basically a problem of how many variables we observe on each individual.

One of the older problems in statistics concerns the extent to which observed relationships can be interpreted causally. This is why in experimental design we try to isolate different effects and render them statistically independent. In many real-life situations we cannot experiment or at least design the experiment in an optimal way, and hence we are faced with a complex of factors which in general are highly correlated. Moreover there are a large number of such factors. One sees this in various fields. In economics, for example, we may easily have thirty or forty variables which influence the behavior of a system: gross national product, production of raw materials and of manufactures, employment rates, prices, wages, bank rates, and so on. Indeed, with a disaggregated model one may have hundreds, and these are observable variables, not merely definitional quantities expressed in terms of others. Again, in a social inquiry we may have answers to a hundred or more questions from each individual. Or again, in a medical diagnosis we may have thirty or forty measurements: height, weight, age, blood count, reactions to various tests, and so on. In each case the variables are highly correlated, and say 100 variables do not give us 100 separate pieces of information.

We then arrive at a series of important questions: Are all these variables necessary? If not, can we discard some of them, and which are the ones to discard? How do we know that in discarding we are not throwing out something of importance from the causational viewpoint? This kind of question is being intensively studied at the present, but we are far from having complete and satisfactory answers. In its acutest form, perhaps, we encounter the problem in economics, for there the situation is bedeviled by a new category of complexity, namely that successive observations even on one variable are correlated in time. I would say that the problem of collinearity in econometrics (as it is described) is the most important unsolved problem we now have to face. Some progress has been made, but it is a very difficult and intractable problem in general.

Incidentally, this problem of discarding redundant information enables me to correct one impression that I may have given in referring to the power of the computer. There are certain classes of problems in multivariate analysis which rapidly outstrip in magnitude the capacity of the largest machine we are ever likely to see. Suppose, for example, we are considering the possibility of relating some variable such as gross national product or the Stock Exchange index number of securities to a number of other variables, perhaps thirty in number. Even if we confine ourselves to linear equations, since in any equation each variable may be either in or out (namely, may occur in two ways), there are 2^{30} possible equations we might want to examine if we were to study the whole range of possibilities exhaustively. This is about 1,000 million equations. A machine which calculated and printed out one equation a second would take about three years to complete the job, and the volume of print-out is frightening to contemplate. We cannot, therefore, expect to solve some of these problems by sheer computing power. There is as much scope for ingenuity in saving the computer's time as there has been in saving a person's time, a reflection which ought to be a comfort and a correction to those who think that the computer is going to take all the fun out of life.

5. Future of Statistics

This brings to mind two general points concerning the future of statistics. In the first place it will be plain that an individual is not capable of tackling many of these complicated problems alone. Some problems require a blend of skills from many different disciplines, over which no one man can have command. We may foresee, then, that teams both of men and of women working together will be most successful in solving practical and theoretical problems. The day of the lone worker may not be gone entirely but is on its way, in statistics as in many other branches of science. I do not want to be misunderstood. The day of the individualist is by no means over and I hope never will be. But the resources required to tackle a major problem are so large that he will find the most rewarding work as part of a team.

Secondly, we seem to have gotten rather a long way from the concept of statistics as the methodology of handling numerical information. What precisely is this subject that we speak of in such wide terms? During this century we have seen the emergence of many subjects with much the same primary motivation, the attempt to bring into human affairs the methods of the biological and natural sciences. Thus we have applied mathematics itself, statistics, organization and method, operations research, cybernetics, management science, psychometrics, biometrics, econometrics, sociometrics, and so on. Some, like statistics, are branches of scientific method, others are concerned with the application of scientific method in particular fields. They have their own societies, their own journals, their own conferences, and usually their own elite. But there is an enormous amount of overlap and, science being what it is, there are just as many theoretical developments generated in the applied subjects as in the more theoretical ones. Only a utopia might have all these different subjects neatly classified as subsets of an overall embracing scientific method.

However, a review of the field of scientific endeavor does enable us to put statistics into some kind of perspective. It does not cover the whole field of scientific method—for example, it does not include deductive logic or pure mathematics, though it draws heavily on both. But apart from that, wherever data arise in the form of counting or measurement, and especially where those data are only a sample from some larger universe of discourse, statistics has a role to play. In that sense it is a branch of scientific method of enormous scope and extent. It has already permeated nearly all the sciences and many of the arts—even theology, with Morton's investigations of the authenticity of the epistles of St. Paul based on the frequency of certain kinds of words in the Greek text.

These widening horizons carry a number of obvious implications: we shall need to sharpen our existing tools, invent new ones, and learn to employ to the full the power which the computer now provides us. It is as though we had gone in twenty years from the spade to the bulldozer. But the operation of digging remains basically the same. There is much more to be done, including the training of the people to do it, but I see no new fundamental revolution in statistical thought over the next fifty years. On the other hand, there are new dimensions of practical problems which will probably engross our attention for at least that length of time.

As time passes, the statistician gets drawn into all kinds of peripheral activities which have nothing much to do with theoretical statistics but very much affect his life and work. He has always had to be something of a diplomat to persuade management that figures really cannot be made to prove anything. Now he has to consider ethical standards for professional statisticians and educational facilities (because his subject is particularly suitable for visual presentation). In his learned societies he must face in an acute form the problems being experienced by such societies all over the world: the tendency to fragmentation, the plethora of publications, the sheer difficulty of keeping abreast of the subject, etc. On the whole I think this is a very good thing. Whatever one may think about the future of statistics, the statistician of the future (outside of a few ivory towers) is going to have to deal with a great many human administrative problems even in his own field. He is, in short, going to be much more human than in the past.

If we look back over the history of our subject and consider the strides made in the last fifty years, the computational power now at our disposal, the ever-increasing field of applied statistics, and the enormous number of problems awaiting solution, we may confidently expect that the subject will continue to develop and to flourish for a long time to come. Most of the history of statistics lies in the future.

13

Early Statistics at Iowa State University

JAY L. LUSH

1. The Teaching of Statistics

THE FIRST COURSE in statistics listed in the Iowa State College catalogue was called "Mathematics as applied to social and economic problems" and was offered by the Department of Mathematics in 1914–15. Apparently it dealt mostly with actuarial methods and probability. It is not listed after 1917–18. "Mathematical theory of statistics" was listed for two years beginning in 1915–16, then its name was changed to "Statistical method of interpreting experimental data." "Biometric methods of interpreting agricultural data" also appeared first in 1915–16 and was continued under that name far into the 1920s. The first statistical course offered by the Department of Economic Science was "Rural statistics" in 1915–16.

George W. Snedecor had joined the mathematics staff in 1913. As interest in statistical topics developed, he became responsible for the teaching in that area. Those wanting such teaching and advice were largely teachers, research workers, or graduate students in the agricultural departments. Especially those doing research with farm crops, genetics, and farm animals were often uneasy about the differences in their results when they repeated an experiment. John M. Evvard of the Department of Animal Husbandry was prominent among those. In addition to seeking help from Snedecor, the staff members urged some of their graduate students to take his courses in statistical methods and to consult him about interpreting the data in their theses.

JAY L. LUSH is Professor of Animal Science and C. F. Curtiss Distinguished Professor in Agriculture, Iowa State University, Ames.

211

Of course, some aspects of statistical methods were taught as parts of the lectures and laboratory work in subjects where these methods were being used actively—especially in genetics, plant breeding, and economics. The men who taught the uses of these methods as incidental parts of their main topics had little interest in the mathematical theory of statistics, apart from occasionally being uneasy about whether some statistical method which they were using was really the best available for answering the particular biological, economic, or physical question which interested them at the moment. They merely wanted extra light on such things as Mendelian ratios, the size of the experimental errors in yield tests or feedlot trials, the growth rates of plants and animals, trends of all kinds, curves of supply and demand, the interpretation of a selection experiment, etc.

The subject matter of statistics, as taught in the 1920s, fitted the needs in experimental biology, particularly in agriculture, more closely than the needs in economics. Because of this divergence the statistical methods used in economics and sociology were taught largely by staff members in that area.

2. Sources of Knowledge about Statistical Methods

The theory behind most statistical methods, especially the biometrical ones, nearly all came by way of Karl Pearson and the Galton Laboratory, although through diverse channels; but parts of that theory can be traced back much further to or through other men such as Gauss and Quetelet. However, the workers at the Galton Laboratory had gathered together and developed further such of this as they thought likely to be useful in the study of living beings. Among statistical methods in common use in the early 1920s, only some of those in economics had a pedigree other than through the Galton Laboratory.

A few of the methods used in studying heredity came from Galton himself, although he provided little of the theory. The de Vilmorins in France had already published in a statistical way on progeny-testing in sugar beets, beginning in the 1850s. Yule and Punnett had helped in clarifying the statistical consequences of Mendelism soon after its rediscovery. Wilhelm Johannsen in Copenhagen and the Swedish plant breeders at Svälof devised many new applications of statistical methods to genetics and to the problems of plant breeding and of variety trials.

Biometrical methods came to the United States through diverse channels. Many plant breeders learned about them through H. H. Love and his students at Cornell. Workers on farm animals generally learned of them from E. D. Davenport's textbook on animal breeding and the many publications by Raymond Pearl. Workers in genetics, especially those in human genetics, often used C. B. Davenport's desk manual; a few of them even read the papers of Weinberg.

Many who were working in plant breeding or with fertilizers did actual experiments, such as uniformity trials, to measure the amounts of uncontrolled variation in crop yields, especially variation caused by heterogeneity of the soils in their experimental fields. They did many experiments with various devices for controlling or discounting the variation caused by this factor or by others which could be identified but could not be controlled experimentally. Such workers included J. Arthur Harris, S. C. Salmon, and F. D. Richey in the United States and, of course, "Student" in Britain and several men at Svälof. These experimenters had acquired considerable familiarity with the actual difficulties of interpreting data from small samples, even before 1919 when Fisher began working at Rothamsted on numerous examples.

Yule's was almost the only textbook dealing wholly with statistical methods until far into the 1920s. Most of those who would have profited by understanding that text thoroughly did not have enough mathematical preparation. However, through those who were able to master it, Yule's book had considerable influence, especially in economics.

Most biologists got their first printed introduction to statistical methods from a chapter on variation and its measurement in some book on genetics. Such a chapter was included in nearly every textbook on genetics printed before the middle of the 1930s. (Chapter 17 in *Genetics in Relation to Agriculture* by Babcock and Clausen is an example.) W. Johannsen's textbook on genetics in 1909 can, with only slight exaggeration, be called the first textbook on the genetics of populations.

Complete textbooks on statistical methods began to appear in the 1920s. They were by such men as Mills and Ezekiel in economics, and by Kelley, Spearman, and Thurstone in psychology and education. By 1925 Pearl's *Biometry for Medical Students* and Fisher's *Statistical Methods for Research Workers* appeared, but the former was aimed a little to one side of the interests of most workers in biology and agriculture, while the latter was so condensed—yet covered so much territory—that only those already experienced in statistical work could begin to appreciate it on first reading.

3. Statistical Methods Commonly Used
in the Early 1920s

Most early statistical methods were descriptive rather than analytical. Tests of statistical significance were used widely in verifying Mendelian ratios and in variety and fertilizer tests, but the idea had barely begun to enter economics. The use of $n - 1$, to discount for the mean having been computed from the sample itself, had been known from the time of Gauss (around 1800) but was usually ignored except in work with field plots. There "Peter's formula" or "Bessel's formula" were used frequently. The replications are necessarily few in field plots. Methods to allow correctly for having taken other information, besides the mean, from the sample itself were not available until after 1915 when Fisher generalized the earlier special case of "Student" into the idea now called "degrees of freedom." Extension of this to various methods actually used with field plots began to come quickly after 1919 when Fisher took a permanent post at Rothamsted. Yet only a few workers with plants and almost none of the workers with animals made much use of degrees of freedom until considerably after Fisher's text on statistical methods appeared in 1925.

It is said that a rule-of-thumb at the Galton Laboratory until near 1920 was that if the number of items per subclass were less than 50, biometrical calculations on them were largely wasted effort, since the results would not be dependable. If the number was as large as 50, correcting for the finiteness of the numbers in the sample would make little difference. Certainly common sense supports this viewpoint; yet following it does leave small biases in the computed statistics, even when the numbers are large. Still more important, this rule excludes from biometrical analysis most of the data from planned experiments with field crops and the larger farm animals. It was no accident that most of the leaders in applying the idea of degrees of freedom (once they learned about it) were experimenters whose basic units were field plots or rows.

"Probable error," rather than "standard error" was still being used in the English-speaking countries in the 1920s and far into the 1930s, in spite of the fact that the extra work of multiplying by .6745 adds no information. Much of Galton's original work had been phrased in terms of "quartiles," and the step from quartiles to probable error was a simple one. He and many of the other early workers in biometry had thought that the many persons who (they expected) would be describing and reading about variation would understand "probable error" more easily than standard deviation. "Probable error" was rarely used in Scandinavia or in the German-speaking countries.

The probability that a difference at least as large as the observed one could have occurred just by sampling, when the real difference was zero, was usually expressed as a ratio such as $1:18$, $1:29$, $1:140$, etc., instead of as: $P > .05$, $.05 > P > .01$, $P < .01$. Each method of expression has psychological advantages. In the older method the experimenter wasted some effort in looking up the exact odds, in view of the few courses of action open to him after he had learned those odds. Also the detail in the figures could easily lead him to an exaggerated idea of the exactness of the odds he had computed. On the other hand, the present method of expressing this as a trichotomy (or a tetrachotomy in some places in Europe, where $P = .001$ is often used as another threshold), when in fact the distribution is continuous, throws away some of the information. It also imputes more importance to being just above or just below one of the two (or three) arbitrarily chosen levels of significance than this deserves. What one would actually *do* on the basis of statistical significance really differs little as between $P = .04$ and $P = .06$ or between $P = .011$ and $P = .009$, whereas one's actions would be quite different as between $P = .049$ and $P = .011$! The present custom also makes it easier for the incautious to make the error of thinking that when a difference is not significant statistically, the evidence has indicated that the difference really is zero.

Multiple correlation was in its heyday, especially in economics but also with some biologists. In 1921 Wright had proposed his powerful method of path analysis for exploring the consequences in cases where a plausible "model" of causal relations behind the observed correlations could be hypothesized from evidence external to that sample of statistics. However, this had not yet had much influence outside his own work on systems of mating.

Fitting straight-line regressions was practiced by almost everyone. Many also fitted curved regression lines, e.g., to the changes in milk production of cows with increasing age. Yet those who fitted curves could not discount the subjectivity introduced by looking at the data first in order to choose the type of curve which would fit well.[1] Multiple curvilinear correlation had been pushed far in the early 1920s by Mills and others among the economists, but the methods were full of subjectivity.

---

1. The present method of fitting polynomial curves routinely if curvature is suspected seems to have nothing to recommend it except its objectivity and the simplicity of its calculations. Fitting and testing a second-degree polynomial is useful for answering the question of whether *any* curvature is present, but once that has been answered affirmatively, subjectivity enters as strongly as ever into deciding whether a polynomial, an exponential, a logistic, or still some other type among the almost infinitely many types of curves which are possible would describe best the curvature which is actually present.

Fitting frequency curves of Pearson's "Types" had come nearly to a dead end by 1920, although the theory continued to be taught in many places for more than a decade. The considerable labor of fitting the curves yielded little increase in one's understanding of the biological or physical basis of the problem.

To the nonmathematical worker, beauty and elegance in the theories of statistics were doubtless admirable for esthetic reasons, but were not worth his own time and effort to master except insofar as they might help him choose the method which would illuminate his own problem most dependably. Doubtless this influenced Snedecor to emphasize actual examples in his teaching and, in his consulting, to ponder about what extra statistical knowledge most of his colleagues would *really use* in their work if he gave that information to them. Those of his students who did not intend to become statisticians simply did not have time to learn more than a smattering of statistical theory. He tried to teach from such actual examples as were in the current literature in their own fields. A jest current among the graduate students in the 1930s was that the main reason for taking Snedecor's course in statistical methods was to learn about plant breeding!

4. The Design of Experiments

The idea of trying to hold all the variables constant except the ones which are the object of the study is centuries old, no doubt going back to the very beginnings of the experimental method itself. Some ways of trying to achieve this will invalidate the experimental error unless the experiment can be repeated completely. Designing experiments to make all the comparisons of interest, and yet to get some idea of the reliability of the observed results, had concerned the investigators at Rothamsted even from the beginning of the research there in the 1840s. Yet little progress was made, especially toward the latter part of this goal, until long after 1900.

Several special designs such as split plots and pairing were being used widely in plant breeding and in experiments on fertilizers by 1910. Designs more complicated than pairing the experimental lot or individual with a "control" lot or individual were used little in other areas.

The statistical significance of differences observed in feedlot trials had been an object of inquiry by animal husbandmen from at least as early as 1913, when Mitchell and Grindley published Bulletin 165 from the Illinois experiment station. The topic progressed only moderately in the ensuing 20 years, as may be seen on examining the papers of a planned symposium (pp. 15–40) on the use of statistical methods, held at the 1932 meeting of the American Society of Animal Production. The standard recommendation concerning feedlot experiments was that the whole experiment should be repeated three times before drawing conclusions; yet few investigators actually did this. Nearly always some new question which seemed to need an answer more urgently had arisen before the experiment could be conducted a third time, or even a second time. To most men, duplicating contemporary lots seemed foolishly wasteful since investigating other interesting rations would have to be postponed. Evvard's curiosity did lead him to duplicate his "check lots" (those receiving the "standard" ration) in the 1920s, but he did not venture to replicate the lots on the other rations.

Randomization was rarely being used enough to guard against systematic errors or to make the error terms entirely valid. Complete randomization often appeared to conflict with the goal of controlling other sources of variation as far as possible. Consequently, the need to modify the simpler statistical designs arose frequently, especially in research with crops, fertilizers, and livestock, so that those designs would permit the investigator to control the irrelevant variation more fully than he could if everything were randomized. A simple example of this is in a paper which Snedecor and Culbertson gave in 1932. The statistical principles were not new but this seems to have been the first time they were applied to a feedlot experiment.

When starting a feeding experiment with pigs, steers, or lambs, the husbandman likes to allot the animals so that initially the lots will be equal in their gaining ability, carcass quality, and any other criteria by which he intends to judge the merits or demerits of the treatments. Most husbandmen believe they can foresee some of the differences in the way individual animals will perform during the experiment. Allotting the animals to make the lots average alike initially in these respects is usually called "balancing," which is a part of the husbandman's effort to make the lots alike in everything except the experimental treatments so that the differences among the lots at the end of the experiment will be only those the treatments caused. Unfortunately, this balancing invalidates the intralot variance as a measure of the experimental error. Specifically, to the extent that the husbandman actually can foresee differences in individual performance, the balancing reduces the variance among lot means and inflates the variance within lots.[2]

2. Incidentally, it does this even if the husbandman has things backward and the correlation between his estimate and the animal's actual performance is negative! Only when the correlation between his estimate and the actual outcome is zero will the intralot variance be a valid experimental error if the experimenter balances the lots initially. The reasons for this can be seen by imagining what would happen if the whole experiment were a uniformity trial; i.e., if all lots were given the same treatment.

If these were the only possibilities, the husbandman would have to choose between (1) allotting the animals at random to get a valid error term, but thereby opening the door for the randomizing sometimes to make the lot means differ more than if he had balanced them initially, or (2) balancing the lots to reduce chance differences among the lot means, but thereby inflating the intralot variance by an uncertain amount. No husbandman in his right mind would adopt alternative (1) if he really believed that he could predict individual performance with any noticeable accuracy. His primary aim is to find out *what differences the treatments actually make*. Getting a valid error term is desirable, of course, but not if getting it endangers his primary aim seriously.

Snedecor's solution was the now well-known one of stratified sampling. The husbandman divides the animals into "outcome groups," each group containing t animals if there are t treatments. He uses any criteria he thinks would help him put the t best animals in the first outcome group, the t next best ones in the second outcome group, the t which are third best in the third outcome group, etc. The t animals in the first outcome group are allotted at random among the t treatments, then those in the second outcome group are allotted among the treatments at random, etc. Or the outcome groups might even be related as the rows in a Latin square if the husbandman thinks he can foretell individual differences *within* outcome groups well enough to make that worth the extra trouble.

By using outcome groups in this way, the husbandman can get a valid estimate of error, even though he has made almost full use of his ability to reduce the experimental errors in the differences among their means by balancing the lots. Moreover, the mean square for differences among outcome groups will show him how nearly he actually did foresee the individual differences in performance. This may be a point of interest in itself.[3] The only price he pays for these advantages is that $x - 1$ degrees of freedom, which could otherwise be used for estimating the average intralot variance more accurately, are used in estimating the sum of squares for differences among the means of the x outcome groups.

The validity of the intralot variance as an error term depends also on intralot competition having had no effect on the trait being studied. This is difficult to ascertain. Also, intralot variance cannot be computed for traits which are observed only on the lot as a unit. These ordinarily include some

3. The correlation between forecast and fact is about .53 in the example which Snedecor and Culbertson used. General experience is that husbandmen can forsee the differences in rate of gain among pigs to an extent worth considering, especially if the pigs vary considerably in age when the experiment begins, as often happens. They have almost no success in forecasting individual differences in rate of gain among steers, but they can foresee many of the differences in carcass merit which will exist among those steers when the experiment is completed.

important ones such as feed consumption and feed conversion. Hence, getting valid error terms for all traits likely to be of considerable interest to the husbandman seems to require repeating the experiment.

The device of outcome groups is a special form of older and more general methods, as often happens when a method is devised for a special purpose. If only two treatments are being compared, the method of outcome groups is the same as "Student's" method of using paired differences. It also can be considered as using the correlation between the husbandman's prediction and the actual performance of the individual animal. That correlation is interclass if only two lots are involved. If the treatments number three or more, the correlation is intraclass, each outcome group being a class and the variance due to treatment means having been removed. Or each outcome group can be considered as a replication of an experiment, the replications being conducted at the same time and place.

5. The Wallace Lectures

An outburst of statistical activity in the Bureau of Agricultural Economics in Washington began about 1919 and was fueled for many years by the economic dislocations of agriculture which became apparent at the end of World War I. Naturally, much of this centered on costs of production and on techniques for predicting trends in prices and production. Multiple correlation and regression occupied the center of the stage in descriptive statistics during the few years before 1920 and far into that decade.

Henry A. Wallace was then an editor of *Wallaces Farmer* in Des Moines. He was also highly interested in research, especially in the fields of plant and animal breeding and economics. His short paper, "What Is in the Corn Judge's Mind?" (Wallace, 1923) will illustrate some of his interests, his enthusiasm for new ideas, and the proficiency he already had in multiple correlation.

Wallace's father, Henry C. Wallace, was Secretary of Agriculture from 1921 until his death in 1924. Henry A. visited Washington frequently during this time, both to see his father and as part of his editorial interest in "seeing what was cooking" in agricultural research and policy. He became enthusiastic about the statistical methods he saw being used so actively in the Bureau of Agricultural Economics and, to a lesser extent, in other bureaus. He was especially interested in the laborsaving potentialities in computing with punched-card machines. Those who began their statistical apprenticeship later than the early 1920s can scarcely imagine the many hours previously spent by high-powered research men in the drudgery of computing correlation coefficients one by one. Only the most tenacious and devoted could come through those ordeals with enough enthusiasm and time left to do the real work of finding the basic biological, economic, or physical mechanisms which were causing their data to behave as they did.

Wallace, as a loyal alumnus of Iowa State, was most eager that the faculty of his alma mater should know these newer statistical methods and use them wherever appropriate. Some of the faculty members challenged him to show them what was really new in this area. He accepted this challenge and spent his Saturdays in the spring of 1924 at Ames, explaining and illustrating to a class of some twenty faculty members and graduate students the things which seemed to him new and useful, especially those which promised to reduce the drudgery of computation. C. F. Sarle, who was then doing graduate work in statistics at Drake Univeristy, came with him on some of the trips. Several lessons near the end of the series dealt with using punched-card machinery for computing. For these lessons Wallace and Sarle borrowed some card-handling equipment from an insurance company in Des Moines, hauling the machines back and forth each Saturday as they were needed.

Wallace was determined that these efforts should grow into something permanent if that would fill a real need at the university. One step in this campaign was the publication of *Correlation and Machine Calculation* (1925). The illustrative data were from Sarle's M.S. thesis. Wallace wrote most of the first draft but Snedecor did the final work, helping especially with keeping it straightforward mathematically. He had helped thus during the whole series of lessons, although Wallace had full responsibility for planning and conducting the course. A revised edition appeared in June 1931. Both editions, but especially the first, were models of lucid writing and were widely influential. Many of the statements were simplified for brevity and some qualifications were omitted for clarity, especially in the first edition. This publication did much to popularize statistical methods and to raise hopes about what could be learned by using them. Many research workers who were using correlations but were not mathematicians themselves kept *Correlation and Machine Calculation* on their desks as an indispensable manual. Some even entrusted their computing to clerks, letting them use this publication as a desk guide.

Snedecor had a card punch and verifier in his possession when A. E. Brandt joined the staff in 1924. The two of them started helping colleagues analyze research data. They would take the punched cards to Des Moines on Saturdays and use a sorter and tabulator at one of the insurance companies. The success and the inconvenience of this led to establishing the Mathematics Statistical Service in 1927 with Snedecor and Brandt in charge. As part of this service the university installed IBM card-sorting and tabulating machines for the first time. The machines were first located at the Physics Building where direct current was available. In the second year they were moved to the third floor at Beardshear Hall, with a generator of direct current installed especially for them.

The Mathematics Statistical Service was rearranged in 1933 to be the Statistical Laboratory. This was responsible directly to the president, rather than going through the Department of Mathematics. The laboratory and the machines, now increased and replaced with 80-column models, were established on the ground floor of the Old Office Building. There they remained until moved in 1939 to the Service Building, now Snedecor Hall.

6. R. A. Fisher's First Visit

The Graduate College was always alert for opportunities to get outstanding scientists to visit and give lectures on their work. This helped keep the local staff abreast of promising developments at other research centers. Largely due to Dean R. E. Buchanan of the Graduate College and to Professor E. W. Lindstrom of genetics, it was the regular custom through the 1930s and far into the 1940s to invite an outstanding scientist as a visiting professor for six weeks each summer. A different visitor, generally with a different field of specialization, was secured each year. The visitor's area of specialization had to be of major interest to several departments. These would join in requesting a particular man and would share in the campus chores of making announcements, arranging rooms, handling registrations, scheduling consultations, etc. The Graduate College or related sources provided the expenses and honorarium of the visitor.

Fisher was a natural choice in 1931 because he had recently published two books of wide interest, besides numerous earlier technical papers, and his subject matter area was keenly interesting to the departments of mathematics and genetics and to many individuals in agronomy and animal husbandry, as well as to some in other areas. The first of these two books was *Statistical Methods for Research Workers*, published in 1925 and in its third edition at the time of his visit. This work was beginning to be appreciated by those biologists who already had experience with statistics. Fisher's other book, then attracting wide attention among geneticists and others interested in evolution, was *The Genetical Theory of Natural Selection*, first published in 1930. Fisher lectured nearly equally on these two topics during that summer, though perhaps a bit more on statistics. Three lectures per week followed rather closely his book on statistical methods. The other three concerned either the theory of statistics or his *Genetical Theory of Natural Selection*.

Members of the local staff, or occasionally a visitor, took turns in presenting at seminars (about two per week) some of their own experimental results which they had analyzed statistically. Then Fisher and the others present would be asked to comment on any errors or ambiguities they had noticed in the procedure or in the speaker's interpretation, whether the question which the experiment answered was really the one the experimenter had intended to ask, what additional inferences might have been drawn, etc. Besides these scheduled or "clinic" sessions, Fisher was available by appointment to individuals singly or by twos or threes, if they wished to inquire more fully into some topic he had mentioned in his lectures, or to ask him about some methodological problem they were encountering in their own work. Brandt looked after this scheduling and did the other campus chores necessary to keep things going smoothly.

Many visitors already experienced in statistical or genetical research attended all or parts of this six-week session. Sometimes these visitors would lecture on their own work or would otherwise contribute formally to the seminars and other discussions.

As consequences of this summer session, the work and interest in statistics continued to expand; experiments were designed better, with the questions in sharper focus; and more instruction in the theory of statistics was given. The first M.S. degree with a major in statistics was granted to Gertrude Cox in 1931. The real problems encountered in local research still furnished most of the examples for teaching and raised most of the questions of theory which became the objects of further research. This was a consequence of the approach taken by both Snedecor and Fisher and of the clinical nature of the seminars.

Tests of *statistical significance* became overemphasized, with consequently less attention to ways of using statistical methods for *describing* populations. For example, all too frequently the first draft of an M.S. thesis, or occasionally a Ph.D. thesis, might inform the astonished major professor that the difference between treatments A and B was statistically significant with $P < .01$ but would neglect to state *how large* the observed difference was! Even more frequently it was implied that if a difference were not significant statistically, the evidence therefore indicated that its true value was zero! These things cure themselves in time, of course; meanwhile they obscure the main object of the research, which is nearly always to learn what the difference *most probably is* or (if one wishes to incorporate the idea of sampling error) the confidence interval which includes it. Even that is usually only a first step toward the real goal of learning more about the biological or other forces or mechanisms which make that difference what it really is. In a very real sense, sampling errors and statistical significance are troublesome artifacts which arise only because the number of units is finitely small and we do not succeed in controlling all the sources of variation which are irrelevant to the question we are trying to answer.

The lectures and discussions of the 1931 summer session were never published as a unit but affected in many ways the subsequent research and writings of those who had heard them. Snedecor published a small book in 1934, *Calculation and Interpretation of Analysis of Variance and Covariance*, but this draws in part from publications subsequent to 1931 and from another two years of experience in teaching and consulting. In the summer of 1934 John Wishart from Cambridge lectured on some problems in the analysis of covariance.

Snedecor was appointed as statistician to the Agricultural Experiment Station in 1933. The general arrangement of his duties was that each formal proposal and outline for a new project was submitted for his opinion as to whether it would yield valid error terms and for suggestions as to how its design might be improved. He did not have veto power; the director reserved that for himself, but Snedecor's advice and suggestions carried weight. Brandt, like Snedecor, had a joint appointment in the Agricultural Experiment Station and in the Department of Mathematics and assisted in these station duties. When the results of an experiment were offered for publication as a bulletin, either Snedecor or Brandt was likely to be named as a member of the examining committee if the manuscript contained more than a little statistical treatment.

A separate statistical section in the Agricultural Experiment Station was organized in 1935. Through this mechanism, research into statistical theory and statistical methods could be supported financially if money was available and the director approved.

7. Fisher's Second Visit

Fisher moved from Rothamsted to the Chair of Eugenics in the Galton Laboratory at the University of London in 1933. Brandt spent parts of 1934 and 1935 studying with him there. Fisher's first visit to Iowa State University had generated so much enthusiasm about statistics and had opened up so many vistas of theory, which needed clarifying to keep the methods from being applied to situations they did not really fit, that the university staff rejoiced when Fisher could accept the invitation to return for another six weeks in the summer of 1936.

About half his lectures concerned the design of experiments; he had published his book with that title in 1935. Nearly all the rest of his lectures concerned topics on the theory of statistics. He paid less attention to genetics than in 1931, although he remained keenly interested and referred to its problems often in the "clinic" sessions. What are known now as "components of variance and covariance" were just beginning to be used widely, often by people who were somewhat uncertain about how closely the computed figures measured the biological or physical realities which their names suggested.

After 1936 the staff increased, slowly at first but then rapidly. Brandt left in 1937 and C. P. Winsor came in 1938. F. Yates and W. G. Cochran were visiting lecturers for short periods in 1937 and 1938. In 1939 Cochran became a regular staff member. Gertrude Cox left in 1940. Some men active in statistics came on the staffs of other departments or divisions. For example, Gerhard Tintner came in 1937 in economics and mathematics to lead the work in econometrics. John W. Gowen, who edited the biometrics section of *Biological Abstracts* for many years, came as professor of genetics in 1937. Others were located officially in such areas as plant and animal breeding, psychology, and economics.

Extensive formal cooperation with the USDA began in 1938 in the areas of sampling and survey methods, opinion polls, market surveys, taking censuses by sampling, etc. At one time subsequently the USDA had as many as seven resident collaborators stationed in Ames as members of the statistics staff. Some further increase in the staff resulted in the 1940s when the Bureau of the Census joined in this cooperative research.

The increased attention to theory and the expansion of interest to survey methods and other fields, especially after 1938, left relatively less room for teaching principles by clinical methods, i.e., by examples. In statistics, as in all sciences, theory and applications are related through a feedback mechanism. Questions of statistical theory are often suggested first by an actual experience in applying a presently known method to a real problem and finding that method not to fit perfectly. Solving the newly arisen questions of theory leads often to new applications and to new methods which, in turn, are found not to fit perfectly some of the problems for which at first they seem ideal. This is the classical cycle of scientific discovery; applications suggest new problems and the solutions of those problems suggest new applications. To paraphrase Walt Whitman: Theory without applications becomes fantasy, while applications without theory become chaos! The almost explosive expansion in the subject matter of statistics (as in most other sciences) has made this problem acute. Solving it by offering separate courses in the theory of statistics and in statistical methods has been only partially successful.

Snedecor had always sought to teach his statistical courses in such a way that anyone who knew high school algebra well could understand them. Some thought it inappropriate for the mathematics department to allow graduate credit for courses which did not even require calculus as a pre-requisite. Yet these courses in statistical methods were extremely useful to many graduate students, especially those in agriculture or biology. Few of these students had had any formal classes in mathematics past high school geometry. As a compromise, through the 1920s and 1930s the courses in statistical methods carried only minor graduate credit. If a student's major interest was in statistics, he did his graduate work under the auspices and rules of the Department of Mathematics. The first Ph.D. with a major in statistics was granted in 1940 under this arrangement to H. C. Fryer, who used data from genetic experiments he had conducted in the laboratory of John W. Gowen. In his thesis he explored the reliability of the chi-square test when the classes are numerous but the expected numbers in many of them are small, perhaps even less than one. This arrangement about graduate credit for work in statistics became increasingly unsatisfactory as the scope of the graduate work in statistics widened, especially after 1938. This was among several considerations which led in 1947 to establishing a Department of Statistics which was independent of the Department of Mathematics. Other considerations included the need for flexibility and quick decisions if the large volume and the varied kinds of work in the Statistical Laboratory were to be done quickly and efficiently.

8. Snedecor's Textbook

Statistical Methods appeared first in 1937. It found worldwide acceptance quickly among the thousands who wanted a desk manual or "cook book" where they could read what to do but would not be bothered with theory, except for a few bits which might help them estimate better whether a particular method fitted the case they were wanting to analyze. More than 100,000 copies had been sold before 1970, not counting translations which had been published in Spanish, Japanese, Rumanian, and perhaps other languages. As a "how to" book containing also a little "why," it nicely complements Fisher's *Statistical Methods for Research Workers*.[4]

The latter contains much of "why" in an extremely condensed form, although with examples to illustrate each topic. The ease with which Snedecor's book can be understood had much to do with its worldwide and long-lasting popularity.

4. This comparison is epitomized by a remark a European agricultural research worker, who had himself published a few simple statistical analyses, made to me in 1939. On learning that I was on the staff of the same university as Snedecor, he said: "When you see Snedecor again, tell him that over here we say, 'Thank God for Snedecor; *now* we can understand Fisher!' "

Snedecor reached the age for retirement from administrative duties June 30, 1947. Consequently, he never was head of the new department officially, although he had guided its development and shaped it to the moment of its launching. He remained on its staff, teaching and consulting on a part-time basis until 1958. When not at Iowa State University during those eleven years, he was usually a visiting professor or in other ways was helping in the local statistical work at universities or other institutions in such places as North Carolina, Alabama, Florida, Virginia, and Brazil. He had guided the work in statistics at Iowa State from its beginning, when it was only a part-time effort by one man, through ten years when he had only one major colleague, and through its explosive expansion from 1938 until the middle 1940s. To the "old-timers" around Iowa State, he will remain "Mr. Statistics."

References

American Society of Animal Production. 1933. Symposium on statistical methods in experiments in animal husbandry. Proceedings of annual meeting, 1932, pp. 15–31.

De Vilmorin, Louis. 1856. Note sur la création d'une nouvelle race de betteraves à sucre. Considerations sur l'hérédité dans les végétaux. *Compt. Rend. Acad. Sci.* 43:871–74.

Fisher, R. A. 1925. *Statistical methods for research workers.* Edinburgh: Oliver and Boyd.

———. 1930. The genetical theory of natural selection. London: Oxford Univ. Press.

Johannsen, Wilhelm. 1909. Elemente der exaktan Erblichkeitsleihre. Jena: Gustav Fischer.

Mitchell, H. H., and H. S. Grindley. 1913. The element of uncertainty in the interpretation of feeding trials. Ill. Agr. Exp. Sta. Bull. 165.

Snedecor, George W. 1934. *Calculation and interpretation of analysis of variance and covariance.* Ames: Iowa State College Press.

———. 1937. *Statistical methods.* Ames: Iowa State Univ. Press.

Snedecor, George W., and C. C. Culbertson. 1933. An improved design for experiments with groups of animals whose outcome may be estimated. Proceedings of the Twenty-fifth Annual Meeting of the American Society of Animal Production, 1932, pp. 25–28.

Wallace, H. A. 1923. What is in the corn judge's mind? *J. Am. Soc. Agron.* 15:300–304.

Wallace, H. A., and George W. Snedecor. 1925. Correlation and machine calculation. Iowa State College official publication 30, no. 4.

14

On the Variance
of the Mean of
Systematically Selected Samples

M. RAY MICKEY

1. Introduction

THE THEORY of systematic sampling developed quite rapidly with the publication of the initial study of Madow and Madow (1944). The Madows formulated the problem in terms of sampling from a finite population and gave expressions for the variance of estimates of the population mean for both unstratified and stratified systematic sampling. The investigation included comparisons between systematic and random sampling. Results were expressed in terms of the (finite) population variances and serial correlations. Cochran (1946) formulated the problem in terms of populations whose elements have a (wide sense) stationary distribution. He presented instructive expressions for the expected value of the variance of the mean from systematic and from stratified random sampling. Results were expressed in terms of the correlation function, and sufficient conditions were given for systematic sampling to be more efficient (in the expected value sense) than stratified sampling. Yates (1948) reported an extensive investigation of one-dimensional systematic sampling, which included numerical studies as well as contributions to the theory. From these and other contributions the topic rapidly became well developed, as indicated by an early review article of systematic sampling literature

M. RAY MICKEY is Research Statistician, Department of Biomathematics, University of California, Los Angeles.

Research for this paper was supported by NIH grant RR-3. Laboratory work on which section 4 was based was supported by a grant from NIH, NICHHD.

The author expresses appreciation to Mrs. Maureen Motola, UCLA Health Sciences Computing Facility, for her very competent assistance with the numerical work.

prepared by Buckland (1951). Some of the subsequent developments were summarized by Dalenius (1962) in his review of the broader topic of survey sampling.

The population characterization underlying the theoretical development of systematic sampling has been the correlation function. An exception is found in Hannan (1962) in which use is made of the spectrum of the presumed underlying stationary process. The present development is more along these lines. The problem is formulated in terms of estimating the average of a function $y(x)$ defined over a finite interval. A random point in the interval is chosen, and the value of y is observed at the selected point and at successive intervals of length L/n, where L is the length of the interval and n is the size of the sample. The observed Y may include an independent error in addition to the function value $y(x)$, but the main interest here is in the variance attributable to the systematic sampling of y. Instead of characterizing y in terms of a correlation function, we choose the representation in terms of the Fourier coefficients of y. This change of reference leads to some general theoretical results. It also leads to a different type of numerical analysis in the study of preliminary pilot data.

The principal results to be presented are: (1) an expression for the variance of sample average in terms of the Fourier coefficients of the function $y(t)$ for the given interval

$$\mathrm{Var}(\bar{y}_n) = \frac{1}{2} \sum_{k=1}^{\infty} b_{kn}^2, \tag{1.1}$$

where

$$y(t) = b_0 + \sum_{m=1}^{\infty} b_m \cos(2\pi mt + \theta m), \quad 0 < t < 1$$

(there is, of course, no loss of generality in taking the fixed interval to be $[0, 1]$) and (2) a bound on the variance:

$$\mathrm{Var}(\bar{y}_n) \leq \frac{(\text{total variation of } y)^2}{12n^2} \tag{1.2}$$

which is established given the condition that $y(x)$ is of bounded variation over the interval. "Bounded variation" is a mathematical idea and may be expressed as implying that the given function is the difference of two nondecreasing functions. The total variation is defined as the sum of the increases of the two functions over the interval. It needs to be kept in mind that for our purposes the values of the function at the end points of the interval are redefined, if necessary, so as to be equal.

The main result is more informative than may appear at first glance. Since for sample size n one sums only every nth term of the series b_m^2, it would be expected that in most cases the variance would be no greater than $1/n$th the variance of a sample of size one, the result for random sampling. But in addition, if the curve is fairly smooth, the harmonic coefficients tend to decrease fairly rapidly, so that the variance of the sample mean can decrease much more rapidly than $1/n$. If data is available for estimating the harmonic coefficients b_m, (1.1) provides a convenient basis for estimating the variance.

The bound (1.2) is interesting in that it shows that the variance will decrease at least as $1/n^2$ rather than $1/n$. Two seeming contradictions that may be noted are the following. Consider first the function $y(x)$ $y(x) = \cos(2\pi M x)$, where M is a positive integer. The result (1.1) shows that $\mathrm{Var}(\bar{y}_n) = 1/2$ if n divides M and is 0, otherwise. In particular a sample of size M has the same variance as a sample of size one. But the total variation in this case is $V = 4M$, so that the bound of (1.2) is $16M^2/12n^2$. For $n = M$ the bound has the value $4/3$. Since one does not ordinarily know the value of the total variation, the bound does not appear overly helpful. Nevertheless it is significant that the bound shows that a decrease at least as fast as $1/n^2$ is assured since bounded variation may be taken as given in applied problems. A second case to be considered is that in which the curve is essentially noise. In this case, for small sample sizes at least, the variance will decrease as $1/n$. Again there is no genuine contradiction since in these cases the total variation will be very large. On the other hand, if the curve is "smooth" and slowly varying, the total variation may be relatively small so that the rapid decrease with n will occur for small n as well as for large. In the applied problems we have had occasion to consider, the decrease of variance with increasing sample size has been impressively rapid.

The applied problem leading to the development of these results concerned estimation of total tissue components from serial sections. For example, Hale et al. (1968) were interested in determining the volumes of various tissue components, e.g., muscle, in a bronchus segment. The area of muscle tissue in a cross section could be estimated from a histologically prepared section. The area at a distance x along the segment defines a function $y(x)$, and the desired volume could be computed by multiplying the segment length by the average value of $y(x)$ along the segment. The observed area of a section could thus be represented as $Y(x) = y(x) + \varepsilon$, in which ε is a measurement error. The immediate questions were: Would the more conveniently taken systematic sample of serial sections lead to greater precision than a random sample of sections? And, if so, how many such sections should be assessed? The approach to the problem was based on data obtained from assessment of several serial sections from a single specimen bronchus segment for each of a number of tissue components. Fourier coefficients were estimated from the relatively abundant data base and formed a basis for judging the effects of varying sample size. The indications were that if the sample size were large enough to adequately control the average measurement error, the error arising from the systematic sampling could probably be neglected. A similar study was done in estimation of the content of various tissue components in rat ovaries (Stern and Mickey, 1970). This study will be considered as an illustration (section 4), since in this case the laboratory work was planned to supply estimates of error which could be compared with the calculated predicted values.

2. Results and Derivations

Let $y(x)$ be given a function defined over the interval $0 \le x \le 1$ and extended as a periodic function $y(x + j) = y(x)$, $j = \pm 1, \pm 2, \ldots$. Consider the systematic sample of size n,

$$x_1 = u, \quad x_2 = u + 1/n, \ldots,$$

$$x_j = u + \frac{(j-1)}{n}, \ldots, x_n = u + \frac{n-1}{n},$$

where u is a uniformly distributed random variable on the interval $(0, 1/n)$. Consider the estimate of the mean of $y(x)$

$$T_n = \frac{1}{n} \sum_{j=0}^{n-1} y(u + j/n) = \sum_{k=0}^{n-1} y\left(u + \frac{j}{n} + \frac{k}{n}\right) \tag{2.1}$$

in which the periodic nature of $y(x)$ has been used in the second expression. Assume, with no loss of generality, that the average of y over the entire interval is zero. Then $\mathrm{Var}(T_n)$ can be expressed as

$$\mathrm{Var}(T_n) = \frac{1}{n^2} \sum_{k=0}^{n-1} \left[\sum_{j=0}^{n-1} \int_0^{1/n} y\left(u + \frac{j}{n}\right) y\left(u + \frac{j}{n} + \frac{k}{n}\right) \cdot n\,du \right]$$

$$= \frac{1}{n} \sum_{k=0}^{n-1} \int_0^1 y(x)\, y\left(x + \frac{k}{n}\right) dx \tag{2.2}$$

$$= \frac{1}{n} \sum_{k=0}^{n-1} \phi\left(\frac{k}{n}\right)$$

in which

$$\phi(t) = \int_0^1 y(x)\, y(x + t)\, dx. \tag{2.3}$$

Let $y(x)$ have the representation

$$y(x) = \sum_{m=-\infty}^{\infty} a_m e^{2\pi i m x}. \tag{2.4}$$

It then follows that $\phi(t)$ has the representation

$$\phi(t) = 2 \sum_{m=1}^{\infty} |a_m|^2 e^{2\pi i m t} \tag{2.5}$$

since it is assumed that $a_0 = 0$. In these terms (2.2) becomes

$$\mathrm{Var}(T_n) = \frac{2}{n} \sum_{m=1}^{\infty} |a_m|^2 \sum_{k=0}^{n-1} e^{2\pi i m (k/n)}, \tag{2.6}$$

and from the identity

$$\sum_{k=0}^{n-1} e^{2\pi i (mk/n)} = \begin{cases} n & , \quad m = ln \\[2mm] \dfrac{e^{2\pi i m} - 1}{e^{2\pi i/n} - 1} = 0, & m \neq ln \end{cases} \tag{2.7}$$

where l is an integer, it follows that

$$\mathrm{Var}(T_n) = 2 \sum_{k=1}^{\infty} |a_{kn}|^2. \tag{2.8}$$

If the series representation (2.4) is expressed as

$$y(x) = b_0 + \sum_{m=1}^{\infty} b_m \cos(2\pi m x + \theta_m), \tag{2.9}$$

b_m and $|a_m|$ are related as $|a_m| = b_m/2$ so that (2.8) becomes

$$\mathrm{Var}(T_n) = \frac{1}{2} \sum_{k=1}^{\infty} b_{kn}^2. \tag{2.10}$$

The result can be stated formally as the following theorem.

THEOREM 2.1. If $y(x)$ is a real valued function defined over a finite interval (x_0, x_1) such that $\mathrm{Var}[y(u)]$ exists, where u is a random variable uniformly distributed over the interval of definition, the variance of the average of a systematic sample of size n is given as

$$\text{Var}(T_n) = \frac{1}{2} \sum_{k=1}^{\infty} b_{kn}^2,$$

where b_m is defined as in (2.9), or

$$b_m = 2 \left| \int_{x_0}^{x_1} e^{(-2\pi imx)/(x_1 - x_0)} y(x) \, dx \right|. \tag{2.11}$$

PROOF. The integrals of (2.11) exist by virtue of the assumption that $\text{Var}[y(u)]$ exists. The remainder of the proof is contained in the preceding derivations.

The expression of (2.10) does not immediately convey much of an idea of the decrease in variance with increasing sample size n. This is in part because the rate of decrease depends very much on the function $y(x)$. The relevant characteristics of the function are summarized by its Fourier coefficients. Some general conclusions can be drawn, however. Since only every nth term of the series b_m^2 is retained, it would seem reasonable that the variance should decrease at least as fast as $1/n$. But, in fact, by imposing the very mild restriction that $y(x)$ be of bounded variation, a more definite result can be obtained. We will show that the variance decreases at least as fast as $1/n^2$. Depending upon the order of differentiability of $y(x)$ (extended as a periodic function), it is possible for the variance to decrease much more rapidly. We develop the $1/n^2$ result by considering first the more transparent case in which $y(x)$ is differentiable. We then remove the restriction.

Suppose, then, that the (periodic) function $y(x)$ is differentiable. Upon integrating by parts one obtains

$$a_m = \int_0^1 e^{-2\pi imx} y(x) \, dx$$

$$= -\frac{1}{2\pi im} \left[y(1) - y(0) - \int_0^1 e^{-2\pi imx} y'(x) \, dx \right], \tag{2.12}$$

so that, since under the restrictive assumption of differentiability $y(1) = y(0)$, there follows the inequality

$$|a_m| \leq \frac{1}{2\pi m} \int_0^1 |y'(x)| \, dx = \frac{V}{2\pi m}, \tag{2.13}$$

where the value of the integral of (2.13) is the total variation V of y over a period. Then

$$\mathrm{Var}(T_n) = 2 \sum_{k=1}^{\infty} |a_{kn}|^2$$

$$\leq 2 \sum_{k=1}^{\infty} \frac{V^2}{4\pi^2 k^2 n^2} = \frac{V^2}{2\pi^2 n^2} \sum_{k=1}^{\infty} k^{1/2}$$

$$= \frac{V^2}{2\pi^2 n^2} \frac{\pi^2}{6},$$

so that

$$\mathrm{Var}(T_n) \leq \frac{V^2}{12n^2}. \tag{2.14}$$

The restriction that $y(x)$ be differentiable is unduly severe and (2.14) holds provided that $y(x)$ is a bounded variation. The result is stated formally as the following theorem.

THEOREM 2.2. If $y(x)$ is of bounded variation over the interval $[x_0, x_1]$ and T_n is the average of a systematic sample of size n over the interval, then

$$\mathrm{Var}(T_n) \leq \frac{V^2}{12n^2}, \tag{2.15}$$

where V is the total variation over a single period of the function $y_1(x)$,

$$y_1(x) = \begin{cases} \frac{1}{2}[y(x_0) + y(x_1)], & x = x_0 \quad \text{or} \quad x = x_1 \\ y(x) & x_0 < x < x_1 \end{cases}$$

PROOF. There is no loss in supposing $x_0 = 0$, $x_1 = 1$. Since $y_1(x)$ is of bounded variation, there exist two nondecreasing functions f_1 and f_2 such that $y_1(x) = f_1(x) - f_2(x)$. Then, upon integrating by parts,

$$a_m = \frac{1}{2\pi i m} \int_0^1 e^{-2\pi i m x}[df_1(x) - df_2(x)]$$

$$|a_m| \leq \frac{1}{2\pi m}\left[\int_0^1 df_1(x) + \int_0^1 df_2(x)\right]; \tag{2.16}$$

and since the total variation is expressed as

$$V = f_1(1) - f_1(0) + f_2(1) - f_2(0)$$

$$= \int_0^1 df_1(x) + \int_0^1 df_2(x),$$

(2.16) becomes (2.13) and the remainder of the preceding development applies to complete the argument.

A special case of interest is that in which the population consists of N ordered elements with associated values $y_k i = 1, \ldots, N$. Define the function $y(x)$ as

$$y(x) = y_i, \qquad \frac{i-1}{N} \le x < \frac{i}{N}, \qquad i = 1, \ldots, N. \qquad (2.17)$$

The total variation then has the value

$$V = |y_N - y_1| + \sum_{i=2}^{N} |y_i - y_{i-1}|. \qquad (2.18)$$

If contiguous elements have quite similar values, the bound (2.15) shows directly that systematic samples of modest size can lead to fairly small standard errors of the estimated mean.

3. A Formal Example

To illustrate the mathematical ideas, consider the particular function y

$$y = x \qquad 0 < x < 1. \qquad (3.1)$$

In this case, the statistic T_n can be expressed as

$$T_n = \frac{1}{n} \sum_{j=0}^{n-1} y(u + j/n)$$

$$= u + \frac{n(n-1)}{2n^2} = u + \frac{n-1}{2n}, \qquad (3.2)$$

where u is uniformly distributed over the interval $(0, 1/n)$. One obtains directly that

$$\text{Var}(T_n) = \frac{1}{12n^2}. \qquad (3.3)$$

The Fourier coefficients are obtained as

$$a_m = \int_0^1 e^{-2\pi i m x} x \, dx$$

$$= \frac{i}{2\pi m}, \qquad m \ne 0. \qquad (3.4)$$

Application of Theorem 2.1 yields

$$\text{Var}(T_n) = 2 \sum_{k=1}^{\infty} |a_{kn}|^2 = \frac{2}{4\pi^2 n^2} \sum_{k=1}^{\infty} \frac{1}{k^2}$$

$$= \frac{2}{4\pi^2 n^2} \cdot \frac{\pi^2}{6} = \frac{1}{12n^2}, \qquad (3.5)$$

which agrees with (3.3), as of course it should. The total variation has the value $V = 2$. Accordingly, the bound of Theorem 2.2 becomes

$$\text{Var}(T_n) \le \frac{(2)^2}{12n^2} = \frac{1}{3n^2},$$

and the example shows that it is possible for the bound to be close enough to be of interest.

4. A Laboratory Example

As part of a laboratory investigation, Stern and Mickey (1970) wished to assess quantitatively some of the tissue components of rat ovaries. The problem was similar in nature to that of Hale et al. (1968), and the same assessment method was used. Among the questions to be answered were those of how many sections should be assayed per animal and what magnitude of error variation was likely to result from the sampling and assay procedures. Rat ovaries are small, of the order of a few millimeters in the short axis, but even so, many 6 μ-thick serial sections can be obtained. The material available consisted of ovary halves, the other half being used for other purposes. A single well-formed specimen was selected for pilot study, and every fifth section was assayed for the tissue characteristics of interest. A total of 41 sections were examined.

The area of each tissue component was estimated for each of the selected sections. Results for the component corpora lutea are shown in Figure 14.1. The curve is reasonably smooth and carries the suggestion that systematic sampling would be quite effective. Since the assay error is not negligible with the method used, the component attributed to systematic sampling only may be less than that indicated by Figure 14.1.

On the basis of the theoretical results presented in section 2, estimates can be constructed of the approximate magnitude of the variance of the sample mean for small sample sizes. This can be done by using the pilot data to estimate the Fourier coefficients. If we denote the distances along the axis perpendicular to the plane of the sections by $x_j, j = 1, 2, \ldots, N$ and the corresponding area estimates by y_j, then the desired coefficient estimates can be obtained from the regression problem

$$y_j = C_0 + C_{11} \cos 2\pi t_j + C_{12} \sin 2\pi t_j + \cdots$$

$$+ C_{k1} \cos 2\pi k t_j + C_{k2} \sin 2\pi k t_j + \cdots$$

$$+ C_{m1} \cos 2\pi k m t_j + C_{m2} \sin 2\pi m t_j + \varepsilon_j \qquad (4.1)$$

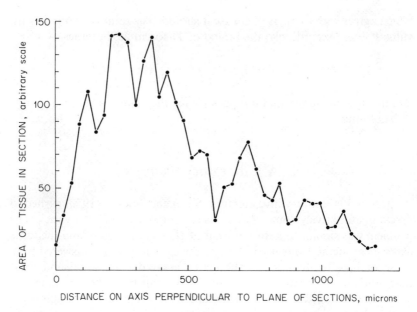

Fig. 14.1 Area of corpora lutea in cross sections of rat ovary as measured from serial sections of a half-ovary.

in which

$$t_j = \frac{x_j - x_0}{x_{N+1} - x_0} \tag{4.2}$$

$$m = \begin{cases} \dfrac{N-1}{2}, & N \text{ odd} \\[2mm] \dfrac{N-2}{2}, & N \text{ even} \end{cases} \tag{4.3}$$

and (x_0, x_{N+1}) is the interval of interest for the function $y(x)$. Results for the corpora lutea, obtained using the regression analysis computer program BMDO2R (Dixon, 1970), are given as Table 14.1. In addition to $\hat{C}_{m,1}$ and $\hat{C}_{m,2}$ the value $4|\hat{a}_m|^2 = \hat{C}_{m,1}^2 + \hat{C}_{m,2}^2$ is tabulated. Estimates of variance of the sample mean for systematic sampling can then be obtained easily by "decimated" summing of the entries $2|A_m|^2$. Thus, for samples of size one, the estimate is one-half the sum of all entries; for samples of size two, sum the 2nd, 4th, 6th, . . . entries; for samples of size three, sum the 3rd, 6th, 9th, . . . entries, etc. The results are tabulated in Table 14.2, and plotted in Figure 14.2. Figure 14.2 suggests that the relation $\sigma_n^2 = \sigma_1^2/n^2$ may be a reasonable approximation for small n for this tissue component. Very similar results were obtained for the other components.

Table 14.1
Regression Estimates of Fourier Coefficients for the Ovary Component Corpora Lutea Ave $= C_0 = 66.66$

m	$C_{m,1}$	$C_{m,2}$	$C^2_{m,1} + C^2_{m,2}$	m	$C_{m,1}$	$C_{m,2}$	$C^2_{m,1} + C^2_{m,2}$
1	−10.68	43.83	2034.84	11	3.74	.62	14.30
2	−21.66	9.98	568.97	12	−1.22	4.77	24.20
3	−4.41	−3.18	29.61	13	−.16	−2.58	6.70
4	−7.52	5.11	82.78	14	−.70	3.20	10.74
5	2.23	−1.30	6.68	15	.22	2.17	4.77
6	−3.89	1.77	18.31	16	−2.45	.51	6.21
7	−1.55	3.75	16.46	17	.00	−5.99	35.92
8	−7.19	−1.49	53.93	18	1.15	2.45	7.34
9	1.51	−3.26	12.94	19	−.77	−.26	.66
10	2.64	−7.24	59.32	20	−1.02	.26	1.12

Table 14.2
Estimated Variance of $\tilde{\sigma}^2_n$ of Average Corpora Lutea Measurement for Sample Sizes $n = 1, \ldots, 10$

n	$\tilde{\sigma}^2_n$	n	$\tilde{\sigma}^2_n$
1	1497.9	6	24.9
2	416.5	7	13.6
3	48.6	8	30.1
4	84.1	9	10.1
5	35.9	10	30.2

Example: $\tilde{\sigma}^2{}_5 = 1/2(6.68 + 59.32 + 4.77 + 1.12) = 71 \cdot 89/2 = 35.9$, in which the numbers summed are from Table 14.1 for $m = 5, 10, 15,$ and 20.

Among the defects of the foregoing procedure is the confounding of the Fourier coefficients. The coefficients $\hat{C}_{m,i}$ estimates $C_{m,i}$ plus the sum of $C_{kN \pm m, i} \, k = 1, 2, \ldots$, where it is assumed that $t_j = (j - 1)/N$. If the function is fairly smooth, however, the higher harmonic coefficients can be expected to decrease rapidly enough that the "aliasing" may be relatively unimportant for at least the first few terms.

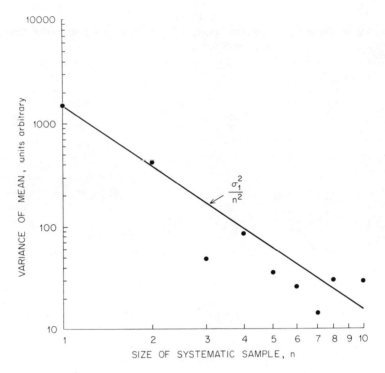

Fig. 14.2 Decrease of variance of systematic sample mean with increasing sample size. Estimates are for rat ovary component and corpora lutea and are based on data shown in Figure 14.1. See Tables 14.1 and 14.2 for illustration of calculation.

A second consideration is that the $\hat{C}^2_{m,i}$ tend to overestimate the "aliased" $C^2_{m,i}$ as a consequence of the measurement error of the y. This will not ordinarily present a problem provided one does not take the resulting estimated variances too seriously.

A third consideration is that the specimens selected for a pilot study may turn out to be unrepresentative of the cases of interest. In some cases it may be feasible to take a random sample of specimens as a pilot study for purposes of estimating variances, but it seems likely that in most cases this would represent an unjustifiable diversion of resources from the primary study. We suggest that the relative values of the variance estimates are likely to be more stable than the values themselves for small sample sizes.

In the study of Stern, sampling was done so that each of two independent systematic samples of size three were taken from each of the experimental units studied. The differences between the two sample means provide a basis for estimating the sampling variance attributable to the

Table 14.3
**Comparison of Predicted Variance with that Estimated from
Paired Independent Samples from 20 Experimental Units**

Tissue Components	Pilot Study	Paired Samples
Corpora lutea	48.6	54.9
Follicles	10.9	18.6
Interstitial tissue	132.7	72.2

systematic sampling. The results are given in Table 14.3 for the components: corpora lutea, follicles, and interstitial tissue. No broad conclusions are warranted on the basis of this single case and one sample size. The agreement in the case of corpora lutea is excellent. In the case of follicles and interstitial tissue the comparison is more difficult to evaluate. The paired sample estimate has 20 degrees of freedom. The pilot study estimate has 8 degrees of freedom but would be approximated using a noncentral rather than a central χ^2. At the extreme of neglecting the variability of the pilot study estimates, the ratios are significant at about the 1% point. Nevertheless, the pilot study estimates are of the approximate same general magnitude as those from the paired samples.

In experiments of the type considered in the example, there may not be particular interest in estimating the systematic sampling component of the total variation other than for planning purposes. For example, it may be the case that the variance of ultimate interest can be roughly expressed as

$$\sigma^2 = \frac{A}{m} + \frac{B}{mn} + \frac{C}{mn^2} \tag{4.4}$$

in which m is the number of experimental units assigned to a treatment and n is the size of the systematic sample involved in assessing each unit. For a fixed number of total determinations (i.e., $mn = M$) σ^2 is minimized by taking $n = \sqrt{C/A}$. Different cost considerations may lead to other allocations, but it is suggested that the method presented for estimating the systematic sampling component—represented in (4.4) by the term C/n^2— may be useful in arriving at the allocation and the provisional estimate of σ^2.

In other cases, such as in the clinical applications envisioned by Hale et al. (1968), verification of variance estimates may be required. The approach of paired independent samples, such as that used in the study of Stern and Mickey (1970), is a straightforward way of accomplishing this.

5. Discussion

It appears that systematic sampling can lead to large gains in precision, as compared with simple random sampling, in sampling from a continuous curve or from a situation in which contiguous elements are relatively similar. In some cases the measurement error may dominate the systematic sampling component—e.g., Hale et al. (1968)—and in this case the problems of deciding upon matters of sample size are much easier.

The rate at which the variance decreases with increasing sample size depends largely on the continuity and differentiability characteristics of the curve sampled. If there are jump discontinuities (including the end points, since we "circularize" the curve), the variance cannot decrease faster than $1/n^2$. If the curve is absolutely continuous (again, circularized) and if its derivative is of bounded variation, then a continuation of the derivation of Theorem 2 shows that the variance decreases as $1/n^4$.

The curve sampled may be continuous and possibly differentiable except at the end points. In this case a slight modification to what might be termed "reflected systematic sampling" can greatly increase the rate at which the variance decreases with sample size.

By reflected systematic sampling we will mean element selection in which u is a random number $(0, 2/n)$ and sample points are taken as follows. Let s_j be defined as

$$s_j = u + \frac{2(j-1)}{n}, \qquad j = 1, \ldots, n.$$

Then the sample points are taken at points $x_j, j = 1, \ldots, n$, where

$$x_j = \begin{cases} s_j & \text{if } s_j \le 1 \\ 2 - s_j & \text{if } s_j > 1. \end{cases} \tag{5.1}$$

If one thinks of the selection as starting from the left (i.e., at $s = 0$), moving to the right and stopping at the random start point and at intervals $2/n$ thereafter, when the boundary $x = 1$ is reached, one simply reverses direction and continues back toward the start. An equivalent scheme is to choose a random u in the interval $(0, 1/n)$ and to take samples at $u, 2/n - u$, and at intervals of $2/n$ after each of these two starting points.

We will restrict consideration to the case in which n is even. Consider for example the illustration used in section 3, $y = x$, $0 \le x \le 1$, and let $n = 2m$. Then the sample points are at

$$u + \frac{(j-1)}{m}, \qquad \frac{1}{m} - u + \frac{j-1}{m}, \qquad j = 1, \ldots, m \tag{5.2}$$

the sum of which is .5 independently of u so that the variance is zero.

A general result can be obtained along the lines of Theorem 2.1. Define $z(x)$ as

$$z(x) = \begin{cases} y(2x) & 0 \le x \le 1/2 \\ y(2 - 2x) & 1/2 \le x \le 1. \end{cases} \tag{5.3}$$

Then the reflected systematic sampling of $y(x)$ is equivalent to the usual systematic sampling of $z(x)$, and Theorem 2.1 applies. The Fourier coefficients of $z(x)$ can be expressed in the terms of those of $y(x)$, if n is even, as follows:

$$a'_k = \int_0^1 e^{-2\pi ikx} z(x)\, dx$$

$$= \int_0^{1/2} e^{-2\pi ikx} y(2x)\, dx + \int_{1/2}^1 e^{-2\pi ikx} y(2 - 2x)\, dx.$$

By making the change of variables $t = 2x$ and $t = 2 - 2x$ in the integrals of the second line, one obtains

$$a'_k = \frac{1}{2} \int_0^1 e^{-2\pi i(k/2)t} y(t)\, dt + \frac{1}{2} \int_0^1 e^{2\pi i(k/2)t} y(t)\, dt \tag{5.4}$$

so that

$$a'_{2k} = \frac{1}{2}(a_k + a_{-k}) = \text{real}\,(a_k) \tag{5.5}$$

in which a_k denotes the Fourier coefficients of $y(x)$, $k = \pm 1, \pm 2, \ldots$. In the regression equation context,

$$a'_{2k} = \frac{1}{2} C_{k,1} \qquad k = 1, 2, \ldots$$

in which $C_{k,1}$ is the coefficient of $\cos 2\pi kx$. The variance of the reflected systematic sampling average for even n is, by Theorem 2.1,

$$\text{Var}(\bar{Z}_n) = \frac{1}{2} \sum_{k=1}^{\infty} C^2_{nk/2,1}, \qquad n \text{ an even integer.} \tag{5.6}$$

The calculation of estimates proceeds as in section 4 except that the coefficients for the sine terms are ignored, and one decimates according to half the sample size.

Some numerical comparisons for the example of section 4 are given in Table 14.4. There is no apparent clear advantage to either procedure for this data. The advantage of the reflected procedure would be expected to be pronounced in case there is a substantial difference between $y(0)$ and $y(1)$ and the curve is otherwise fairly smooth; the situation would be described as a trend.

Table 14.4
Comparison of Estimated Variance of Mean for Ordinary and Reflected Systematic Sampling

Sample Size	Corpora Lutea		Follicles		Interstitial Tissue	
	Ordinary	Reflected	Ordinary	Reflected	Ordinary	Reflected
2	416.5	384.5	9.2	32.6	107.4	652.8
4	84.1	305.0	3.5	5.2	44.1	60.3
6	24.9	19.9	3.4	7.1	21.3	12.5
8	30.2	58.4	1.1	1.2	10.1	21.1
10	30.2	6.5	1.4	.8	5.3	4.5

Other modifications to the sampling procedure could be considered that would lead to improved performance for general types of curves. Along the lines of variance estimation used in section 4, it would be appropriate to adjust for the bias resulting from the sampling variability of the coefficients that is attributable to measurement error. The problem is more severe as the sample size increases. However the variance may decrease rapidly enough that the distortion is of no practical interest, depending upon the investigation involved.

The results reported and illustrated here are directed to the problems of forming a notion of sampling variances that might be expected in dealing with material that is similar to that for which preliminary data is available. This data need not come from a systematic sample; it suffices that it supplies the basis for estimating Fourier coefficients, e.g., by a regression analysis. The presented results then provide a direct and simple way of estimating the variance contribution attributable to systematic sampling in the assessment of sampling units or experimental units, and in this way they provide a planning aid. It is perhaps risky to apply the results as standard error estimates. The numerical results for a single instance, section 4, provide a very small degree of assurance that such standard errors are not completely untrustworthy.

References

Buckland, W. R. 1951. A review of the literature of systematic sampling, *J. Roy. Statist. Soc.* Ser. B, 13 : 208–15.

Cochran, W. G. 1946. Relative accuracy of systematic and random samples for a certain class of populations. *Ann. Math. Stat.* 17: 164–77.

Dalenius, Tore. 1962. Recent advances in sample survey theory and method. *Ann. Math. Stat.* 33: 325–49.

Dixon, W. J. 1970. *BMD, biomedical computer programs*. Berkeley: Univ. Calif. Press.

Hale, F. G., R. Olsen, and M. R. Mickey. 1968. The measurement of bronchial wall components. *Am. Rev. Respiratory Diseases* 98 : 978-87.

Hannan, E. J. 1962. Systematic sampling. *Biometrika* 49: 281-84.

Madow, W. G., and L. H. Madow. 1944. On the theory of systematic sampling. *Ann. Math. Stat.* 15: 1-24.

Stern, E., and M. R. Mickey. 1970. An effect of enovid in DMBA-treated rats. Paper presented to 10th International Cancer Congress.

Yates, F. 1948. Systematic sampling. *Phil. Trans. Roy. Soc. London* Ser. A, 241: 345-77.

15

An Empirical Study
of the Distribution of
Primes and Litters of Primes

FREDERICK MOSTELLER

1. The Multinomial Model for Primes

HOW DO PRIMES distribute themselves in relatively short intervals when the
sizes of the integers studied are large? We study this question empirically
for integers up to 92.5 million with the help of tables constructed by others
(Gruenberger and Armerding, 1961; Lehmer, 1956). From the theory of
the distribution of primes it is well known that the number of primes less
than x, when x grows large, tends to $x/\ln x$. This means that the density of
primes in intervals near x tends toward $1/\ln x$, or speaking roughly, the
probability of a prime at the integer x is $1/\ln x$. Thus as x grows large, the
primes become rare, but slowly.

An idea coming from probability and statistics that must have occurred
to many people is that the continuous Poisson distribution is closely
associated with rare events and therefore might produce distributions or
spacings like those of the primes if one allowed for their slowly changing
density and paid no attention to their fine structure. Against this intuition
is the systematic nature of the construction of primes. For example, since
all primes but one are odd, a sampling theory of the distribution of primes
might produce false conclusions if it did not attend to this and other

FREDERICK MOSTELLER is Professor, Department of Statistics, Harvard University,
Cambridge, Massachusetts.
This work was facilitated by Grant GS-2044X from the National Science Foundation.
The author expresses appreciation to D. H. Lehmer, J. Barkeley Rosser, George B.
Thomas, Jr., The RAND Corporation, George W. Armerding, and Thomas C. Leiser for
correspondence and materials that have been of help in assembling this paper and to
Cleo Youtz for her work in computing.

245

restrictions. In a vein similar to the Poisson theory we can think of primes in a given "short" interval as tossed into the interval at random, and we investigate this model.

We investigate this local distribution in several ways. We assess how much more uniformly primes are distributed over short intervals than a random distribution of the primes would allow. On the other hand, from 10^7 to 10^8 twin and triple primes have local distributions with properties more like those of the Poisson. Quadruple primes come even closer.

One way to look at the distribution of primes is to take an interval of the integers and break it into parts of equal length. Provided the length of the interval is small compared to its least value, the density of the primes changes but slightly from one end of the interval to the other; and we make little error in our simple probability model by computing as if there were supposed to be a uniform density over the interval. If we break the interval into k segments of equal length, we can ask whether the numbers of primes in the segments varies as one would expect from an urn scheme. Under the simplest urn scheme, there is some fixed number of balls N corresponding to the number of primes in the whole interval. Balls are assigned to the k segments, or boxes, independently with equal probability. Upon the completion of the experiment the ith segment has $x_i = 0, 1, 2, \ldots, N$ balls—or random pseudoprimes—and if the random sampling theory were correct, the joint distribution of the random variables X_i, $i = 1, 2, \ldots, k$, would be determined by the multinomial distribution with probability function

$$\text{Prob}\,(X_1 = x_1, \ldots, X_k = x_k) = \frac{N!}{x_1!x_2!\ldots x_k!}\,(1/k)^N,$$

where $N = \sum x_i$.

Some restraint upon the multinomial character of the distribution is imposed by the "without replacement" aspect of the actual process. When a ball representing a pseudoprime is assigned to a segment, it uses up one of the integers in that segment, with the result that the next choice has one less integer available to be a pseudoprime. The importance of this feature depends upon the ratio of the number of primes to the length of the interval and, therefore, approximately upon $1/\ln x$.

A general way of thinking about this effect is to recall that the variance of a mean for a sample of size N drawn without replacement from a population of size M is the variance of the mean based upon a sample drawn with replacement multiplied by a factor $1 - f$, where $f \approx N/M$, the relative size of the sample to the size of the population, in our problem essentially $1/\ln x$. If this ratio is small, say 0.10, the reduction of the variance is of the order of 10%. In our problem, values of x larger than 22,026 would have reduced the effect of this sort of sampling without replacement below 10%.

The effect of "without replacement" is heightened if we note that the number of integers available to be primes in an interval is reduced because we cannot use those divisible by 2, 3, and so on.

We plan to extend the use of the multiplier $1 - f$ to give a measure of "pressure toward uniformity" from both the causes of "without replacement" and the special property that primes have no divisors other than themselves. If $f = 1$, the variability would be 0, and we would have primes systematically spaced, that is, uniformly spaced along the line (not distributed according to a uniform frequency distribution)—except for their slowly changing rate. If $f = 0$, then according to the model the variability of the count in a segment would be that of the binomial, or as $1/\ln x$ became small would be like that of the Poisson (see Snedecor, 1938). We shall use the term Poisson to describe the $f = 0$ case and to describe the situation where the mean and variance coincide, without necessarily implying all the distributional properties associated with that name.

2. Chi-Square for the Multinomial

One standard way to measure the departure of the sample of size N from the theoretically expected values is through the chi-square statistic. The measure would be especially appropriate for such an urn scheme because, owing to the near constancy of $1/\ln x$, we would have good theoretical reasons to assume that the segments of equal length have nearly equal expected counts of primes. The chi-square statistic for our problem is

$$\chi^2 = \sum_i (x_i - N/k)^2/(N/k) = \sum \frac{(\text{observed} - \text{expected})^2}{\text{expected}},$$

(see Snedecor and Cochran, 1967). The quantity χ^2 follows approximately the chi-square distribution with $k - 1$ degrees of freedom when the X_i's come from a multinomial distribution, providing N is large compared to k, as it will be in our work.

Insofar as the multinomial distribution may be false in this investigation of primes, one can anticipate that the true distribution will be more uniform than the multinomial predicts because of the systematic character of the construction of primes and because of the sampling without replacement features. If this additional uniformity holds, then the values of χ^2 that we observe should be smaller than appropriate for the multinomial theory.

Usually when the chi-square distribution is used in a goodness-of-fit problem, we anticipate that a poor fit between theory and data will produce a large sum of squares, but our ideas about the distribution of primes go in the opposite direction.

3. The Intervals

We have selected intervals in a special way. Let p and p_* be successive primes; we consider the interval from p^2 to p_*^2. The special homogeneity feature of these intervals is that each of the integers in the interval has to be tested for primeness against the same set of primes: 2, 3, 5, ..., p, and no others. (This feature, while attractive may not actually be worth the bother, since it creates intervals having considerably different lengths and containing substantially different numbers of primes.) For $p > 2$, the difference $p_*^2 - p^2$ is the difference between the squares of two odd primes, which means that the interval length is divisible by 24. This gives a convenient range of possible values of k, the number of segments; we use $k = 4, 8$.

4. Distributions in Four Segments

In Table 15.1 we show, for each homogeneous interval whose left end point is the square of an odd prime of 29 or less, the distribution of the primes into four intervals of equal size. We show for each interval the

Table 15.1
Distribution of Primes into Four Equal Intervals between Successive Squared Primes p^2 and p_*^2

Primes		Intervals $\frac{1}{4}(p_*^2 - p^2)$, for $p^2 \leq x < p_*^2$				Total	
p	p^2	1	2	3	4	N	χ^2
3	9	1	1	2	1	5	.60
5	25	1	1	2	2	6	.67
7	49	3	5	3	4	15	.73
11	121	2	2	2	3	9	.33
13	169	6	4	6	6	22	.55
17	289	1	4	2	4	11	2.45
19	361	7	7	6	7	27	.11
23	529	11	14	11	11	47	.57
29	841	4	4	4	4	16	0
Totals		36	42	38	42	158	6.01

computed value of χ^2, which would be a sample from an approximately chi-square distribution with 3 degrees of freedom, and therefore average 3, if the multinomial distribution applied.

The counts of primes in each row of Table 15.1 are very nearly uniformly distributed—each segment has nearly one-fourth the primes in the row. Note that no value of χ^2 was as large as the mean, 3, and that all but one were much smaller.

It is almost embarrassing that the totals do not decrease from left to right, since $1/\ln x$ is monotonically decreasing.

5. Distribution into Eight Segments

When we have more primes available per interval, as we do for the larger p's, we divide the homogeneous intervals into eighths rather than quarters and looked again at the distribution. Table 15.2 shows this

Table 15.2

Distribution of Primes into Eight Equal Intervals between Successive Squared Primes p^2 and p_*^2

		Intervals $\frac{1}{8}(p_*^2 - p^2)$, for $p^2 \le x < p_*^2$								Total	
		1	2	3	4	5	6	7	8	N	χ^2
$p = 19$	$p^2 = 361$										
		3	4	3	4	4	2	4	3	27	1.15
23	529										
		4	7	7	7	4	7	4	7	47	2.87
29	841										
		1	3	3	1	2	2	2	2	16	2.00
31	961										
		7	9	8	6	6	7	9	5	57	2.09
37	1369										
		3	6	7	6	5	6	7	4	44	2.55
41	1681										
		3	2	3	3	2	3	2	2	20	0.80
43	1849										
		7	5	4	8	5	7	7	3	46	3.74
47	2209										
		10	11	10	8	8	8	15	10	80	3.80
53	2809										
		9	11	8	9	10	9	13	9	78	1.79
59	3481										
		2	6	4	4	4	4	3	5	32	2.50
61	3721										
		10	12	12	12	9	15	9	11	90	2.44

Table 15.2 (*continued*)

	Intervals $\frac{1}{8}(p_*^2 - p^2)$, for $p^2 \le x < p_*^2$								Total	
	1	2	3	4	5	6	7	8	N	χ^2
67										
		4489								
	8	7	10	7	9	4	11	10	66	4.30
71										
		5041								
	2	6	3	4	3	4	4	4	30	2.53
73										
		5329								
	14	13	15	12	17	9	13	13	106	2.83
79										
		6241								
	11	12	7	7	8	11	9	10	75	2.76
83										
		6889								
	16	12	14	11	15	17	15	14	114	1.93
89										
		7921								
	19	21	19	19	25	19	21	20	163	1.47
97										
		9409								
	14	8	12	12	12	8	11	12	89	2.78
101										
		10201								
	4	7	7	4	4	7	3	6	42	3.71
103										
		10609								
	12	9	12	10	12	8	13	11	87	1.92
107										
		11449								
	6	5	4	4	6	3	8	6	42	3.33
109										
		11881								
	15	10	12	12	12	14	14	11	100	1.60
113										
		12769								
	47	38	46	40	47	44	46	46	354	1.84
127										
		16129								
	12	11	12	12	14	11	15	12	99	1.12
131										
		17161								
	20	22	21	20	22	22	22	16	165	1.45
137										
		18769								
	5	4	7	6	6	6	8	7	49	1.78
139										
		19321								
	36	40	39	34	34	34	42	40	299	1.98
149										
		22201								
	5	8	6	7	9	6	9	8	58	2.14
151										
		22801								
	23	23	19	26	25	29	18	19	182	4.64
157										
		24649								
	21	24	24	22	24	23	26	22	186	0.75
163										
		26569								
	20	16	16	13	13	17	12	21	128	4.75
167										
		27889								
	26	20	31	26	23	27	24	21	198	3.54
173										
		29929								
	24	23	26	26	27	23	27	19	195	2.13

Table 15.2 (*continued*)

Intervals

$$\frac{1}{8}(p_*^2 - p^2), \text{ for } p^2 \le x < p_*^2$$

	1	2	3	4	5	6	7	8	Total N	χ^2
179		32041								
	11	9	9	11	9	12	8	7	76	2.11
181		32761								
	49	47	41	49	42	46	41	41	356	2.07
191		36481								
	10	9	11	10	12	9	7	9	77	1.65
193		37249								
	18	23	15	17	16	18	16	21	144	2.89
197		38809								
	9	9	9	11	10	11	7	9	75	1.27
199		36901								
	60	51	60	62	62	58	51	59	463	2.40
211		44521								
	61	57	58	58	62	58	63	62	479	0.65
223		49729								
	23	24	19	20	16	21	19	26	168	3.43
227		51529								
	11	10	12	11	9	11	11	7	82	1.71
229		52441								
	20	22	23	21	18	23	22	17	166	1.71
233		54289								
	39	29	32	31	35	30	36	38	270	2.95
239		57121								
	12	12	9	10	12	13	10	12	90	1.20
241		58081								
	56	57	56	53	55	55	52	54	438	0.36
251		63001								
	31	43	30	30	32	35	38	36	275	4.24
257		66049								
	29	35	39	39	37	26	37	32	274	4.72
263		69169								
	35	29	39	35	39	41	35	39	292	2.79
269		72361								
	10	10	15	10	15	11	8	12	91	3.86
271		73441								
	37	37	36	39	34	39	36	34	292	0.71
277		76729								
	24	26	28	28	21	24	23	25	199	1.64
281		78961								
	8	15	13	10	15	10	17	11	99	5.48
283		80089								
	67	64	71	60	62	61	63	64	512	1.37
293		85849								
	87	97	81	101	88	90	103	88	735	4.45

distribution for p's used to determine the left end point of the interval running through successive primes from 19 to 293. Table 15.3 shows some selected additional intervals, the last of which has a lower boundary of $2953^2 \approx 9 \times 10^6$. The χ^2 values in Tables 15.2 and 15.3 never get as high as 7, the theoretical average for the urn model. The table shows clearly then that the primes, at least the early primes, are distributed much more nearly uniformly than samples from the multinomial with equal probabilities would yield. The χ^2's seem to be rising slightly as p increases, but very slightly.

Table 15.3

Distribution of Primes into Eight Equal Intervals between Selected Successive Squared Primes p^2 and p_*^2

			Intervals							
			$\frac{1}{8}(p_*^2 - p^2)$, for $p^2 \leq x < p_*^2$							
									Total	
1	2	3	4	5	6	7	8		N	χ^2
439	192721									
40	36	38	30	36	39	33	38		290	2.14
727	528529									
80	88	73	90	80	77	81	97		666	5.14
769	591361									
53	62	67	52	66	55	57	58		470	3.87
1097	1203409									
111	116	109	111	122	128	122	115		934	2.67
1483	2199289									
94	98	106	106	98	110	98	119		829	4.59
1667	2778889									
53	55	60	58	51	54	66	53		450	2.98
1987	3948169									
206	183	197	191	186	186	194	202		1545	2.43
2029	4116841									
323	323	325	318	333	325	351	338		2636	2.44
2411	5812921									
235	222	237	217	235	237	217	237		1837	2.60
2659	7070281									
181	165	178	166	169	168	165	190		1382	3.45
2953	8720209									
187	188	181	196	181	178	188	185		1484	1.20

Totals: (Tables 15.2 and 15.3 combined)

2689	2666	2708	2653	2684	2675	2721	2744	21540	168.16

But if the computed χ^2's are ever to match the theoretical chi-square distribution, we can see that a close fit, if there is one, would have to be sought much further out in the integers than we have gone. If the urn model is asymptotically correct, then ás the primes get sparser, the χ^2 values should slowly drift up to an average of 7 per line in an extension of Table 15.3. Tables 15.1, 15.2, and 15.3 were constructed with the aid of the tables of primes by Lehmer (1956).

6. Pressure toward Uniformity

Suppose that the N balls are tossed into M boxes in k sets, where $M = kT$, T an integer. Under free tossing the variance of the number of balls in any specific segment is $N(1/k)(1 - 1/k)$ and the expected value of χ^2 is $(k/N)kN(1/k)(1 - 1/k) = k - 1$, as we already knew. The point of the derivation, though, is to see the role of the binomial variation. If we have sampling without replacement as noted earlier, the effect is to multiply the binomial variance by $1 - f$ where f is the fraction that the sample is of the possible values. Initially, if we attend only to the fact that the same numbers cannot be used as primes more than once, the value of f is N/M. But as soon as we note that even numbers are not allowed, then M is essentially cut in half and $f = 2N/M$. As further integers are disallowed, the multiplier of N/M increases. As we noted earlier, we shall use f as our measure of pressure toward uniformity of spacing of primes. Instead of trying to evaluate the change by theoretical methods, we can find the average value of f over the first 90,000 integers by using the data from Table 15.3. The average χ^2 is about 2.5 instead of 7, implying that f must be about 0.74. In sampling problems we usually have small values of f, but this value is large, indeed much larger than it would be if the "without replacement" feature were the only pressure, for then it would be less than 0.1.

7. Larger Integers

These results suggest that we should move to much larger values of the integers if we are to see f reduce. Fortunately we had available the tables by Gruenberger and Armerding (1961). These tables give the counts of primes in intervals of length 50,000. We took the counts in intervals of length 50,000 starting with 10×10^6. The numbers from 10×10^6 to 12.5×10^6 give 50 such counts. We computed the mean and the variance of such counts. Since $1/\ln x$ does decrease a bit over such an interval, in computing the empirical variance, we computed sums of squares for sets of 10 successive intervals and then pooled them. The result is an estimate of variance based upon 45 degrees of freedom instead of 49, but it should be less affected by the change in mean. (See Snedecor, 1946.) The means and variances appear in Table 15.4 for the first interval of length 2.5×10^6 following whole multiples of 10×10^6, plus an extra line in the middle to give us 10 sets. The mean values of the number of primes in these intervals of length 50,000 is 2,845. The variances vary from 615 to 1,286, about a factor of two, and they average 975. According to discrete Poisson theory the mean would be equal to the variance, so the combined effect of sampling without replacement and of restriction are measured by the value of f that makes $2,845(1 - f) = 975$. This means an average value of $f = 0.66$; thus f has only been reduced by about 10% from 0.74 as we moved from 90×10^3 to 90×10^6. There may be a slight effect here due to the different technology of measurement in the small and the large sets of integers, but it seems doubtful. The main point is that f is still large, and the counts of primes do not vary as much from interval to interval as the urn scheme requires.

8. Prime Litters: Twin, Triple, Quadruple

If primes are to be more nearly Poissonly distributed, apparently to see it one will have to study much larger integers than we have. More attractive would be to show theoretically that the distributions would or would not approximate the urn scheme asymptotically.

Another line of empirical study is suggested by the multiple primes— the twin, triple, and quadruple primes. Instead of going further out in the numbers, by choosing a subset of the primes that is rarer, one might hope to see the Poissonness, if it exists, exert itself earlier in the integers. The tables of Gruenberger and Armerding (1961) give tabulations of these values. Our Table 15.4 shows again for intervals from 10×10^6 to 90×10^6 by intervals of 50,000 (except for twin primes where intervals of 100,000 were used) calculations comparable to those just described for the primes. Twin primes occur when p and $p + 2$ are prime (example: 11 and 13). Triple primes do not exist except for 1, 3, 5 (if you allow 1 as a prime) and 3, 5, 7, and so the name of triple prime is reserved for sets of the form p, $p + 2$, and $p + 6$ all prime, called a 2-4 triple prime, with a 4-2 triple prime being a set of the form p, $p + 4$, $p + 6$, all prime. The counts for quadruple primes given by Gruenberger and Armerding (1961) are for type 2-4-2.

Table 15.4

Means and Variances of Numbers of Primes, Twin Primes, Triple Primes, and Quadruple Primes in Intervals of Length* 50,000 for Integers from 10×10^6 to 92.5×10^6

Interval (millions)		Primes		Twin Primes*	
		Mean	Variance	Mean	Variance
10.0	12.5	3082.48	615.20	493.60	394.08
20.0	22.5	2960.66	1139.40	462.10	275.04
30.0	32.5	2897.80	924.60	441.06	448.18
40.0	42.5	2850.92	1024.79	424.56	265.94
50.0	52.5	2815.86	840.09	415.38	210.02
55.0	57.5	2805.50	1286.49	413.76	347.68
60.0	62.5	2787.44	1125.77	408.28	185.93
70.0	72.5	2766.90	1127.40	400.66	344.43
80.0	82.5	2747.56	867.06	397.44	307.25
90.0	92.5	2730.30	797.02	393.40	261.42
Average		2845	975	425	304

Interval (millions)		Triple Primes 2-4		Triple Primes 4-2		Quad. Primes 2-4-2	
		Mean	Variance	Mean	Variance	Mean	Variance
10.0	12.5	33.38	27.71	33.78	25.44	2.84	2.99
20.0	22.5	29.78	24.27	30.20	19.28	2.62	1.71
30.0	32.5	28.02	23.79	27.60	29.49	2.06	1.74
40.0	42.5	26.40	24.44	26.10	20.91	2.30	2.19
50.0	52.5	25.06	21.56	24.58	25.47	1.94	1.51
55.0	57.5	26.76	31.84	24.88	34.12	2.08	2.44
60.0	62.5	24.40	24.48	25.72	23.24	2.08	1.66
70.0	72.5	24.16	18.28	24.28	16.99	1.96	1.88
80.0	82.5	24.82	13.28	24.24	17.52	2.06	1.48
90.0	92.5	23.36	24.40	22.28	18.73	1.72	1.27
Average		26.6	23.4	26.4	23.1	2.17	1.89

Note: The computations are based on intervals of length 50,000 except for twin primes, where it is 100,000.

*The interval for twin primes is doubled, beginning at same lower bound, for example, 10.0–15.0.

For the twin primes the averages of the means and variances of the counts are 425 and 304 respectively, shown in Table 15.4; therefore, we compute the average f to be $f = 0.28$, which does represent a substantial move in the direction of Poissonness compared to $f = 0.66$ for primes. For triple primes, averaging the results for the two kinds gives an average

Table 15.5
Distributions of Number of Quadruple Primes in Intervals of 50,000

Number of Quadruple Primes in Intervals of 50,000	10,000,000 to 20,000,000	Frequency 50,000,000 to 60,000,000	90,000,000 to 100,000,000
0	13	27	23
1	29	52	63
2	42	49	69
3	48	43	30
4	42	20	6
5	15	5	4
6	8	2	4
7	3	1	1
8	...	1	...
Total	200	200	200
Mean	2.84	2.06	1.83
Variance	2.47	2.11	1.65
Residual variance*	2.43	2.06	1.63
f	0.14	0	0.11

*Adjusted for linear change in mean over the 200 intervals.

$f = 0.12$, a serious further reduction. The quadruple primes gave $f = 0.13$. This last proportion raised the question of whether large sampling error may have entered in. We therefore made further calculations. Table 15.5 gives three different frequency distributions of quadruple primes (one each for intervals starting at 10 million, at 50 million, and at 90 million). After allowing for the effect of the slightly changing means, we computed f for each of these distributions and found the average f to be 0.08, a considerable reduction compared to 0.13 based upon fewer intervals.

9. Conclusion

When short intervals of the integers are broken into segments of equal lengths, the empirically observed numbers of primes in the segments are more nearly equal than a simple multinomial model would predict, for $x \leq 100 \times 10^6$. We quantify this "pressure toward uniformity" by f, by extension of the idea that the factor $1 - f$ gives the reduction in variance for effects of sampling without replacement. If $f = 0$, the sampling variability corresponds to that of an urn model or the Poisson; when $f = 1$, the variability is zero and we get uniform spacing within the interval. For primes up to about 90×10^3, $f = 0.74$, and moving up to 90×10^6, $f = 0.66$. We see that f is decreasing slowly at best. Similar calculations for twin primes between 10×10^6 and 90×10^6 give an average $f = 0.28$. Triple primes in this interval give an average $f = 0.12$, and quadruple primes give $f = 0.08$.

References

Gruenberger, F., and G. Armerding. Oct. 1961. Statistics on the first six million prime numbers. The RAND Corporation, P-2460.

Lehmer, D. N. 1956. *List of prime numbers from 1 to 10,006,721*. New York: Hafner.

Snedecor, George W. 1938. *Statistical methods*, pp. 362-78. Ames: Iowa State College Press.

———. 1946. *Statistical methods*, 4th ed., pp. 214-16. Ames: Iowa State College Press.

Snedecor, George W., and William G. Cochran. 1967. *Statistical methods*, 6th ed., sect. 9.3, pp. 231-33. Ames: Iowa State Univ. Press.

16

A Statistical Appraisal of the Protein Problem

P. V. SUKHATME

1. Introduction

PROTEIN MORE than any other constituent of food has been singled out for attention in recent years. Protein deficiency, if allowed to develop during childhood, can lead to impaired body growth and possibly impaired mental development as well, which cannot be fully rectified in later life.

On a countrywide basis supply of protein is about equal to or even slightly exceeds requirement in most of the developing countries (UN, 1968). Nevertheless, there is ample evidence from clinical surveys showing that a considerable part of the population suffers from protein malnutrition. Previous assessments put this proportion at one-quarter to one-third. More disturbing is the fact that this gap between nutritional requirement and actual consumption of protein by the greater part of the population in the developing countries is widening rapidly. If the situation should worsen, the physical, economic, and social development of the populations involved may become completely arrested.

This warning, given by the UN Advisory Committee on the Application of Science and Technology to Development (UN, 1968), led the UN General Assembly to adopt a specific resolution calling upon countries to increase production and use of edible protein in all forms—conventional, semiconventional, and nonconventional (Resolution 2319, XXII, 1967). However, emphasis on increased production and consumption of edible protein by itself may not go far under all conditions in reducing the

P. V. SUKHATME is Director of the Statistics Division, Food and Agriculture Organization, United Nations, Rome, Italy.

Table 16.1

Daily per Capita Calorie and Protein Supply by Expenditure and Level in Maharashtra State, India, 1958

| Item | Monthly per Capita Expenditure in Rupees | | | | | | | Average |
	0–8	8–11	11–13	13–18	18–24	24–34	34	
Total calories	1,120	1,560	1,850	2,190	2,440	2,530	3,340*	2,100
Total proteins in grams	30.7	45.0	52.8	60.4	66.3	71.7	85.7	59.7
Animal proteins in grams	1.0	1.8	2.3	2.9	6.1	7.1	11.9	4.5
Number of households	76	114	87	82	102	83	349	893

*This value appears unduly high. It is stated that this is partly due to the exclusion from the household size of guests and laborers taking meals.

incidence of protein deficiency in the population. For as long as the diet is inadequate in quantity as measured by its calorie content, the protein in the diet will be partly wasted for energy purposes. Moreover, protein is very unevenly distributed, with the rich taking enough and more and the poor only what they can afford. This is even more true of animal protein (see Table 16.1). A substantial part of the additional production is therefore more likely to reach those who already have enough to eat than those in need.

One can broadly classify the diets into four patterns according to whether they are adequate in calories or proteins or not, as shown below.

proteins

	A	B
	E	F

calories

The dividing lines are placed at the critical limits (later refined) for calories and proteins. Diets A and E will clearly lead to protein deficiency; but diet B, even though not deficient in protein, will also lead to protein deficiency because it does not have the calories needed for the synthesis of protein in the amounts needed by man.

Low body weight, retarded body growth, and low physical activity are the normal reactions of a child brought up on an inadequate diet of type A. In other words, a child brought up on diet A will adapt himself to what has been called the condition of "nutritional dwarfism." Experimental results show that when such children are given a diet increasingly adequate in calories and proteins there is continuous improvement in their health and physique, provided the improved diet is not given too late to influence their development. The pressing need of individuals brought up on diet A is more food to bring not only more protein but also more calories. The addition of protein to their diet without adequate calories will only shift them to diet B without proportionately diminishing their susceptibility to protein deficiency. The larger the shortage of total calories relative to protein in the diet, the larger will be the underutilization of protein. The pressing need of group B is clearly to have more calories. Only individuals brought up on diet E have a pressing need for more protein.

Previous estimates of the incidence of protein deficiency have excluded the contribution from diet B being based on the simple criterion of whether the diet fulfills the requirement of protein regardless of its calorie content. It is clearly important to ascertain this contribution from diet B and its share in the total incidence of protein deficiency, so that one may formulate appropriate measures to close the protein gap. Lack of bivariate distributions of calorie and protein intake is among the principal reasons for not attempting such analysis previously. Lack of methodology to allow for the interdependence between calorie and protein intake was another. The object of this paper will be to develop the method of estimating the incidence of protein deficiency in the population in its relation to the different dietary patterns and to illustrate it by the data which have become available for parts of India.

2. Study Sources

Food Consumption Surveys

The first source from which we have prepared the material for this study is the 1958 Round of the National Sample Survey of India (NSS) for rural Maharashtra. The data relate to food consumption of 590 households. The survey was carried out by the Bureau of Economics and Statistics of the government of Maharashtra. The sample of households was selected by the method of stratified random sampling; the sample in each stratum being divided into two independent subsamples. The one-year survey, commencing in July 1958, was divided into six two-month periods or subrounds. In each subround the survey covered the entire rural area of the state. The method of collecting data was interview, with a reference period of one month. There was small nonresponse amounting to some 3% of the households selected for the survey. Altogether, the sample was uniformly spread geographically, over time and over the household size groups, and can be expected to provide a representative picture for rural Maharashtra.

To prepare the material for this paper, the recorded data on consumption of each foodstuff was converted into calories and proteins, using the nutritive values of Indian foodstuffs tabulated by Aykroyd et al. (1966). The number of nutrition units for calories and proteins were separately calculated for each household, using the known information on age and sex of the members of the household and the latest calorie and protein requirement scale (FAO, 1957, 1965). The nutrition unit for calories represents the requirement of the "reference" man, and the nutrition unit for protein likewise represents the protein requirement of an adult man of twenty to thirty.

The data was tabulated in the form of bivariate distributions of calorie and protein intake in households on a nutrition unit basis. The values of protein tabulated were those of reference protein of (Net Protein Utilization) NPU = 100. In converting dietary protein of each household into reference protein, animal protein in the diet was assumed to have an NPU of 80 and vegetable protein to have an NPU of 50. The former is approximately the value of NPU determined for milk, which accounts for the major part of the animal protein in India. The latter is slightly lower than the value of NPU calculated for vegetable protein in the average diet. These assumptions would probably underestimate the protein intake in households, since no account is taken of any possible amino acid supplementation, and Net Protein Utilization operative (NPU op) derived from the calculated value of the protein score of a diet with amino acid pattern in egg as the reference is biased downward by some 10–15% (Payne, 1969). On the other hand, the use of a fixed value of 80 for all animal protein and 50 for vegetable protein for all households, regardless of the actual composition of their diets, might distort the bivariate distribution. However, the distortion is unlikely to be large compared to the errors of underestimation referred to above. Table 16.2 shows the bivariate distribution. Table 16.3 summarizes the material in Table 16.2 and shows the values of mean and standard deviation for calorie and protein supply as well as values of the coefficient of correlation between calorie and protein supply in households.

Apart from errors introduced in converting food consumed into calories and reference proteins, food consumption data themselves are

Table 16.2

Distribution of Households in Rural Maharashtra by Protein and Calorie Intake at Retail Level

Protein Intake per Nutrition Unit in Terms of Reference Protein (g/day)	Calorie Intake per Nutrition Unit per Day											Total	%
	<900	900–1,300	1,300–1,700	1,700–2,100	2,100–2,500	2,500–2,900	2,900–3,300	3,300–3,700	3,700–4,100	4,100–4,500	>4,500		
0–5	—	1	—	—	—	1	—	—	—	—	—	2	.3
5–10	5	2	—	—	—	—	—	—	1	—	—	8	1.4
10–15	5	7	1	—	—	—	—	—	—	—	—	13	2.2
15–20	1	8	11	2	1	—	—	—	—	—	—	23	3.9
20–25	—	—	15	16	4	2	1	—	—	—	1	39	6.6
25–30	—	—	5	19	9	9	4	1	1	—	—	48	8.1
30–35	—	—	—	13	16	17	8	1	2	1	—	58	9.8
35–40	—	1	—	3	26	36	7	5	1	1	4	84	14.2
40–45	—	—	—	1	6	28	14	12	1	1	—	62	10.5
45–50	1	—	—	—	—	14	26	9	5	2	1	58	9.8
50–55	—	—	—	—	—	—	11	18	1	5	7	44	7.5
55–60	—	—	—	—	—	2	7	8	10	4	5	35	5.9
60–65	—	1	—	—	1	1	2	9	6	4	7	30	5.1
65–70	—	—	—	—	—	—	—	2	4	8	9	24	4.1
70–75	—	—	—	—	—	1	—	2	3	2	12	19	3.2
75–80	—	—	—	—	—	—	—	—	1	—	16	17	2.9
80–85	—	—	—	—	—	—	—	—	—	—	4	5	.9
85–90	1	—	—	—	—	—	—	1	1	1	5	9	1.5
90–95	—	—	—	—	—	1	—	—	—	—	1	2	.3
95–100	—	—	—	—	—	—	—	—	—	1	3	3	.5
100–105	—	1	—	—	—	—	—	1	—	—	4	7	1.3
Total	13	21	32	54	63	112	80	69	37	30	79	590	100.0
%	2.2	3.5	5.4	9.1	10.7	19.0	13.6	11.7	6.3	5.1	13.4	100.0	

Table 16.3

Values of Mean, Standard Deviation, and Co-efficient of Correlation between Calorie and Reference Protein Supply in Households per Nutrition Unit Basis in Rural Maharashtra

Mean protein intake per nutrition unit in terms of reference protein in g/day	43.6
Standard deviation	20.4
Percent coefficient of variation	47
Mean calorie intake per nutrition unit per day	2,940
Standard deviation	1,110
Percent coefficient of variation	37
Coefficient of correlation	0.79

subject to a variety of recall biases. We have discussed these at length in an earlier paper (Sukhatme, 1961). Suffice it to say here that food consumption data are among the most difficult to collect and that the net effect of recall biases on the values of calorie and protein intake is difficult to assess but quite likely is appreciable. The data reported in this paper are likely to be subject to even larger errors than the data on food consumption usually obtained by interview because these were collected in an income-expenditure schedule not precisely designed for making the type of nutritional study envisaged in this paper. For example, information on specification of certain items of food was not available. Conversion factors applicable to a group of items had to be used in converting such items into calories and proteins. This undoubtedly renders the estimates of calorie and protein intake less exact than would have been the case if detailed specification had been available. However, the number of such items was so few and their consumption so small, it is unlikely that lack of detailed specification of certain items would materially change the estimated calorie and protein values in households.

The effect of recall biases may be minimized by supplementing the interview by objective measurement of foodstuffs on a small subsample of households which provides for weighing of foodstuffs by the enumerator upon entering the kitchen, as done in the dietary surveys conducted by the state nutrition units under the guidance of the Indian Council of Medical Research (ICMR) for the last several years. Such surveys naturally demand greater cooperation from the people, with the result that nonresponse is large and data collected largely relate to homogeneous groups of households within different income classes of rural areas. In

Table 16.4

Values of Mean, Standard Deviation, and Coefficient of Correlation for Bivariate Distribution of Calorie and Reference Protein Intake in Households on a Nutrition Unit Basis in Andhra

Income in Rupees	0–50	50–100	100–250	>250	All Households
Number of households	944	903	572	256	2,675
Mean protein intake in g/day	37	38	38	39	38
Standard deviation	13	13	11	13	13
Coefficient of variation	36	33	29	33	33
Mean calorie intake	2,600	2,620	2,710	2,860	2,660
Standard deviation	740	660	620	740	680
Coefficient of variation	28	25	23	26	26
Coefficient of correlation	.72	.71	.69	.80	.73

cooperation with ICMR we have assembled data for Andhra covering the period 1952–62. These data, which relate to food consumption in 2,675 households, form the second source of material for this study, were processed in the same manner as described above for Maharashtra, and are summarized in Table 16.4 under each of the four income classes showing values of mean calorie intake per nutrition unit, mean protein intake per nutrition unit, and coefficient of correlation between calorie and protein intake. Admittedly the data are not representative of the state as a whole for providing an absolute picture of the incidence of protein deficiency in the rural population. Considering, however, that they are collected over several years and different seasons of each year and over several districts, they are believed to provide a reasonably good picture of the consumption patterns in the different income groups in the rural areas of Andhra for purposes of assessing the relative importance of low calorie intake and low protein intake in causing protein deficiency.

Protein Requirements

The requirement levels used in this paper relate to protein and are shown in Table 16.5 along with calorie requirements. They represent the latest recommendations made by ICMR (1968). These are based on the factorial approach adopted by the FAO/WHO Expert Committee (1965) and represent the needs for different functions and growth. Thus, adult protein needs are the sum required to replace obligatory protein losses in urine and feces plus losses through sweat and shedding of hair, skin, and nails. These are grouped under the heading "maintenance." In the case of children a component for growth is added. For pregnancy and lactation

Table 16.5

Recommended Protein Requirements and Allowances Expressed as NDp Cal%

Age	Protein Allowances*	Mean Protein Requirements	Calorie Requirements	NDp Cal% Mean requirements	NDp Cal% Allowances
	(grams)				
0-$\frac{1}{2}$	2.1$^+$/kg	—	120/kg	—	5.7
$\frac{1}{2}$-1	1.7$^+$/kg	—	110/kg	—	5.1
1-3	18	15	1,200	2.5	3.0
4-6	22	19	1,500	2.5	3.0
7-9	33	27	1,800	3.1	3.7
10-12	41	34	2,100	3.2	3.9
13-15 { boys	55	46	2,500	3.7	4.4
13-15 { girls	50	42	2,200	3.7	4.5
16-19 { boys	60	50	3,000	3.3	4.0
16-19 { girls	50	42	2,200	3.7	4.5
Adult male	55	46	2,800	4.2	5.1
Adult female	45	37	2,200	4.4	5.3
Pregnant female	55	46	2,500	4.9	5.9
Lactating female	65	54	2,900	4.9	5.8

*Recommended allowances for children are in terms of protein of NPU = 50 and those for adults of NPU = 65.
†In terms of human milk protein.

appropriate allowances are added to the adult requirements in the normal state.

Table 16.5 shows the protein requirements at two levels, one representing the mean and the other the mean plus 20%. The former is given under the heading "requirement" in the table; the latter is called "allowance." The margin of 20% represents the allowance to cover individual variation. The requirements are given in grams of dietary protein of NPU = 65 for adults and NPU = 50 for children. They are also expressed in terms of calories derived from reference protein as a percentage of the calorie needs, called Net Dietary Protein Calories percent, i.e., NDp Cal%. The latter form is particularly suitable for expressing protein requirements for our purpose because if a diet has a value equal to or more than the NDp Cal% corresponding to the allowance set out in Table 16.5, it implies that protein requirements of almost all individuals will be met whenever the diet is adequate in calories.

Experimental evidence shows there is wide variation in the magnitude of each of the various components of maintenance, even when individuals are of the same age, sex, and body weight and are living in good health under standard conditions. The object of adding a margin of 20% to the average requirement is to provide for this intrinsic variation among individuals. Protein requirements are also influenced by climatic conditions and by the level of physical activity, but experimental evidence to evaluate these effects is not conclusive. One can, however, obtain an idea of the magnitude of the influence of these factors from Table 16.6 by comparing the FAO/WHO scale developed for application to average

Table 16.6
A Comparison of FAO and ICMR Scales on Protein Requirements

Age	Body weight	Cal/kg FAO	Cal/kg ICMR	Protein g/kg Reference FAO	Protein g/kg Reference ICMR	NDp Cal% FAO	NDp Cal% ICMR
1–3	10.5	115	114	0.88	0.73	3.1	2.6
4–6	13.2	119	114	0.81	0.69	2.7	2.4
7–9	20.8	93	87	0.77	0.66	3.3	3.0
10–12	27.7	84	76	0.72	0.62	3.4	3.3
13–15 { Boys	38.2	76	65	0.70	0.59	3.7	3.7
13–15 { Girls	35.7	67	62	0.70	0.59	4.2	3.7
16–19 { Boys	45.1	63	67	0.64	0.54	4.1	3.3
16–19 { Girls	39.4	50	56	0.64	0.54	5.1	3.7
Adult male	55.0	46	51	0.59	0.54	5.1	4.2
Adult female	45.0	43	49	0.59	0.54	5.0	4.4

Table 16.7

Computation of Adult Protein Requirement
(g/kg/day)

Data	Protein Equivalent of Daily Nitrogen Loss			Total Requirement in Terms of Reference Protein
	Fecal	Urinary	Cutaneous	
Indian	0.125	0.240	0.125	0.490
FAO/WHO	0.125	0.284	0.125	0.534

healthy individuals under average conditions, with the ICMR scale developed for application in India. In fact, in developing its recommendations, the ICMR used the FAO/WHO recommendations as a starting point and brought the latter up to date in the light of experimental evidence available in the country. Where the Indian data was inconclusive, ICMR merely adopted the figures used by FAO/WHO. Table 16.7 illustrates the procedure adopted by ICMR. As the table shows, the difference in the recommended values for protein requirements arises from the lower amount of nitrogen loss through urine reported by Indian research workers. The Indian data confirms the nitrogen figure for loss through feces adopted by FAO/WHO, but Indian evidence on cutaneous losses was considered incomplete; for this reason ICMR merely retained the figure used by FAO/WHO, pending further collection of data in India.

The same approach was used by ICMR in updating FAO/WHO recommendations on protein requirements for children. In general, whenever the experimental evidence in the country was conclusive, as in the determination of nitrogen losses, ICMR adopted it in place of FAO/WHO figures. Where it was inadequate, FAO/WHO figures were used. Although the procedure has resulted in a downward revision of FAO/WHO recommendations, there is nothing to suggest that in their application to India these represent an upper limit. For example, it is likely that the component due to losses of nitrogen through sweat among adults may turn out to be higher under Indian conditions than shown in the FAO/WHO scale, which rates losses through sweat under average conditions in the world as a whole. It is not unlikely that FAO/WHO recommendations themselves may undergo upward revision as more data on nitrogen losses become available. As an example, the joint FAO/WHO Expert Committee on Nutrition at its seventh session recognized that the requirement for children of the age group one to three, i.e., .88/kg of body weight, is probably on the low side and needs to be revised slightly upward in the light of recent evidence. The possibility thus exists that both ICMR and FAO/WHO scales may undergo revision with additional evidence.

This is particularly likely to happen in the case of children because the available information is admittedly meager. In time, both FAO/WHO as well as ICMR scales will therefore change, though it seems unlikely, judging from the magnitude of changes already suggested by additional evidence on nitrogen losses, that any substantial modifications will be made.

Comparing ICMR with FAO/WHO scales shown in Table 16.6, we see that ICMR recommendations expressed as gram of protein per kilogram of body weight are lower than the corresponding FAO/WHO figures, in the case of all age groups, by 10–20% and on the average lower by about 13%. This difference in the two scales can be taken to mean that the use of the FAO/WHO scale for interpreting Indian intake data when the ICMR scale based on that data is the more appropriate to use, may overestimate the size of the protein gap by 13% on the average. In other words, the determination of average requirements can be considered to have a degree of uncertainty amounting to some 10–15%. Little significance can be attached to such small differences between the average intake and requirement as we find in the case of a number of countries. These differences appear even less significant when we consider the errors involved in the determination of NPUop. Thus, available data indicate that net utilization of protein of the type generally consumed by adults may be anywhere between 50 and 80, depending upon the level of intake and the protein source. The range observed in children is 40–66. They show that the assumption that NPU based on minimum requirement for protein is 65 for adults and 50 for children can introduce an appreciable error. It is thus not unlikely that what is shown as an overall deficit using the FAO/WHO scale may well turn out to be an overall excess by about the same or larger amount when the local scales developed for application in the countries are used in place of FAO/WHO scales and vice versa.

The differences between the ICMR and FAO/WHO scales become larger when the requirements are expressed in terms of NDp Cal%. This is due to the differences in calories needed per kilogram of body weight according to the two scales, as can be seen from columns 3 and 4 of Table 16.8. FAO/WHO recommendations for calories per kilogram of body weight are seen to be higher in the case of children up to fifteen but are lower for adolescents and adults. This change of sign in the differences between the two scales for calories explains the erratic and large variation in the values of NDp Cal% for FAO/WHO and ICMR scales. The differences in the values of NDp Cal% become still larger when body weights used for applying the FAO/WHO scale under Indian conditions differ from those used in the ICMR scale, but we shall not elaborate further here on the influence of errors in body weight on the two scales.

To sum up, protein requirements are subject to several sources of error whose net effect can be comparable with if not larger than the error in determining protein intake. On the other hand, the protein requirements as drawn up are based on the best available knowledge. Therefore we need not hesitate to use them as long as we recognize and allow for their limitations in interpreting the data presented here. Since the data in this paper relate to India, it is only appropriate that we should use the scale drawn up by ICMR.

Notwithstanding the differences in the two scales, certain broad trends emerge which we ought to note. Thus, requirements expressed as gram of protein per kilogram of body weight are seen to be high at birth and decline rapidly with increasing age. The same trend is observed for calories needed per kilogram of body weight which, however, decline more rapidly than protein with increase in age. This is reflected in the values of NDp Cal%. As will be seen, NDp Cal% is larger for adults than for children, infants excepted. As we shall show later, this observation is of considerable importance because it shows that the belief that children need foods which are much more concentrated in protein than those normally eaten by adults has no firm basis in the latest recommendations on protein and calorie requirements.

3. Overall Size and Incidence of the Protein Gap

The Size of the Gap

An estimate of the overall size of the protein gap for India is given by the difference between the per capita supply and the corresponding requirement and is set out in Table 16.8. Along with the estimate of the protein gap that of the calorie gap is also shown. Estimates of supply are taken

Table 16.8

Calorie and Protein Supplies per Capita (1963-65) and Corresponding Requirements' Based on ICMR Recommendations (1968) for India

Level	Calorie Supply	Calorie Requirement	Dietary Protein Supply	Protein Supply in Terms of Ref. Protein of NPU = 100	Reference Protein Requirement
				(grams)	
Physiological	...	2,080	...	...	21.2
Retail	2,050	2,280	53.4	28.6	23.3

from the food balance sheet for 1963–65, and those of requirement are based on the latest recommendations of ICMR (1968). It will be seen that supply of protein exceeds requirement by over 20%. Unlike calories, whose supply falls short of requirement by over 10%, the supply of protein is seen to exceed the allowance obtained by adding a safety margin of 20% to the mean requirement and should therefore meet the needs of most individuals in the population. The conclusion is that the average Indian diet, while inadequate in calories, appears to have adequate protein to meet the physiological needs.

The Incidence of the Gap

In practice there is ample evidence, clinical and subclinical, to indicate that an appreciable part of the population does not get adequate calories and proteins to meet physiological needs. As Table 16.1 shows, calories are unevenly distributed in the population and protein even more so. Table 16.1 by itself, however, does not make it possible to estimate the proportions of the population which are calorie and protein deficient. To assess the incidence one must turn to the distributions of calorie and protein intakes relative to the respective requirements. Given such distributions $f(x/y)$ for calorie intake x and requirement y and $f(u/v)$ for protein intake u and requirement v, we can express the incidence of calorie and protein deficiency formally thus

$$I = \int f(x/y)\, d(x/y) \qquad (3.1)$$
$$x/y < 1$$

and

$$I' = \int f(u/v)\, d(u/v) \qquad (3.2)$$
$$u/v < 1.$$

The evaluation of (3.1) and (3.2) presents a difficulty since there is almost a complete lack of information on the two frequency functions. However, the household distributions $g(\bar{X})$ and $h(\bar{U})$ of calorie and protein intake per nutrition unit basis can be prepared, as shown in Table 16.2. Therefore, the most one can do is to estimate the proportions of households in which calorie and protein deficiencies exist. Given such distributions it can be shown that (3.1) can be approximated by

$$\int g(\bar{X})\, d(\bar{X}), \qquad (3.3)$$

the integral being evaluated over the range $\bar{X} < C - 3\sigma_{\bar{Y}}$ (Sukhatme, 1961), where

$\bar{X} = \dfrac{X}{R/C}$ = calorie intake of a household per nutrition unit.

$\bar{Y} = \dfrac{Y}{R/C}$ = calorie needs of a household per nutrition unit.

R = calorie requirement of a household on the assumption that the physical activity corresponds to that of the nutrition unit.

C = calorie requirement of the reference nutrition unit.

$\sigma_{\bar{Y}}$ = standard deviation of $\bar{Y}$.

Ordinarily, in an adequately nourished population with no one calorie deficient and assuming normal distribution, most households can be expected to have their calorie intake per nutrition unit higher than the critical limit given by $C = 3\sigma_{\bar{Y}}$, i.e., the average requirement of the "reference" man minus three times the standard error. It follows that in any observed distribution the proportion of households with calorie intake per nutrition unit below $C - 3\sigma_{\bar{Y}}$ may be taken to provide an estimate of the incidence of calorie gap in the population.

In a similar way it can be shown that (3.2) can be approximated by

$$I' = \int g(\bar{U})\,d(\bar{U}), \qquad \bar{U} < C' - 3\sigma_{\bar{V}}, \tag{3.4}$$

where

$\bar{U} = \dfrac{U}{R'/C'}$ = protein intake of a household per nutrition unit.

$\bar{V} = \dfrac{V}{R'/C'}$ = protein needs of a household per nutrition unit.

R' = protein requirement of a household on the assumption that the level of physical activity corresponds to that of the nutrition unit for protein.

C' = protein requirement of the reference nutrition unit.

$\sigma_{\bar{V}}$ = the standard deviation of $\bar{V}$.

It follows too that in any observed distribution the proportion of households with protein intake per nutrition unit falling below $C' - 3\sigma_{\bar{V}}$ may be taken to provide an estimate of the incidence of protein deficiency in the population.

To estimate the critical limit we recall that the requirement of the reference man for calories is 2,780 according to the ICMR scale and 2,620 according to the FAO/WHO scale adjusted for application to India. The difference is rather wide. Considering the errors of determination, we may not be far out if we adopt a figure for requirement midway between the two, i.e., 2,700. The standard deviation of energy expenditure among healthy active adults is known to be around 400. Assuming households on the average to have four nutrition units for calories, the standard error may be placed at 200. It follows that the critical limit for calories may be placed at 2,700 − (3 × 200), or 2,100 at the physiological level or 2,300 at the retail level. Table 16.2 shows that about one-fourth the people in rural Maharashtra are calorie deficient or undernourished.

In a like manner we can work out the value of the critical limit for proteins. The requirement of the reference man is 30 grams of protein of NPU = 100. There is not adequate experimental data to estimate the standard deviation of protein requirement of healthy active individuals of the reference type, but the Expert Committee on Protein Requirements has hazarded a guess, based on observed variation in requirement for basal metabolism, that the coefficient of variation is of the order of 10%. Assuming then, as in the case of calories, that households consist of four nutrition units for protein (actually the size will be slightly smaller), this implies a standard error per household per nutrition unit of the order of 1.5 grams of reference protein. It follows that the critical limit for assessing the proportion of people with inadequate intake of protein can be placed at 30 − (3 × 1.5), or 25.5 grams. However, in view of the fact that protein distribution is more skew than calorie distribution, we may place the cutoff point at twice the standard error, thereby giving for the critical limit the value of 30 − (2 × 1.6) = 26.8 ≅ 27 grams of reference protein at the physiological level, or roughly 29–30 grams at the retail level. Table 16.2 shows that approximately one-fifth of the households in rural Maharashtra have a protein supply less than the critical value. In other words, some one-fifth of the people can be considered to be protein deficient.

We note that the incidence of protein deficiency is smaller than the incidence of calorie deficiency in the population. It needs to be stressed, however, that the incidence of protein deficiency as estimated above represents the proportion of persons with diets inadequate in protein but excludes people who although having adequate protein in the diet are not able to utilize it for lack of calories. In other words, the incidence of protein deficiency as estimated above does not represent the total incidence of protein deficiency in the population. It should be added that the existence of calorie or protein deficiencies in the households does not mean that every member of the household is necessarily calorie or protein deficient. Food is not distributed in proportion to needs, especially in poor households. For example, earners might take an adequate share of food, leaving children to exist on much less than their needs. The same may also hold, though to a somewhat smaller extent, in households with a supply above the critical limit. For lack of information on individual intake it has been assumed that the two groups broadly counterbalance each other, and in consequence the proportions of calorie and protein deficient households may be equated to the incidence of calorie and protein deficiencies in the population.

Incidence of Both Calorie and Protein Deficiencies

We can further extend the above analysis to estimate the proportion of people whose diet is deficient in both calories and proteins. This is done by calculating the proportion of households with calorie and protein intakes below the respective critical limits for both. From Table 16.2 we find that this proportion is about one-sixth.

Table 16.9

Percent Incidence of Protein and Calorie Deficiency in Rural Maharashtra

	PD	NPD	% Sub-total
CD	17	8	25
NCD	4	71	75
%Subtotal	21	79	100

CD = calorie deficient.
NCD = not calorie deficient.
PD = protein deficient.
NPD = not protein deficient.

The results of the analysis presented in the last three sections are summarized in Table 16.9. The majority of the protein-deficient households are also calorie deficient. This high association between protein deficiency and calorie deficiency is to be expected since ordinarily as calorie intake increases, so does the intake of protein.

4. Incidence of Protein Deficiency Associated with Inadequate Calorie Intake

In the analysis presented above it has been assumed that the utilizable protein in the diet is independent of the calorie intake. This assumption is not valid, for when the diet is inadequate in calories, the protein in the diet is partly diverted from its primary function to the provision of energy. This is best brought out in Table 16.10, based chiefly on the evidence

Table 16.10

Estimated Protein Loss (g/day) by an Adult Taking a Diet Restricted in Protein or Calories, or Both

Protein Intake (g/day)	Calorie Intake per Day	
	900	2,800
20	− 30	− 20
40	− 30	0
60	− 30	0

Source: Calloway and Spector (1954), *Am. J. Clin. Nutr.* 2:405.

presented by Calloway and Spector (1954). When the diet is restricted to a calorie level of about 900, protein intake in excess of 10 grams is not retained by the body. Therefore, no matter how much protein a man ate, it would be wasted as long as calorie intake was restricted to about 900. As the calorie intake is increased, the loss of protein is decreased. With an intake of about 2,000 calories the protein loss is reduced to about 10 grams a day. With further increases in calorie intake the loss becomes progressively smaller. Only when the calorie intake is adequate and above the critical limit is a man seen to utilize the protein fully to meet his body needs. But here again excess protein over and above body needs is diverted to providing energy. In other words, intake of dietary protein higher than requirement does not appear to leave any benefit on protein balance. The data show that in any assessment of the incidence of protein deficiency in the population we cannot be guided by the inadequacy of protein intake alone but must also take into account adequacy of calories as well. People who take adequate or more than adequate proteins but are not able to utilize them for lack of adequate calories in the diet, obviously form a part of the protein deficient in the population.

Miller and Payne (1961) have shown that on the average a man will need a little over 6 calories for the synthesis of 1 calorie of protein. Since a reference man for India on the average needs 1,400 calories for basal metabolism, it follows that the additional energy for anabolizing needed protein will be roughly $120 \times 6.2 = 750$ calories. In other words, a reference man must have a minimum of 2,150 calories at the physiological level or the equivalent of approximately 2,300 calories at the retail level so that he may be able to utilize fully the protein in his diet. This minimum value is precisely the critical limit below which a man can be said to be calorie deficient.

Given then a bivariate distribution $g(\bar{X}, \bar{U})$ for calorie and protein intake in households on a nutrition unit basis, we can express the incidence of protein deficiency I formally thus

$$I = \int_{\bar{U} < C' - 2\sigma_{\bar{V}}} g(\bar{U}) \, d\bar{U} + \int\int_{\substack{\bar{X} < C - 3\sigma_{\bar{Y}} \\ \bar{U} > C' - 2\sigma_{\bar{V}},}} g(\bar{X}, \bar{U}) \, d\bar{X} \, d\bar{U}$$

where

$$g(\bar{U}) = \int_{\text{all } \bar{X}} h(\bar{X}, \bar{U}) \, d\bar{X}. \tag{4.1}$$

Using the notation of Fig. 16.1 we may denote the second term of (3.1) by B and write

$$I = \int_{\bar{U} < C' - 2\sigma_{\bar{V}}.} g(\bar{U}) \, d\bar{U} + B \tag{4.2}$$

In Table 16.8 we have given values for the first and second terms of (4.2). Substituting from the table we get $I = 21 + 8 = 29$. We thus see that whereas the incidence of protein deficiency when based on protein intake alone was 21%, it is increased to 29% when the interrelationship between protein and calories is taken into account. It will be seen that nearly 10% of the total households, although receiving adequate protein, are not able to meet their protein needs for lack of adequate calories in the diet. To leave these out in estimating the incidence of protein deficiency is to underestimate the incidence by some 25%.

5. The Relative Shares of Different Diets in the Incidence of Protein Deficiency

We are now in a position to estimate the relative shares of different diets in the incidence of protein deficiency. We start by expressing the first term in the expression for I given in (4.1) as the sum of two integrals as follows:

$$\int_{\bar{U} < C' - 2\sigma_{\bar{V}}} g(\bar{U}) \, d\bar{U} = \int\int_{\substack{\bar{X} < C - 3\sigma_{\bar{Y}} \\ \bar{U} < C' - 2\sigma_{\bar{V}}}} g(\bar{X}, \bar{U}) \, d\bar{X} \, d\bar{U} + \int\int_{\substack{\bar{X} > C - 3\sigma_{\bar{Y}} \\ \bar{U} < C' - 2\sigma_{\bar{V}}}} g(\bar{X}, \bar{U}) \, d\bar{X} \, d\bar{U}. \quad (5.1)$$

The two terms clearly correspond to the cells A and E of Fig. 16.1. Substituting from (5.1) in (4.2), interchanging terms, and using notation of Fig. 16.1, we have

$$I = A + B + E. \quad (5.2)$$

Table 16.9 shows the contribution to the total incidence made by the three terms representing the three cells. By far the largest contribution to the incidence of protein deficiency is seen to come from cell A. In other words, by far the greater number of protein-deficient households are deficient in both protein and calories. As the table shows, these households represent some 60% of the protein-deficient households in the population. Next in order of magnitude comes the contribution of cell B with households which are calorie deficient and are not able to utilize their protein fully. They represent some 30% of the total number of protein-deficient households in the population. The remaining 10% are households predominantly deficient in protein with adequate or more than adequate calories; they fall in cell E.

There is yet another way we can express the incidence of protein deficiency in the population. This is done by combining the first and the third terms of (5.2) into a single term comprised of all calorie-deficient households. We write

$$I = \int_{\bar{X} < C - 3\sigma_{\bar{Y}}} g(\bar{X}) \, d\bar{X} + \int\int_{\substack{\bar{X} > C - 3\sigma_{\bar{Y}} \\ \bar{U} < C' - 2\sigma_{\bar{V}}}} g(\bar{X}, \bar{U}) \, d\bar{X} \, d\bar{U}$$

$$= \int_{\bar{X} < C - 3\sigma_{\bar{Y}}} g(\bar{X}) \, d\bar{X} + E. \quad (5.3)$$

Expression (5.3) holds special interest for nutrition workers because the first term represents the contribution of a diet which is known to lead to marasmus, while the second term represents the contribution of a diet which is known to lead to kwashiorkor. From the statistician's point of view the merit of (5.3) lies in showing the contribution of calorie deficiency to the total incidence of protein malnutrition in the population. Substituting values from Table 16.9 in (5.3) we have $I = 25 + 4 = 29$. We see that over 85% of the households which are protein deficient are also calorie deficient. In other words, most protein-deficient households have an inadequate diet.

6. Analysis of Data from Andhra and Other States

Table 16.11 shows the results of the analysis for Andhra. In analyzing the data, we have considered that the waste of edible food was small in keeping with the method of measurement adopted in the surveys, i.e., entering the kitchen to weigh food before cooking and after meals. This is a reasonable assumption, considering that most households covered came from the poor-income strata. Accordingly, the critical limits used in analyzing data were 2,200 for calories and 27 grams of protein. The table shows that the incidence of protein deficiency decreases as the income increases. The incidence is about one-third in the lowest-income group and decreases to about one-fifth in the high-income group of the rural area. For the rural area as a whole the incidence is about 30%.

The data for Andhra are seen to confirm the main conclusions for rural Maharashtra; i.e., the incidence of the protein deficiency is about 30%, and most of the protein-deficient households are also calorie deficient. On the other hand, there appear to be significant though small differences in the relative share of the dietary patterns in the total incidence of protein deficiency. Thus, it will be seen that the cell E contributes a relatively larger share of one in every five to the total incidence of protein deficiency in Andhra. The table also shows that there is a slightly larger proportion of households in Andhra in cell B and a somewhat smaller proportion in cell A, probably arising from the use by poor-income households of somewhat larger amounts of low-priced pulses in place of cereals compared to that used in rural Maharashtra. For purposes of implications for measures for closing the gap, the picture in essence is the same as for Maharashtra.

Estimating Incidence for Other States and India as a Whole

It will be some time before we obtain data for other states of India. Pending their collection it will be interesting to determine if the incidence of protein deficiency and the relative shares of the different dietary patterns in causing it can be estimated from the form of distributions for Andhra and Maharashtra, using the available data to estimate parameters. In Tables 16.12 and 16.13 we have made such an attempt.

Table 16.12 shows the computed values of the incidence of protein deficiency in the different cells on the assumption that the bivariate distribution for Andhra is of the normal bivariate form; Table 16.13 shows the calculations for Maharashtra. The agreement with the observed values is fairly close. The comparison indicates that where the data are not tabulated in the form of bivariate distributions, computed values on the assumption of normality can broadly serve the purpose of assessing the picture of protein deficiency in the population.

Table 16.11

Observed Values of Percent Incidence of Protein, Calorie, and Protein-Calorie Deficiency in Andhra

Expenditure in Rupees	0–50		%	50–100		%	100–250		%	>250		%	All Households		%
	PD	NPD	Subtotal	PD	NPD	Subtotal	PD	NPD	Subtotal	PD	NPD	Subtotal	PD	NPD	Subtotal
CD	15	14	29	11	13	24	9	10	19	8	7	15	11	13	24
NCD	5	66	71	6	70	76	5	76	81	6	79	85	6	70	76
% Subtotal	20	80	100	17	83	100	14	86	100	14	86	100	17	83	100

CD = households with calorie supply less than 2200 per nutrition unit/day.
NCD = households with calorie supply above 2200 per nutrition unit/day.
PD = households with protein supply less than 27 g per nutrition unit/day.
NPD = households with protein supply above 27 g per nutrition unit/day.

Table 16.12

Computed Values of Percent Incidence of Protein, Calorie, and Protein-Calorie Deficiency in Andhra

Income in Rupees	0–50		%	50–100		%	100–250		%	>250		%	All Households		%
	PD	NPD	Subtotal	PD	NPD	Subtotal	PD	NPD	Subtotal	PD	NPD	Subtotal	PD	NPD	Subtotal
CD	15	15	30	13	13	26	10	10	20	11	7	18	14	12	26
NCD	7	63	70	7	67	74	6	74	80	6	76	82	6	68	74
% Subtotal	22	78	100	20	80	100	16	84	100	17	83	100	20	80	100

Table 16.13

Comparison of Observed (O) with Expected (E) Percent Frequencies of Protein-Calorie Deficiency in Rural Maharashtra

		PD	NPD	% Subtotal
CD	O	17	8	25
	E	18	7	
NCD	O	4	71	75
	E	3	72	
% Subtotal		21	79	

This is an important finding in that it shows the way to estimation of the incidence of protein deficiency in the population, given the values of the mean calorie and protein intake, the coefficients of variability, and the coefficient of correlation between the calorie and protein intake. Thus, using information on the variability and on the coefficient of correlation from the data for Andhra, assuming in particular that the coefficient of variation for calories is 25% and that for protein is 33% and that the coefficient of correlation is .75, and applying it to India as a whole, we obtain a picture as shown in Table 16.14. Some 30% of the population in

Table 16.14

Computed Value of Percent Incidence of Calorie and Protein Deficiency in India

	PD	NPD	% Subtotal
CD	16	13	29
NCD	5	66	71
% Subtotal	21	79	100

the country is calorie deficient, about one-fifth is protein deficient in the sense that their diet does not meet the protein requirements, and about one-sixth are both calorie and protein deficient. The table shows that about one-third the population is protein deficient, including individuals who although having adequate protein are not able to use it for lack of adequate calories in the diet.

We are currently examining available data for other states and countries. If our examination should indicate the possibility of estimating parameters with reasonable confidence, we may be able to extend the above method, using food balance sheet data to obtain estimates of the incidence of protein deficiency and also to ascertain the relative importance of low calorie and protein intake in causing it. Pending this exercise, we consider that the broad picture we obtain for India probably holds for other countries with dietary patterns similar to that in Andhra and Maharashtra.

7. Nutritive Values of Diets

We have already seen that for a reference man taking his exact requirements of calories and proteins, NDp Cal needed to *cover individual variation* is 5.1% and that needed to cover household variation is 4.7%. How do these values compare with the nutritive value of the protein in the various diets which cause protein deficiency? This is the next natural question.

In Table 16.15 we have tabulated for each of the different cells values of NDp Cal% for Maharashtra and Andhra. The diet represented by *A*, although deficient in both calories and proteins, has NDp Cal exceeding the allowance of 4.7% needed to cover variation among households. The

Table 16.15
Values of NDp Cal% for Different Diets

	Rural Maharashtra		Andhra (all households)	
	PD	NPD	PD	NPD
CD	5.0	6.1	4.8	6.0
NCD	3.1	5.3	4.2	5.5

value of NDp Cal% for diet *B* is very much higher than needed, as one would expect. It is only for diet *E* that the value for NDp Cal is significantly lower than the allowance of 4.7% needed by households. This cell accounts for about 5% of the total number of households for whom the real need is more protein. For the vast remaining majority, the need is for more food of the type they are eating today.

That common diets have a nutritive value adequate to cover the protein needs expressed in terms of NDp Cal% can also be seen from Table 16.16, which shows the calculated values of chemical scores and of NDp Cal% for principal foodstuffs. Wheat and pulse have NDp Cal of 7%, which is more than adequate to cover the needs of man at all ages, including those of pregnant and lactating women. Rice and pulse also cover needs of man at all ages; the only exceptions are pregnant and lactating women who are seen to require a slightly higher concentration of utilizable protein than is present in the rice-pulse diet. Infants are also exceptions since they cannot be given a solid diet if they are weaned during the first year. If, however, a

Table 16.16

NDp Cal% in Different Foods (mixtures providing 2,800 calories)

Staple	A/T*	A/T† Revised	% Calorie from Protein	NDp Cal%
Wheat	39	43	12.6	5.0
Wheat + pulse (12:1)	51	56	14.1	7.0
Rice	56	62	7.6	4.6
Rice + pulse (25:1)	60	66	9.0	5.6

*A/T represents the chemical score with amino-acid composition of egg as the reference pattern.

†A/T revised, allows for underestimation of the chemical score based on Payne's relation between NPU standardized and the chemical score.

Table 16.17

NDp Cal% in Different Foods (mixtures providing 2,200 calories)

Staple Alone	Rice (4.9)	Wheat (5.7)	Maize (4.8)
Staple + 30g legume	6.2	7.0	6.3
Staple + 60g legume	6.7	8.3	6.5
Staple + 30g skim milk powder	7.1	8.0	7.0
Staple + 45g legumes + 15g skim milk powder	6.8	8.8	6.9

small amount of milk were added to a rice-pulse diet, the protein concentration would adequately improve to cover the special needs of pregnant and lactating women, as can be seen from Table 16.17 reproduced from the work of Aykroyd and Doughty (1964).

8. Implications

The analysis given in the preceding section has far-reaching implications for policy measures to close the protein gap. A detailed discussion of this aspect is outside the scope of this paper. We only briefly shall indicate some of the more important points which emerge from the results of the previous sections.

About one-half the individuals who are protein deficient have a diet deficient in both proteins and calories. The protein value of their diet appears adequate to meet the allowances. It follows that the primary need of these individuals is to have an adequate quantity of food. To give more protein to households of this cell without ensuring that this protein brings proportionately more calories with it will be a wasteful way of meeting their needs.

Between one-fourth to one-third of the protein-deficient individuals have a diet adequate in protein but deficient in calories. They have a diet which more than meets the protein requirements. Their primary need is for more calories so that they may utilize the protein in the diet fully.

About one-sixth to one-eighth of the protein-deficient individuals have a diet adequate in calories but inadequate in protein. The pressing need of this group is clearly for more protein. This type of dietary pattern is said to lead to subclinical signs of kwashiorkor. The condition occurs during infancy soon after weaning. An infant is at the breast during the first six months or so of its life and usually receives adequate calories and protein from mother's milk. But after that time breast milk is no longer able to provide the needed amounts of calories and protein for satisfactory body growth. At this stage a child receives supplements of starchy foods which can usually provide the needed additional calories but not the protein. Consequently, the protein value of the mixed diet falls below the value needed to ensure satisfactory growth. As the infant grows, protein deficiency develops and eventually becomes acute, affecting the liver of the child. In due course edema makes its appearance. The child feels averse to food. The result is increased emaciation accompanied by increased weight loss and starvation, often leading to death.

The widespread incidence of protein deficiency associated with this pattern of diet can be understood under such conditions as exist in parts of Africa where the staple food is starchy root, like cassava or plantain, with poor content of protein and NDp Cal less than 2%; although even here a typical poor diet consists of staples taken with legumes and spices and is found to have NDp Cal exceeding 5% (Payne, 1969). In India, unlike Africa (except in parts of Kerala among low-income households), starchy roots do not form a principal part of the diet. The principal food is some cereal or other, e.g., rice, wheat, jowar, etc. Together with a small amount of pulse and vegetables this provides the diet for most people in the country. Irrespective of what cereal a man takes, this diet is adequate to provide the needed protein if enough is taken to satisfy energy needs. Much the same holds for other countries in the Far East with cereal-pulse-based diets.

It follows that the widespread incidence of protein deficiency we find today is mostly the result of inadequate quantity of food. The critical factor is calorie intake, not protein.

How then does one explain the call to produce more protein in all forms irrespective of whether they are high or low in calories. This call must be traced to the widespread belief that the principal deficiency in the diet is protein, not total calories, and that infants and children need foods much more concentrated in protein than foods normally eaten by adults. The former is indeed true of conditions where the staple food is starchy root and infants do not get all the protein they need until they begin to take a solid diet. The need for protein-rich foods is also in line with the old scale of protein requirements (ICMR 1958, FAO 1957). However, with revised recommendations children's needs for protein can be met with the cereal-pulse-based diet normally eaten by adults provided they take enough of it. There is therefore no particular danger of protein deficiency occurring once an infant begins to take solid food and can tolerate the associated bulk without impairing his digestive capacity. Protein-rich foods of high biological value, such as milk and eggs, nevertheless have a vital role to play in ensuring a smooth transition in the diet of recently weaned infants and even more in providing other nutrients (vitamins and minerals) needed during infancy and the states of pregnancy and lactation.

References

Aykroyd, W. R., and J. Doughty. 1964. FAO Nutr. Stud. 19, Rome.

Aykroyd, W. R., C. Gopalan, and S. C. Balsubramanian. 1966. Spec. Rept. Ser. Indian Council Med. Res. 42, Delhi.

Calloway, D. H., and H. Spector. 1954. *Am. J. Clin. Nutr.* 2:405.

Food and Agriculture Organization. 1957. FAO Nutr. Stud. 16, Rome.

————. 1965. FAO Nutr. Meeting Rept. Ser. 37, Rome.

Indian Council of Medical Research. 1968. Recommended dietary allowances for Indians, ICMR, Delhi.

Miller, D. S., and P. R. Payne. 1961. *J. Nutr.* 75: 225.

Payne, P. R. 1969. *Overdruck, VIT, Voeding* 30:182.

Sukhatme, P. V. 1961. *J. Roy. Statist. Soc.* 124:463.

World Health Organization. 1965. Tech. Rept. Ser. 301, Geneva.

United Nations. 1968. International action to avert the impending protein crisis. Rept. of UN Advisory Committee on the Application of Science and Technology to Development, New York.

17

Use of Transformations and Statistical Estimations of Long-Term Population Trends

GERHARD TINTNER, GOPAL KADEKODI,
and STURE THOMPSON

LOGISTIC AND GOMPERTZ FUNCTIONS are considered frequently as possible trends of long-term population development (also as trends of other economically important variables) in the older literature (Davis, 1941a, b). Some work on the estimation of a logistic function (Tintner, 1960), inspired by earlier work by Hotelling (1927), suggests the following model: Let X_t be population, define a transformation:

$$Y_t^{(k)} = X_t^k, \qquad (1.1)$$

where k is a real number. This idea is related to the earlier work of Lord Keynes (1948). See also Box and Cox (1964) and Zarembka (1968). A simple first-order linear stochastic difference equation is

$$Y_{t+1}^{(k)} = A + BY_t^{(k)} + u, \qquad (1.2)$$

where u_t is a random variable.

One of the authors previously treated the logistic ($k = -1$), and large sample approximations of the standard errors of the estimates have been derived (Tintner, 1960, Tintner et al., 1961). We also would like to say

GERHARD TINTNER is Distinguished Professor of Economics and Mathematics, Economics Department, University of Southern California, Los Angeles.

GOPAL KADEKODI is Teaching Assistant, Department of Economics, University of Southern California, Los Angeles.

STURE THOMPSON is in Teaching and Research, Department of Economics, University of Goteborg, Sweden.

Research for this paper was supported by the National Science Foundation, Washington, D.C. We are much obliged to N. Keyfitz (University of California, Berkeley) for comments and criticism.

something about the small sample distributions. Recent work indicates the desirability of generalizing our results for the Gompertz curve $k = 0$ (Chow, 1967) or perhaps for other values of k.

The problem then is to estimate the constants A, B, and k. If we assume that the u_t are independent and have a normal distribution with mean zero and constant variance σ^2, the maximum likelihood is applicable.

If we do not make the assumption that u_t are independently distributed but only that they themselves are the result of a stationary process, we might use modern methods of spectral analysis (Granger and Hatanaka, 1964; Hannan, 1962). This work is related to the model of Yaglom (1955) (see also Whittle, 1963) and to some earlier work connected with the variate difference method (Tintner, 1940; Rao and Tintner, 1962, 1963).

If $Y_t^{(k)}$ is interpreted as a vector variable, methods suggested by Quenouille (1957) ought to be explored (see also Tintner, 1968). This would permit an investigation of the relation between population trends and other economic variables.

The transformation (1.1) is continuous for all k, except for $k = 0$. However, a linear transformation of (1.1) of the type

$$Y_t^{(k)} = \frac{X_t^k - 1}{k} \tag{1.3}$$

is continuous for all k and for $k = 0$; in the limit the transformation is $Y_t^{(0)} = \log X_t$. For such a transformation we can rewrite the linear stochastic difference equation (1.2) as

$$Y_t^{(k)} = A' + B Y_t^{(k)} + u_t'. \tag{1.4}$$

Unfortunately, for the estimability of either logistic ($k = -1$) or Gompertz ($k = 0$) curves, the transformation (1.3), and hence the stochastic difference equation (1.4) are not readily useful. Moreover, because of linear relationships between (1.2) and (1.3) the maximum likelihood estimates of k, B, and the maximum likelihood will not be different for the two relations.

We have fitted the relations (1.2) and (1.4) to both yearly U.S. population from 1789 to 1967 (Table 17.1) and to Swedish population from 1749 to 1966 (Table 17.2). For the different values of k we tabulate the approximate maximum log likelihood. They are:

$$L_{\max}(k) = -\frac{T}{2} \log s^2(k) + T \log |k| + \sum (k - 1) \log X_t \tag{1.5}$$

for the transformation (1.1) and,

$$L_{\max}(k) = -\frac{T}{2} \log s^2(k) + (k - 1) \sum_{t=1}^{T} \log X_t \tag{1.6}$$

for the transformation (1.3)

where
$$s^2(k) = \sum_{t=1}^{T} \hat{u}_t^2 / T - 2 \qquad (1.7)$$

From Tables 17.1 and 17.2 we see that the maximum of the log likelihood is not attained either at $k = 0$ or at $k = -1$, $k \, \varepsilon \, R$. Around both $k = -1$ and $k = 0$ the L max is increasing continuously. The maximum for Sweden is attained at $k = 0.92$ and for the United States at $k = 1.81$. In an earlier paper (Tintner et al., 1969) the authors and (in a comment) Keyfitz erred in estimating L_{max}, i.e., instead of estimating approximate L_{max} by (1.5) which is more accurate and close to the true estimate of L_{max} (not neglecting the terms involving k), (1.6) was used for both the transformations.

Table 17.1

Fitting of $Y_{t+1}^{(k)} = A + B Y_t^{(k)} + U_{t+1}$ to U.S. Data* on Population for Both Transformations Treating $Y^{(0)} = \log X$

k	$\hat{B}$	$L_{max}(k)$
2.00	1.02858	-1258.369
1.90	1.02698	-1254.262
1.84	1.02601	-1253.178
1.83	1.02585	-1253.097
1.82	1.02569	-1253.044
1.81	1.02552	-1253.019 ⟵
1.80	1.02536	-1253.021
1.79	1.02520	-1253.051
1.78	1.02504	-1253.107
1.77	1.02488	-1253.191
1.76	1.02472	-1253.302
1.70	1.02375	-1254.511
1.60	1.02212	-1258.463
1.50	1.02050	-1264.528
1.40	1.01884	-1272.333
1.30	1.01718	-1281.529
1.20	1.01550	-1291.812
1.10	1.01380	-1302.936
1.00	1.01272	-1314.734
0.90	1.01028	-1326.982
0.80	1.00864	-1339.651
0.70	1.00660	-1352.638
0.60	1.00466	-1365.885
0.50	1.00267	-1379.354
0.40	1.00062	-1393.016

Table 17.1 (*continued*)

k	$\hat{B}$	$L_{\max}(k)$
0.30	0.99849	-1406.851
0.20	0.99630	-1420.846
0.10	0.99404	-1434.992
0.09	0.99381	-1436.415
0.08	0.99358	-1437.839
0.07	0.99335	-1439.264
0.06	0.99312	-1440.691
0.05	0.99289	-1442.119
0.04	0.99266	-1443.549
0.03	0.99242	-1444.980
0.02	0.99219	-1446.412
0.01	0.99196	-1447.837
0.00	0.99172	-1449.280
-0.01	0.99149	-1450.710
-0.02	0.99125	-1452.154
-0.03	0.99102	-1453.588
-0.04	0.99078	-1455.039
-0.05	0.99054	-1456.481
-0.06	0.99031	-1457.925
-0.07	0.99007	-1459.371
-0.08	0.98983	-1460.817
-0.09	0.98959	-1462.265
-0.91	0.96897	-1585.767
-0.92	0.96871	-1587.333
-0.93	0.96845	-1588.900
-0.94	0.96819	-1590.469
-0.95	0.96793	-1592.039
-0.96	0.96767	-1593.611
-0.97	0.06741	-1595.180
-0.98	0.96714	-1596.747
-0.99	0.96688	-1598.332
-1.00	0.96662	-1598.912
-1.01	0.96636	-1601.487
-1.02	0.96609	-1603.070
-1.03	0.96583	-1604.655
-1.04	0.96557	-1604.241
-1.05	0.96530	-1607.823
-1.06	0.96504	-1609.413
-1.07	0.96478	-1611.002
-1.08	0.96451	-1612.592
-1.09	0.96425	-1614.185

Source: Historical Statistics of the United States Colonial Times to 1957, U.S. Bureau of the Census and also Statistical Abstract of the United States, U.S. Bureau of the Census.
*Yearly data from 1789 to 1967 in thousands.

Table 17.2
Fitting of $Y_{t+1}^{(k)} = A + BY_t^{(k)} + U_{t+1}$ to Swedish Data* on Population for Both Transformations Treating $Y^{(0)} = \log X$.

k	$\hat{B}$	$L_{max}(k)$
1.50	1.00979	-609.316
1.40	1.00912	-606.112
1.30	1.00846	-603.455
1.20	1.00780	-601.376
1.10	1.00715	-599.902
1.00	1.00652	-599.057
0.99	1.00644	-599.011
0.98	1.00638	-598.969
0.97	1.00632	-598.933
0.96	1.00625	-598.905
0.95	1.00619	-598.883
0.94	1.00613	-598.867
0.93	1.00606	-598.858
0.92	1.00600	-598.856 ⟵
0.91	1.00593	-598.861
0.90	1.00587	-598.872
0.89	1.00581	-598.891
0.88	1.00574	-598.916
0.87	1.00568	-598.946
0.86	1.00562	-598.986
0.85	1.00555	-599.031
0.84	1.00549	-599.083
0.83	1.00543	-599.142
0.82	1.00536	-599.208
0.81	1.00530	-599.281
0.80	1.00524	-599.361
0.70	1.00461	-600.542
0.60	1.00400	-602.426
0.50	1.00337	-605.020
0.40	1.00276	-608.325
0.30	1.00215	-612.337
0.20	1.00155	-617.045
0.10	1.00095	-622.434
0.09	1.00089	-623.010
0.08	1.00083	-623.593
0.07	1.00077	-624.182
0.06	1.00071	-624.777
0.05	1.00065	-625.379
0.04	1.00059	-625.987
0.03	1.00053	-626.600
0.02	1.00047	-627.219
0.10	1.00041	-627.851
0.00	1.00036	-628.480

Table 17.2 (*continued*)

k	$\hat{B}$	$L_{\max}(k)$
−0.01	1.00030	−629.118
−0.02	1.00024	−629.773
−0.03	1.00018	−630.420
−0.10	0.99976	−635.175
−0.20	0.99917	−642.476
−0.30	0.99859	−650.358
−0.40	0.99800	−658.791
−0.50	0.99742	−667.740
−0.60	0.99683	−667.175
−0.70	0.99625	−687.062
−0.80	0.99567	−697.368
−0.90	0.99509	−708.063
−0.91	0.99503	−709.142
−0.92	0.99497	−710.240
−0.93	0.99491	−711.342
−0.94	0.99486	−712.446
−0.95	0.99480	−713.548
−0.96	0.99474	−714.654
−0.97	0.99468	−715.760
−0.98	0.99462	−716.882
−0.99	0.99456	−717.991
−1.00	0.99451	−719.117
−1.01	0.99445	−720.230
−1.02	0.99439	−721.375
−1.03	0.99433	−722.488
−1.04	0.99427	−723.634
−1.05	0.99421	−724.759

Source: Statistisk Arsbok for Sverige 1945 and 1966, Statistika Centralbyran, Stockholm.
*Yearly data from 1749 to 1966 in thousands.

For Sweden $\hat{B} = 1.0060$ and $\hat{k} = 0.92$ and hence the solution of the stochastic difference equation (1.2) is of the form

$$X_t = [\hat{\alpha} + \hat{\beta}(1.006)^t]\frac{1}{0.92}, \qquad \alpha < 0, \beta > 0, \qquad (1.8)$$

and the same for the United States is of the form

$$X_t = [\hat{\alpha}' + \hat{\beta}'(1.0252)^t]\frac{1}{1.81} \qquad (1.9)$$

From these we see that approximate annual growth rates for Sweden and the United States are 0.65% and 1.36% respectively.

We may define approximate 95% confidence intervals for our estimates of k, i.e.,

$$L_{max}(\hat{k}) - L_{max}(U) < \frac{1}{2} X_1^2(0.05) = 1.92 \qquad (1.10)$$

$$L_{max}(\hat{k}) - L_{max}(L) < \frac{1}{2} X_1^2(0.05) = 1.92 \qquad (1.11)$$

where U and L are the upper and lower limits of the interval. For Sweden, we then have $L_{max}(U) = L_{max}(L) = -600.776$. Linear approximates of the upper and lower limits for Sweden are 1.1593 and 0.6867 respectively. The confidence interval includes 1.0. Hence, for Sweden the stochastic difference relationship is almost linear. However, this is not the case for the United States. The approximate (linear interpolation) upper and lower limits for the United States are 1.9165 and 1.6894 respectively and do not include 1.0; hence the relation is much different from linear.

From tables 17.1 and 17.2 we see that the L_{max} curve is quite flat within the confidence bounds; therefore, we do not attach the uniqueness of the estimate of $\hat{k}$ lying in that region. The flatness of the L_{max} estimates in the neighborhood of $\hat{k}$ cautions us that the approximations (1.5) and (1.6) may not be sufficient to locate the exact maximum. However, the continuity in the neighborhood and the change of direction of L_{max} around $\hat{k}$ are sufficient to judge the existence of the maximum.

Thus, in this short analysis we conclude that there is not sufficient evidence to assume that the long-term population trends of the United States and Sweden are either logistic or Gompertz in form.

References

Box, G. E. P., and D. R. Cox. 1964. An analysis of transformation. *J. Roy. Statist. Soc.* Ser. B, 26:211–43.

Chow, G. C. 1967. Technological change and the demand for computers. *Am. Econ. Rev.* 57:1117–30.

Davis, H. T. 1941a. *The theory of econometrics.* Bloomington, Ind.: Principia Press.

——. 1941b. *The analysis of economic time series.* Bloomington, Ind.: Principia Press.

Granger, C. W. J., and M. Hatanaka. 1964. *Spectral analysis of economic time series.* Princeton: Princeton Univ. Press.

Hotelling, H. 1927. Differential equations subject to errors and population estimates. *J. Am. Statist. Assoc.* 22:283.

Hannan, E. J. 1962. *Time series analysis.* New York: Wiley.

Keyfitz, N. 1968. *Introduction to mathematics of population reading.* Reading, Mass.: Addison-Wesley.

Keynes, J. H. 1948. *A treatise on probability*. London: Macmillan.

Quenouille, M. H. 1957. *The analysis of multiple time series*. New York: Hafner.

Rao, J. N. K., and G. Tintner. 1962. The distribution of the ratio of the variances of variate differences in the circular case. *Sankhya* Ser. A, 24: 385–94.

———. 1963. On the variate difference method. *Australian J. Stat.* 5: 106–16.

Tintner, G. 1940. *The variate difference method*. Bloomington, Ind.: Principia Press.

———. 1960 *Handbuch der Oekonometrie*. Berlin: Springer.

———. 1968. Time series (general). *International encyclopedia of the social sciences*, ed. D. L. Sills, vol. 16, pp. 47–59. New York: Macmillan.

Tintner, G., G. V. L. Narasimham, L. Patil, and N. S. Raghavan. 1961. A logistic trend for Indian agricultural income. *Indian J. Econ.* 42: 80–85.

Whittle, P. 1963. *Prediction and regulation*. London: English Univ. Press.

Yaglom, A. M. 1955. The correlation theory of processes whose nth differences constitute a stationary process. *Matem. Sbornik* 37: 141–96.

Zarembka, P. 1968. Functional form in the demand for money. *J. Am. Statist. Assoc.* 63: 508–11.

18

Some Graphic and Semigraphic Displays

JOHN W. TUKEY

1. Introduction

GRAPHS and semigraphic displays are made for purposes. Different purposes usually call for different graphs (or displays), although they do not always get them. In order of increasing importance come three broad classes:

- Graphs from which numbers are to be read off—substitutes for tables.
- Graphs intended to show the reader what has already been learned (by some other technique)—these we shall sometimes impolitely call propaganda graphs.
- Graphs intended to let us see what may be happening over and above what we have already described—these are the analytical graphs that are our main topic.

Five directions of innovation concern us:

1. Displays that lie between the conventional graph and the conventional table offer real opportunities. The thought that numbers should participate in an exhibit that is at least partly graphical has been too

JOHN W. TUKEY is Professor of Statistics, Princeton University, Princeton, New Jersey, and Associate Executive Director of Research, Bell Telephone Laboratories, Inc., Murray Hill, New Jersey.

This paper has been prepared in part in connection with research at Princeton sponsored by the Army Research Office, Durham, and is based on a paper presented at the annual meetings of the American Statistical Association, Institute of Mathematical Statistics, and ENAR of the Biometric Society, August 1969.

heretical to mention, though the conventional typewriter and a variety of multilithlike reproduction processes long ago made such separation uncalled for. The computer has now made separation unbearable—both for its printed output and for its visual displays.

2. Data that repeat periodically by their very nature (climate provides many examples) offer a challenge to visual representation that we seem to have neglected. It can be met.

3. We need to learn condensation in plotting, especially when dealing with relatively unstructured bodies of data. Simple techniques can attract our attention to what is most often valuable for us to see.

4. Additive fits to two-way tables underlie much of the analysis of variance, and thus very many of the applications to which George Snedecor has contributed so much. We have lacked convenient ways to look at such fits. They are now available.

5. Histograms have been with us for generations without any real change. Yet all we know about the behavior of counts, the relative importance of tail values, and the effectiveness of different visual comparisons has called for change. We can now do better.

Earlier references seem to be lacking on all but the last of these, where two papers [1, 9] are likely to have escaped wide notice.

In one way or another, as one would expect of significant innovations for familiar problems, these five advances all have the flavor of heresy. In (1) we merge the inseparables, in (2) and (4) we use a coordinate which has a meaning we are to forget (or better never learn), in (3) we treat different observations quite differently, and in (5) we give up the idea of area proportional to count. This is as it must be if we are to make progress.

The most institutionalized of all has been the separation of "table" and "graph," involving as it has special technical skills and the division of labor. Any exhibit containing numbers had to be set in letterpress by a printer, who could not be expected to understand what was to be made clear and therefore had little choice but to make sure only that his table could yield its facts, if not its insights, to those skilled in the archeology of numbers. Anything graphic was to be drawn by a draftsman, who equally could not be expected to understand what was to be made clear and thus had little choice but to draw for the eyes of an unperceptive viewer whose thoughts were not to be stimulated.

As we move through an era of photographic and xerographic reproduction toward an era of computer-controlled composition, we have the opportunity of taking back into the analyst's hand and mind the control of what is to be shown and how the key points are to be emphasized. The instances that follow do not show this being done, but they may help us to see what sorts of things we can begin to do. If our simple examples encourage others to undertake rewarding explorations of still further possibilities, especially in computer output, they will have served a second important function.

Many of the approaches illustrated have been developed in connection with various evolving versions of the author's *Exploratory Data Analysis* (to be published by Addison-Wesley). All are directed at doing a better job of *looking* at data.

2. Position and Numerical Text as Coarse and Fine

We are all used to working with numbers. All not-small numbers with which we are familiar, whether in figures or words, are written so that we can look at either a part, telling us the broad story, or at the whole, telling us the details. Thus

$$137 = 100 + 37$$

is in the one hundreds, and is 137.

$$MCMLXX = MCM + LXX$$

is in the twentieth century, and is 1970; while George Snedecor's birth year of

eighteen hundred and eighty-one

is in the nineteenth century, and is 1881.

How do we apply the same principle in a semigraphic way? By giving the coarse information by position and by giving the detailed information by the character or characters, the numerical text, that are positioned. Sometimes the characters can repeat the coarse information; sometimes they should not.

Stem-and-Leaf Displays

A *batch* of numbers is by definition a collection of values that we may wish to look at or do something to together. Our first requirements are likely to be:

- To collect a batch together from some source.
- To write down its values in a form that is appropriately compact and easy to look over.

The latter requirement faces us squarely with the issue of adequate detail. We all ought to be aware of $h^2/12$, the approximate increase in variance due to grouping in cells of width h. This reaches 2% of the initial variance when $h = \sigma/2$, thus supporting the classical suggestion that frequency distributions with 10 to 20 occupied cells are adequate for most purposes.

However, tallying values into frequency distributions is wasteful. If we are to make a mark, it may as well be a meaningful one. The simplest—and most useful—meaningful mark is a digit. By taking certain digits as the stem and one more as the leaf, we can stack up the leaves on the stem. Thus, 39, 31, and 33 combine to yield

$$3|913$$

which is compact, quickly written, and easily scanned.

Figure 18.1 shows an application of this technique (stem-and-leaf display) to a simple example. Extensions and modifications to meet a variety of problems are easily possible.

```
 to
   8      0 | 98766562                    0 | 9 = 900 feet
  16      1 | 97719630
  39      2 | 69987766544422211009850
  57      3 | 876655412099551426
  79      4 | 99988443319294333361107
 102      5 | 97666666554422210097731
 (18)     6 | 898665441077761065
  98      7 | 98855431100652108073
  78      8 | 653322122937
  66      9 | 377655421000493
  51     10 | 0984433165212
  38     11 | 4963201631
  28     12 | 45421164
  20     13 | 47830
  15     14 | 00
  13     15 | 676
  10     16 | 52
   8     17 | 92
   6     18 | 5
   5     19 | 39730                       19 | 3 = 19,300 feet
```

Fig. 18.1 Stem-and-leaf displays: heights of 218 volcanoes, unit 100 feet. Source: *The World Almanac, 1966* (New York: The New York World-Telegram and The Sun, 1966), pp. 282–83.

The "to" column, given to the left of the main display, accumulates counts and makes it easy to "count in" to any desired depth from either end. Thus the "hinges," at depth 55 in any batch of 218, lie at 38 and 96 (at 3,800 and 9,600 feet) in this batch, while the median, at depth 109.5 in any batch of 218, lies at 66 (at 6,600 feet).

Another use of this idea comes when we are giving comparative numbers. If these are naturally spread from left to right, we can usually afford a small up and down spreading—enough to make some important aspects clear. Figure 18.2, showing New York City rainfall, makes broad swings clearly visible, and yet gives detailed results.

Slightly Graphic Lists

	1890 −95	1900 −05	1910 −15	1920 −25	1930 −35	1940 −45	1950 −55
(44 or 45)		45.5				45.5	44.0
(42 or 43)	42.5						
(40 or 41)							
(38 or 39)			39.0				
(36 or 37)					37.5	37.5	

Fig. 18.2 Slightly graphical comparisons of median annual rainfalls in New York City for six-year periods.

Figure 18.3 appropriately uses annual mean temperatures to illustrate a similar technique in comparing the individuals of two groups in a way reminiscent of a thermometer. Again broad information comes by direct vision; yet details are available.

Carthago Delenda Est

While undoubtedly brought forth by almost unbearable examples of unnecessary repetition in detail, the editorial principle that nothing should be given both graphically and in tabular form has to become unacceptable. The great usefulness of graphs is their portrayal of the gross and easily visible. They should not be used for detail. If both the coarse and the fine are needed, we need both graphical and tabular aspects. As we have now seen, we may be able to combine the two aspects in one display. When this is so, the editor's strain toward condensation will make less difficulty. When it is not, we will have to fight harder—whenever both the coarse and the fine are really needed.

3. Cyclical Behavior as a Challenge

We have rarely faced up to the challenge of plotting truly cyclical data— data where exact repetition occurs every cycle. The average hourly demand for electricity in New York City during the day (for some particular period of years) repeats exactly every 24 hours, day after day. Mean monthly

Thermometer Comparisons

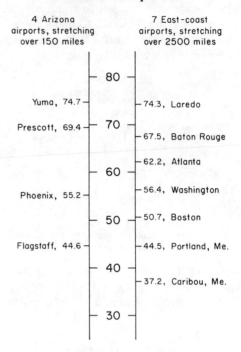

Fig. 18.3 Annual means of monthly mean temperatures (°F).

temperatures repeat every 12 months, year after year. And there are many others.

The usual, almost wholly inadequate, approach to plotting such data is to plot one cycle, say 12 monthly values from January to December. When this is done, we can *look* at the summer, but not at the winter. (Who can put the two end quarters together visually?)

Doing this is ignoring the challenge, not meeting it.

Two Cycles Can Help

As Figure 18.4 illustrates, making our plot extend over two full cycles lets us have a clear look at any part of the cycle. Indeed, we can look at any 12-month stretch, although some come so close to the ends of the plot that we could gain still more from a version extending for at least two and one-half cycles.

The data plotted in Figure 18.4 come from the analysis of a 4 × 12 table of mean monthly temperatures for the four Arizona airports

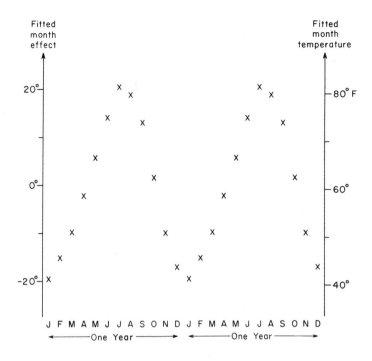

Fig. 18.4 Mean monthly temperatures for four Arizona airports.

appearing in Figure 18.3. The scales show two alternative ways of looking at each month:

- In terms of a fitted temperature.
- In terms of a fitted effect—of the deviation of the fitted temperature from the fitted grand mean temperature.

Two-scale plotting is often useful after two-way analysis.

Oval Plots Do Even Better

To show the cyclic behavior going on from left to right for two, two and one-half, or three cycles helps. We can look at whatever we need to, but the cyclic nature of the situation is not emphasized enough. It is an essential; it should be a ground bass to the music of the picture, playing an essential supporting role, whether or not we notice it as such.

Figure 18.5 shows how this can be done by letting the months run around and around, instead of only up and down. The data come from the other margin of the analysis of the same 4 × 12 table.

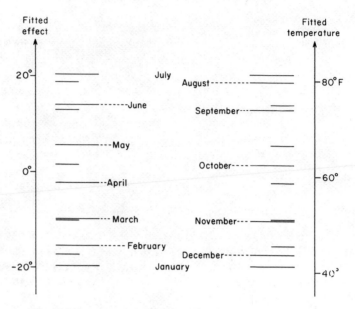

Fig. 18.5 Temperature by month, mean of four Arizona airports.

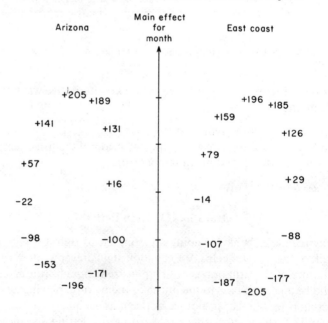

Fig. 18.6 Monthly mean effects (monthly fits minus grand fit) for four Arizona airports and seven east-coast airports (unit 0.1°F). January was the lowest and July was the highest in both cases.

This is an introductory example of "forget it" plotting, in which one coordinate has a meaning that no one should bother with—where in fact we are much better off to forget not only its meaning but often even its existence.

Actually, the horizontal coordinates in Figure 18.5 are proportional to the horizontal projections of 12 points, one for each month, equally spaced on a circle. Clearly we did not need precision, since the months are labeled by words without localized markers. A little experimentation shows that we can change shape either slightly or moderately with little loss.

Comparisons Can Apply Both Ideas

Figure 18.6 shows a comparison of margins of two analyses—one of the 4 × 12 table we have mentioned and one of a similar 7 × 12 table for the same seven east-coast airports that appeared in Figure 18.4. Here the graphic aspect emphasizes the similarity of the two results, while the numerical aspect allows their detailed comparison.

4. The Need for Condensation

One of the simplest errors in graphing, particularly for those fortunate enough to have convenient graphical output from their computers, is to plot *too many points*. The idea that if 20 or perhaps 50 points are good, then 200 or 500 are better is almost always wrong. Once upon a time the price of the draftsman, in effort if not in money, held down the number of points. Now that only a small increment in the cost of computations is involved, there is a need for expanding use of a variety of schemes for condensing data to levels of detail that are reasonable for plotting.

Schematic Plots

The need for condensation arises even when dealing with batches. Once we have more than about half a dozen observations, they are not all likely to need individual attention and labeling. If they do not, then they do not deserve individual plotting.

For those who find hinges (or other approximate quartiles) useful as indications of location and spread, it is natural to make these the central two of six equally spaced values: "corner," "side," "hinge," "hinge," "side," "corner." (The spacing is defined to be one midspread.) It may help

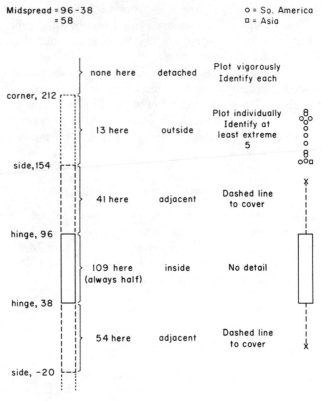

Fig. 18.7 Many elements of a schematic plot illustrated on the data of Figure 18.1.

us to know that for samples from Gaussian distributions, something like 5% of sample points are "outside" (lie beyond the sides), while only 1 or 2 in 1,000 are "detached" (lie beyond the corners).

Figure 18.7 shows the hinges, sides, and one corner (the other is negative) for the 218 heights of volcanoes listed in Figure 18.1. It also shows the coding routinely used for observations of the four kinds: inside, adjacent, outside, and detached.

The final schematic plot, in addition to the elements shown in Figure 18.7 (box between hinges, dashed lines to cover adjacent points, individual characters for outside or detached points), often helpfully shows one or both of: median (shown by a horizontal bar) and the trimean (shown by a round blob). Here

$$\text{trimean} = \frac{\text{hinge} + 2(\text{median}) + \text{hinge}}{4}.$$

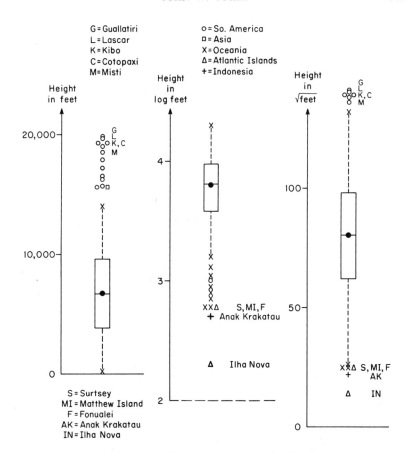

Fig. 18.8 Comparison of symmetry for three expressions of the heights of 218 volcanoes (see Figures 18.1, 18.2).

Figure 18.8 shows the use of this type of plot, with both medians and trimeans shown, to compare the relative symmetry of the heights of the 218 volcanoes when expressed in (1) raw feet, (2) log feet, (3) square roots of feet. The long tails of the first two expressions of the batch (toward high values for raw feet, toward low values for log feet) are very obvious, as is the near symmetry when heights are expressed in square roots.

Schematic Plots for *x, y* Batches

It is possible to modify and generalize the scheme just illustrated to produce quite useful two-dimensional schematic plots for batches, each of whose members is an *x, y* pair. A description of the techniques used to

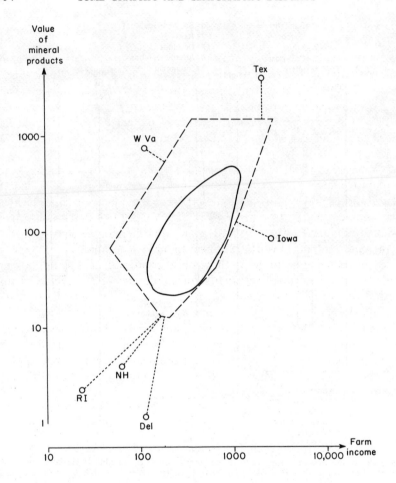

Fig. 18.9 Values of farm and mineral products by state, 1958 (scales in millions of dollars per year). Source: *The World Almanac, 1966* (New York: The New York World-Telegram and The Sun, 1966), pp. 660, 700.

determine what is plotted is a little too lengthy for inclusion here. An example will fit in, however.

Figure 18.9 shows the 1958 values by state of farm income and mineral products production. Six states deserve identification. We are not surprised that Iowa shows an unusually high farm income in comparison with value of mineral products or that West Virginia shows the reverse. The other four states are clearly unusual in their geographic size.

5. Two-Way-Fit Plotting

We have made additive fits to two-way tables of responses for many decades. During the last third of a century we have most often done this as part of an analysis of variance. Even when we are making only one-way fits, we have had to work hard stressing the need of looking at the means as well as at the significance test. (As Snedecor said in 1946 [7], " . . . means, unnecessary in the computations, should always be recorded in the table.")

We have been able to get information about one-way fits across, both to ourselves and to others, by drawing simple pictures. For two-way fits we have suffered under a severe handicap. While Egon Pearson [5, 6] tackled this problem about 35 years ago, as did Welch [10], and showed that something could be done, we must admit that there has been no pictorial representation judged worth the effort to make it routinely.

By introducing a "forget-it" horizontal coordinate and again, looking only at vertical positions, we can make a simple and effective picture of any additive two-way fit, one that brings a reality to linear models that we have sadly lacked.

Figure 18.10 shows the additive fit to the 4×12 table of Arizona airport temperatures we have already mentioned. Given a plot such as this, even cross-comparisons (such as that of Yuma in April with Phoenix in July) are easy and clear. We now have no excuse for not looking at our two-way fits to see what they are really like.

Construction of Plots

The construction of these plots is very simple, with two provisos:

- That we accept an initial plot on graph paper followed by a tracing on tracing paper as *the* natural process.
- That we are prepared to turn the graph paper through 45° before making the tracing.

We are all used to various forms for the additive fit in a two-way table, including that two-term form which in our example reads

(place effect) + (month effect + common term).

(We have already given the numerical values for these two parentheses in our example. See Figures 18.3 and 18.6.) If we take the values of the parentheses as x and y, the lines of constant sum will be 45° lines of the form

$$x + y = \text{constant}.$$

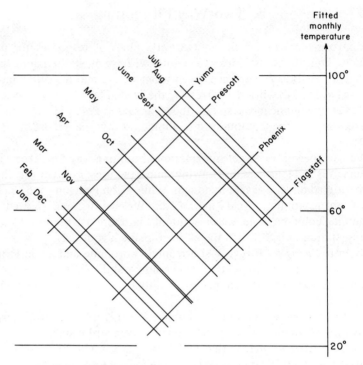

The fitted temperature for any combination of month and place
is shown by the vertical coordinate of the intersection
of the corresponding lines

Fig. 18.10 A two-way-fit plot (compare Figures 18.3, 18.5, 18.6).

For each place there is a horizontal line, for each month a vertical. Adding a few 45° lines of constant sum is easy. The result is Figure 18.11. If we turn through 45° and trace, we have Figure 18.10.

Picturing of Residuals

Given a good clear picture of the fit and recalling the basic relation

$$\text{data} = \text{fit} + \text{residuals}$$

it is very natural to want to show the residuals as radiating from the intersections that describe the fit, and thus to show everything—data, fit, and residuals—on a single picture. After all, do we not do this when fitting curves?

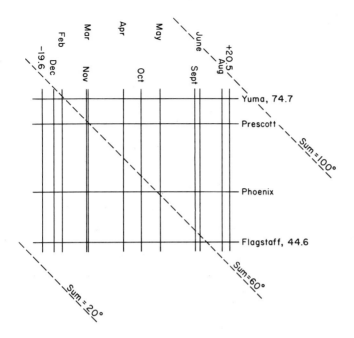

Fig. 18.11 Construction plot for Figure 18.10.

On the one hand, no scheme so far tried for including residuals with the two-way fit is useful for other than very small tables (3 × 3, yes; 3 × 4, maybe; 4 × 4, doubtful). Indeed, even plausibly small steps in this direction seem to degrade the value of the picture of the fit more than we might expect.

On the other hand, the sort of picture of curve fitting that shows the fitted curve nosing its way through the haze of points is a propaganda picture, not an analytical one. After curve fitting, the picture which is analytical (looking forward for us, not back) is the plot of the residuals themselves against x. While such a picture may need to be assisted by a picture of the fit, possibly in propaganda form, the residual plot is itself the main analytical step.

We ought then to expect to show our two-way table residuals similarly: (1) in a separate picture and (2) placed to show their relation to rows and columns (and perhaps even to fitted values). We could do this if we could let the two-way plot of the fit determine where we plot information about the residual. The result is hard to look at unless we streamline the description of the residual's value very considerably.

Classification of value*	Character used in plot	In example†
Detached high	Large heavy "X" or "+"	(above 1.5°F)
Outside high	Large "X" or "+"	(1.0 to 1.3°F)
Adjacent high	Small "x" or "+"	(0.7 to 0.9°F)
Inside	Inconspicuous dot	(−0.3 to 0.3°F)
Adjacent low	Small "o"	(−0.9 to −0.4°F)
Outside low	Large "O"	(−1.5 to −1.0°F)
Detached low	Large heavy "O"	(below −1.5°F)

*See section 4.1 and Figure 18.7 for definition and example.
†Ranges of values applicable to Figure 18.13.

Fig. 18.12 Coding of residuals

If we think of the residuals as a batch, the breakdown illustrated in Figure 18.7 works quite well, especially with the coding given in Figure 18.12. It is important to use clearly antithetic characters for high residuals and low ones. (Plus and minus signs just do not look different enough.) Returning to our 4 × 12 table of temperatures, the residual plot appears as in Figure 18.13.

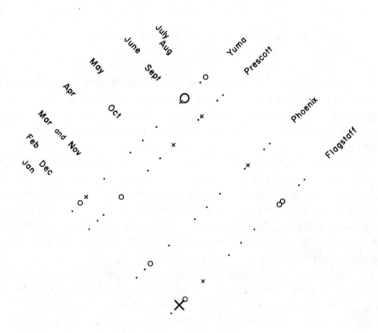

Fig. 18.13 Residuals plotted in the pattern of the two-way-fit plot.

Since the characters describing the residuals are located at the intersections of the lines of the two-way fit, this picture serves some of the purposes served by the earlier plot of the two-way fit. Any attempt to ask it to do this better, as by adding vertical scales and tenuous horizontal lines, seems to significantly decrease its value as a way of looking hard at the residuals. For the analyst who needs above all to inform himself and his immediate colleagues, the two plots serve complementary purposes, not competing ones. Since the two-way-fit plot is needed in construction of the two-way-residual plot, he may as well look at both routinely.

Is Our Grouping Too Coarse?

Some will feel, no doubt, that using merely seven (usually only five) categories for our residuals is throwing away much information. One simple fact may give them pause: If y has a Gaussian distribution and z is defined as follows:

$$z = \begin{cases} 2.4, & \text{when } y \text{ is high outside or high detached} \\ 1.2, & \text{when } y \text{ is high adjacent,} \\ 0 \ , & \text{when } y \text{ is inside,} \\ -1.2, & \text{when } y \text{ is low adjacent,} \\ -2.4, & \text{when } y \text{ is low outside or low detached,} \end{cases}$$

the square of the correlation coefficient between y and z is larger than .86. Thus the efficiency of z as a substitute for y in assessing location is more than 86%. Not very much has been thrown away. A moderate improvement in our ability to see patterns because of the simplicity of the coding can gain us much more than we lose.

Showing Something About Residuals on the Two-Way-Fit Plot

We note that it is both possible and useful to give some information about size of residuals on the two-way-fit plot. Available techniques work; but since better ones are hoped for, we shall not attempt an example.

Extended Fits

The usefulness of changes of expression of the response in cleaning up two-way analyses was emphasized in the late 1940s [8]. The alternative of adding a term of the form

$$(\text{constant}) \cdot \frac{(\text{row effect}) \times (\text{column effect})}{(\text{common term})}$$

or equivalently,

$$- \frac{(\text{row effect}) \times (\text{column effect})}{(\text{saturation change})}$$

(as well as certain generalizations) has been emphasized in the interim, especially by John Mandel [2, 3, 4].

A conversation with Frank Anscombe developed the understanding that any fit of the form

$$\text{fit} = \text{common term} + \text{row effect}$$
$$+ \text{column effect} - \frac{(\text{row effect} \times \text{column effect})}{(\text{saturation change})}$$

can be plotted almost as simply as a fit involving only the first three terms. Figure 18.14 shows how the result appears for the 7×12 table of mean monthly temperatures at seven east-coast airports.

The denominator of the last term in the extended fit is called the "saturation change" because if either row effect or column effect is equal to this value, changes in the other do not affect the fit. This is shown in a picture

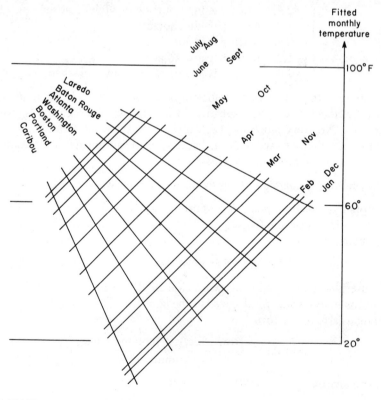

Fig. 18.14 Extended fit to mean monthly temperature at seven east-coast airports (saturation change = $+56.1°F$).

by all lines of one family meeting in a single "vanishing point," one of them being horizontal. In the construction plot, the coordinates of this common vanishing point are (in terms of effects, i.e., with the common term set aside):

$$(0, \text{saturated change}).$$

Since the added term vanishes when either effect does, these radiating lines cut the line

$$(\text{any}, 0)$$

in the same places that the corresponding lines did for the three-term fit.

To convert the construction plot for a simple two-way fit into one for the extended fit, then, we have only to: (1) add one zero-effect line; (2) add the vanishing point; and (3) rotate the lines of one family so that while passing through the same points on the zero-effect line as before, they now pass through the vanishing point. Figure 18.15 illustrates this for the east-coast airport monthly temperature.

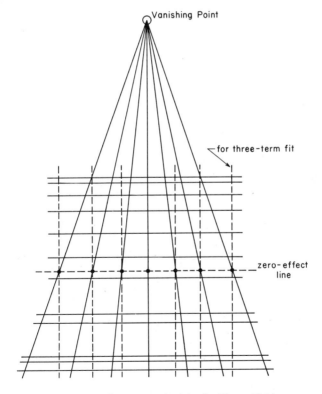

Fig. 18.15 Making the construction plot for Figure 18.14.

6. Rootograms

As statisticians we are probably unduly conscious of sampling fluctuations. Intellectually and numerically we expect larger counts to be more variable than smaller counts. Most of us know that square roots of counts are likely to be almost uniformly perturbed. Yet when we make our most common plots of counts, we pay no heed to this knowledge.

The idea that a histogram must have area proportional to count seems to be deeply ingrained. Why? There seem to be a few clear answers. The argument is far from impregnable that (1) impact is proportional to area and (2) impact ought to be proportional to count. We all know that one more case in the tails has far greater importance than one more in the middle.

Once we have tried plotting "rootograms" with height proportional to the square root of count (in general, square root of count per unit breadth), we realize that such plots have great advantages. Figures 18.16 and 18.17 show rootograms for the heights of 218 volcanoes used in Figures 18.1,

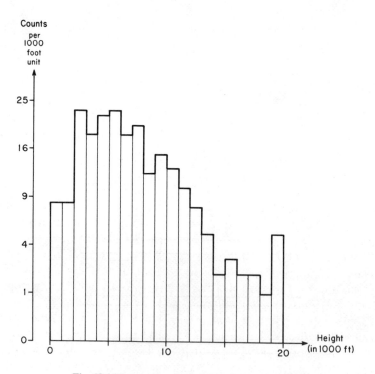

Fig. 18.16 Rootogram for heights of 218 volcanoes.

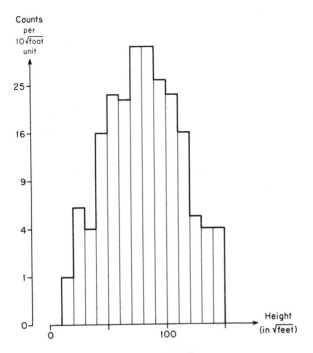

Fig. 18.17 Rootogram for $\sqrt{\text{height}}$ of 218 volcanoes.

18.7, and 18.8. One is for heights in raw feet and shows substantial lack of symmetry. The other is for heights in square roots of feet and shows reasonable symmetry. The interested reader can use counts from Figure 18.1 to plot a histogram for comparison.

Hanging and Suspended Rootograms

Once we see Figure 18.17, it is natural to want to compare it with a "normal curve." At first, the use of square roots of counts makes this seem purposeless; but then someone points out that the square root of a Gaussian density is just a multiple of another Gaussian density, a fact that makes the comparison quite sensible—at least in the following way: If the rootogram fits a Gaussian density closely, the corresponding histogram fits another Gaussian density closely, and vice versa.

From another point of view, having fitted "a normal curve" to the pattern of Figure 18.17, we ought not to be happy to then draw the curve on top of the rootogram. This would leave us making a comparison that is graphically poor, since the only comparisons that are graphically good are comparisons with straight lines.

We ought to make a plot which compares with a straight line, preferably a horizontal one. To do this will also mean making some sort of a plot of residuals, always a good thing.

We have fixed the widths and heights of the blocks that represent the individual counts in our picture. Their spacing from left to right is naturally fixed. We are still free, however, to slide these blocks up and down. Classically, we put one end on the straight baseline and tried to compare the other end with the normal curve. Why not put one end on the normal curve and compare the other end with the straight baseline? Figure 18.18 shows what happens to Figure 18.17 when this is done. Now we can see the deviations from fit.

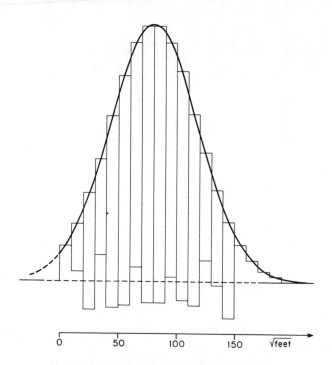

Fig. 18.18 A hanging rootogram for 218 volcanoes.

Some of us like this plot turned over, with the blocks above a down-bowing normal curve. While we are turning the plot over, we can drop excessive detail, shade excesses one way and deficits the other, and add warning limits at ± 1 and control limits at ± 1.5. Figure 18.19 shows what happens to Figure 18.18 when this is done. It is now clear that the fit of the normal curve to the square roots of these volcano heights is very good indeed.

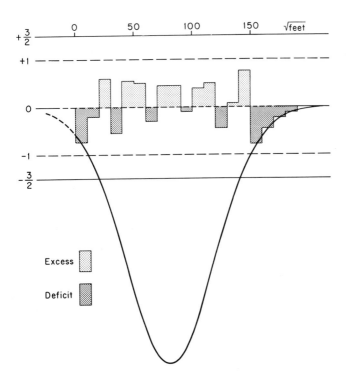

Fig. 18.19 Suspended histogram of the data in Figure 18.18.

References

[1] M. J. R. Healy. 1968. Disciplining of medical data. *Brit. Med. Bull.* 24: 210–14.
[2] John Mandel. 1956. Non-additivity in two-way analysis of variance. *J. Am. Statist. Assoc.* 56: 878–88.
[3] ———. 1964. *The statistical analysis of experimental data,* esp. pp. 261–66. New York: Interscience Publ. (Wiley).
[4] ———. 1969. A method for fitting empirical surfaces to physical or chemical data. *Technometrics* 11: 411–29.
[5] E. S. Pearson. 1936. Discussion of Messrs. Jennett's and Dudding's paper. *Suppl. J. Roy. Statist. Soc.* 3: 12–17, esp. pp. 14, 16.
[6] ———. 1936. Discussion on Mr. Gould's and Dr. Hampton's paper. *Suppl. J. Roy. Statist. Soc.* 3: 156–62, esp. 157.
[7] George W. Snedecor. 1946. *Statistical methods,* 4th ed. p. 217. Ames: Iowa State Coll. Press.

[8] John W. Tukey. 1949. One degree of freedom for non-additivity. *Biometrics* 5: 232–42.

[9] ——. 1965. The future of processes of data analysis. Proceedings of the Tenth Conference on the Design of Experiment in Army Research Development and Testing, pp. 691–729. (ARO-D Report 65-3) U.S. Army Res. Off., Durham.

[10] B. L. Welch. 1936. Discussion on Mr. Gould's and Dr. Hampton's paper. *Suppl. J. Roy. Statist. Soc.* 3: 162–65, esp. 163.

19

Sensitivity:
An Alternative Representation
of the Power Function of a
Test of Significance

FRANK YATES

1. Introduction

THE MEASURE OF SENSITIVITY of a test of significance here described was
originally devised for use in a Monte Carlo investigation into the efficiency
of the combination of probabilities or probability integral test proposed
by Fisher in an early edition of *Statistical Methods for Research Workers*.
The investigation was completed in 1955, but because of the rapid advances in electronic computers and attendant advances in Monte Carlo
methods the paper reporting these results was set aside.

At the time it occurred to me that sensitivity provided a useful way
of looking at the classic problem of balancing the loss to the investigator
of having relatively few degrees of freedom available for a t test, when
using, say, a Latin square instead of a randomized block design, against
the probable gain in accuracy. It never seemed to me that the solution
given by Fisher in *The Design of Experiments* (sect. 74) provided a satisfactory answer to this problem. Fisher concluded that the loss of information with a t test based on n degrees of freedom relative to the corresponding normal test is $2/(n + 3)$, but I doubt whether the loss to the investigator
can be adequately summarized by a single measure of "loss of information" applicable in all circumstances.

This paper contains the results then worked out for the t test and also
puts on record the results of the Monte Carlo investigation, without

FRANK YATES is former Head of the Statistics Department, Rothamsted Experimental
Station, Harpenden, Herts, United Kingdom.

going into the details of the somewhat complicated sampling methods by which they were obtained.

Most of the numerical calculations required for these investigations were done by Averil M. Sharp (née Munns). She also supervised the computer work, a not entirely simple task on the somewhat primitive computer we then had at Rothamsted.

2. The Power Function of a Test of Significance

A test of significance is essentially a test of whether a given set of observations conforms with or is at variance with a particular hypothesis (the null hypothesis). Although only a single hypothesis is directly under test, all tests of significance necessarily involve consideration of the likely types of departure from the null hypothesis. A test will be so chosen that it will be sensitive to the particular types of departures that are of interest. In many cases (e.g., in most experimental work) the types of departure that are of interest are clearly defined without reference to the actual observations. In other cases the type of departure to be tested will be suggested by the observations themselves.

In order to measure the sensitivity of a test to a particular type of departure, the concept of power was introduced. The power P of a test is defined as the proportion of samples that will on the average deviate significantly from the null hypothesis H_0 at some chosen significance level α when repeated samples are taken from a population for which some alternative hypothesis H_1 holds. P is clearly a function of α as well as of the departure of H_1 from H_0. In the common case in which the hypothesis to be tested is that a parameter μ has some particular value, μ_0 say, we may write the power function in the form

$$P = f(\mu, \alpha). \tag{2.1}$$

We may put $\mu_0 = 0$ without loss of generality; μ will then represent the departure of H_1 from H_0. It should be noted that we are not in fact concerned with a unique alternative H_1; the whole permissible range of alternatives is under test. If we are testing H_0 against a single alternative H_1, there is no logical distinction between H_0 and H_1, and there are four alternative results for a given significance level:

1. H_0 is confirmed, H_1 is contradicted.
2. H_0 is contradicted, H_1 is confirmed.
3. Neither H_0 nor H_1 are significantly contradicted.
4. Both H_0 and H_1 are significantly contradicted.

If it is objected, if (4) occurs, that we *know* that either H_0 or H_1 is true, it can only be said that an exceptional chance has occurred, or that there is something wrong with the observations, or that our belief, however confident, that either H_0 or H_1 is true is incorrect. Note also that for (3) or (4) the conclusions are at variance with those given by the "errors of the second kind" approach, which for (3) would be "accept H_0," for (4) "accept H_1."

The behavior of a test of significance when there is some departure from the null hypothesis is completely defined by the power function, but power does not in itself give a very satisfactory measure of sensitivity. The power in a two-sided test, for example, varies from α when the null hypothesis is satisfied ($\mu = 0$) to unity as μ becomes large in either direction. With a one-sided test the power becomes unity when μ is large in the test direction and zero when it is large in the opposite direction. The relative power of two alternative tests varies in a similar manner. In a pair of two-sided tests, for example, the ratio of the powers will tend to unity when μ tends to zero or becomes large in either direction, and consequently cannot have any element of constancy.

3. Definition of Sensitivity

Sensitivity provides an alternative way of specifying the power function of a test of significance. With the standard graphical representation of a power function F for significance level α shown in Figure 19.1, sensitivity for the power A of the test is given by $(AB/AC)^2$, B being a point on the power function S for significance level α of the test of a normal deviate with known unit standard deviation. The main advantage of this form of specification is that for many types of test the sensitivity for a given α does not vary greatly over the range of P, and thus provides a compact way of summarizing the performance of a given test.

Sensitivity is derivable from the power function in the following manner. If the null hypothesis is such that a parameter μ has the value 0, the actual value of μ will measure the departure from the hypothesis, and the power P will be a function of μ and the test level α. Consequently we may write

$$\mu = \phi(P, \alpha). \tag{3.1}$$

This equation gives the magnitude of the departure from the null hypothesis required for a proportion P of all samples to be judged significant at test level α. If now

$$\mu_N = \phi_N(P, \alpha) \tag{3.2}$$

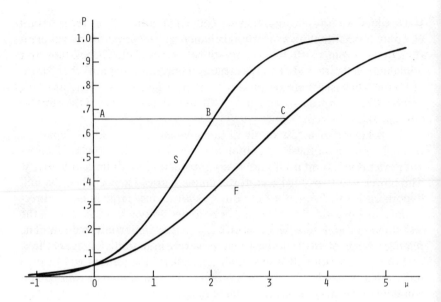

Fig. 19.1 Sensitivity of a test with power function F. A one-sided test at significance level $\alpha = .05$ is illustrated. S is the power function of the corresponding unit normal deviate test. The sensitivity for value A of P is $(AB/AC)^2$. If, as here, F is a normal deviate test, the sensitivity is constant for all P and all α.

is the value of μ required for a proportion P to be significant at test level α when the test is based on an error function which is normally distributed with (known) unit variance (which we may term the *unit normal test*), the sensitivity S is defined as

$$S(P, \alpha) = \mu_N^2/\mu^2. \tag{3.3}$$

If τ_θ denotes the deviate of the unit normal curve corresponding to the probability level (single tail) of θ, we may note that

$$\mu_N = \tau_\alpha - \tau_P. \tag{3.4}$$

If the test under consideration is itself a normal test—i.e., is based on a normally distributed error with known variance σ^2—it can easily be shown that the sensitivity is independent of P and α, having everywhere the value

$$S = 1/\sigma^2. \tag{3.5}$$

Thus for a normal test the sensitivity is equivalent to the amount of information in the corresponding estimate. In nonnormal tests the sensitivity will vary with P and α, but in tests which approximate to normal tests the amount of variation will be small. More generally, if the sensitivity for a given P and α is S, we may say that the test will behave at these values of P and α (or at equivalent values of μ and α) as will a normal test with variance $1/S$. Moreover fiducial statements concerning a departure μ at fiducial probability α will be the same as they would be if based on a normal estimate with variance $1/S$.

The concept of sensitivity can be immediately extended to the relative sensitivity of two tests, which may be defined as

$$S_1/S_2. \tag{3.6}$$

This will in general be a function of P and α, but if both tests are normal with (known) variances σ_1^2 and σ_2^2,

$$S_1/S_2 = \sigma_2^2/\sigma_1^2 \tag{3.7}$$

for all values of P and α. If the second test is based on an efficient estimate and the first on an inefficient estimate, the relative sensitivity under these circumstances is equivalent to the efficiency of the first estimate.

It should be noted that the sensitivity can be tabulated as a function of P and α or of μ and α as convenient, since every P and α defines a μ. On the other hand, in the case of the relative sensitivity, the μ's of the two tests for a given P and α will be different; therefore, if the relative sensitivity is tabulated against μ, the test to which the μ refers must be specified.

4. Derivation of the Relative Sensitivity of the t Test from the Noncentral t Distribution

The power function of the t test with $\sigma = 1$ is equivalent to the noncentral t distribution and can be written in the form, corresponding to form (3.4) of μ_N,

$$\mu = t_\alpha - \lambda A, \tag{4.1}$$

where t_α is the value of t for significance level α (single tail) and n degrees of freedom,

$$A = \sqrt{\left(1 + \frac{t_\alpha^2}{2n}\right)}, \tag{4.2}$$

and λ is a function of n, t_α, and P which was tabulated by Johnson and Welch (1939).

Using these tables (4.1) may be calculated for the required values of n, P and α. Equations (3.3) and (3.4) then give the sensitivity. The relative sensitivity of the t test and the corresponding normal test when $\sigma = 1$ is known is equal to this sensitivity since the sensitivity of the normal test is unity when $\sigma = 1$.

5. Values of the Sensitivity of the t Test

Table 19.1 shows the values of the sensitivity calculated in the above manner for $n = 4, 6, 9, 16$, and a range of values of P and α.

The main features of this table are:

1. The relatively small amount of variation in the sensitivity for variations of P (and consequently also for variations in μ). The central line of the table $P = .50$ may in fact be taken as reasonably representative.
2. The marked falling off in sensitivity as the test level decreases. This implies that a greater increase in accuracy is required to compensate for the estimation of the error variance when a more stringent level of significance is demanded. Thus for 6 degrees of freedom and a test level of 1 in 10 (single tail) the error variance must be decreased by a factor of 0.87 ($P = .50$) to compensate for its estimation, whereas at a test level of 1 in 40 (single tail) the factor is 0.70 and for 1 in 200 is 0.53. This result is, of course, not unexpected and can be deduced qualitatively from the t table itself.

Table 19.1
Values of the Sensitivity of the t Test for $n = 4, 6, 9, 16$, and a Range of P and α

$P \backslash \alpha$	.25	.10	.025	.005	.0005	.25	.10	.025	.005	.0005
			$n = 4$					$n = 6$		
.01	.95	.83	.64	.48	.28	.97	.87	.76	.60	.42
.10	.95	.81	.61	.42	.23	.96	.87	.72	.56	.38
.30	.94	.80	.59	.39	.19	.96	.87	.71	.55	.36
.50	.94	.80	.58	.37	.17	.96	.87	.70	.53	.34
.70	.94	.79	.56	.35	.16	.96	.86	.70	.52	.32
.90	.94	.78	.54	.33	.14	.96	.86	.69	.50	.31
.99	.94	.77	.52	.30	.12	.96	.85	.67	.48	.28
			$n = 9$					$n = 16$		
.01	.98	.92	.81	.70	.56	.99	.95	.89	.81	.71
.10	.97	.91	.80	.69	.53	.99	.95	.89	.81	.71
.30	.97	.91	.80	.68	.52	.99	.95	.89	.81	.70
.50	.97	.91	.80	.67	.51	.99	.95	.88	.80	.70
.70	.97	.91	.79	.66	.50	.99	.95	.88	.80	.69
.90	.97	.91	.79	.65	.48	.99	.95	.88	.80	.68
.99	.97	.90	.78	.64	.46	.99	.95	.88	.79	.68

Table 19.1 may also be used to determine the relative sensitivity of t tests based on different numbers of degrees of freedom by taking the ratio of the corresponding values. Thus for tests 6 and 9 degrees of freedom respectively, the relative sensitivity $P = .50$ is $.70/.80 = 0.88$ for $\alpha = .025$ and $.53/.67 = 0.79$ for $\alpha = .005$. A 4×4 Latin square has 6 degrees of freedom for error, whereas an experiment with 4 treatments in 4 randomized blocks has 9 degrees of freedom for error. Both designs require 16 plots, and choice of one or the other will therefore depend on how much the Latin square may be expected to decrease the error variance. As far as tests of significance are concerned, the Latin square will give better results at the .025 level if the error variance of the Latin square is less than 0.88 of that for randomized blocks. For the .005 level the factor is 0.79.

The advantage of sensitivity over power as a measure of performance will be apparent from a comparison of Table 19.1, (which gives a meaning-ful and easily remembered conspectus of the performance of the t test over a range of n, P, α) and the two tables for $\alpha = .05$ and $\alpha = .01$ of the "Probability of second kind errors" (equivalent to $1 - P$ tabulated against n and μ) given by Neyman (1935, pp. 131–2), or the similar tables given by Neyman and Tokarska (1936) for values of μ tabulated against n and P.

If for any reason a table or graph of the power function is required, this can be constructed from Table 19.1 and a table of the normal probability integral, using (3.3) and (3.4) to obtain μ for the required values of P and α. Thus for $n = 4$, $P = .70$, and $\alpha = .025$ we obtain

$$\mu = (1.960 + 0.524)/\sqrt{0.56} = 3.3.$$

Cox and Stuart (1955) made the useful suggestion that it is much better to plot power functions on probability paper. A family of curves for a normal test is then represented by parallel straight lines, and the lines for most nonnormal tests will only have slight curvature.

6. Sensitivity for Small Deviations

The power function cannot be used to determine the sensitivity when $P = \alpha$, since in the limit when $P \to \alpha$, $\mu \to 0$ and $\mu_N \to 0$ also. It can be shown, however, that as $\mu \to 0$ the sensitivity tends $[\text{to } 0(\mu^2)]$ to the value

$$\exp{(\tau_\alpha^2)}/(1 + t_\alpha^2/n)^n. \tag{6.1}$$

The values given by this formula for $\alpha = .10$, .025, and .005 and various n are shown in Table 19.2. They will be found to agree with the relevant interpolates of Table 19.1.

Table 19.2
Values of the Sensitivity of the t Test for $P = \alpha$

P, α	4	6	9	16
0.1	.813	.870	.912	.949
0.025	.635	.733	.811	.887
0.005	.483	.600	.703	.816

(column group header: n)

7. Relation of Sensitivity to Dixon's Power Efficiency Function

Dixon (1953) proposed what he termed the power efficiency function, which he later (1954) defined as "the ratio of the sample size of t-test which results in equal power for a given alternative, δ [my μ], to the sample size of the nonparametric test under consideration."

Dixon's power efficiency function is similar to Walsh's power efficiency (1949a, b, c), the difference being that Walsh has adopted an averaging process over all μ so as to obtain a single value for the power efficiency for a given α.

Leaving aside questions of discontinuity which affect nonparametric tests but which will not occur with other tests to which the power efficiency function might be applied, we may note that knowledge of the sensitivity of the given alternative test relative to the t test for a given n, μ, α, and the sensitivities of the t test enable the power efficiency to be calculated for that n, μ, α, (or equivalent P).

As an example take the case of a test with sensitivity of 75% of that of the t test when $n = 4$, $P = 0.5$, $\alpha = 0.025$. We need to find the value of n for which the sensitivity of the t test at $P = 0.5$, $\alpha = 0.025$ is equal to $2.32/.75 = 3.09$. When $\sigma = 1$, the sensitivity of the t test is (Table 19.1) $0.58n = 0.58 \times 4 = 2.32$. When $n = 6$, the sensitivity equals $0.70 \times 6 = 4.20$; and linear interpolation (which is sufficiently accurate for our purpose) gives $n = 4.82$ for a sensitivity of 3.09. The power efficiency is therefore $4/4.82 = 83\%$.

The power efficiency is greater than the relative sensitivity because increase in sample size increases the sensitivity more than proportionately to n because of increase in the degrees of freedom. The use of sensitivity instead of the power efficiency function keeps this effect separate from the relative performance of the two tests for a given n. Quite apart from the avoidance of the fiction of fractional degrees of freedom, the sensitivity approach is preferable because it leaves open what action is to be taken when confronted with the choice between two tests of differing sensitivity. We may in fact:

1. Accept the loss of sensitivity.
2. Use the more sensitive test.
3. Increase the number of observations to compensate for the loss.
4. Reduce the variance of individual observations to compensate for the loss.

The power efficiency function is only relevant to (3). Sensitivity enables all alternatives to be rationally considered.

In many experiments the estimates are more important than tests of their significance. For a single experiment, however, sensitivity is still relevant, as it indicates the spread of the fiducial limits for different probability levels. When a combined estimate from several experiments is envisaged, the importance of the estimates of error for the individual experiments depends on the way the different estimates are combined. If there is an additional component of variance from experiment to experiment, the loss due to inaccurate estimates of error is less than when a fully weighted mean is required.

8. The Combination of Probabilities Test

Monte Carlo methods were used to investigate the performance of the combination of probabilities test on a set of experiments giving quantitative estimates with normally distributed errors and known error variance. The weighted mean then provides an efficient combined test of significance. Sets of 2, 4, 8, and 16 experiments were taken, with standard errors (assigned in equal numbers to the different experiments) in the ratios (a) $1:1$, (b) $3:4:5:6$, (c) $1:2$, (d) $1:2:3:4$, and (e) $1:4$; (b) and (d), therefore, are not relevant when there are only two experiments.

Sampling distributions of logarithms of the probabilities given by the combination of probabilities test were generated for three values—μ_1, μ_2, μ_3 of μ (the "true" response)—the values chosen being those which would give 50% significant results with the efficient test at probability levels 0.1, 0.01, and 0.001 respectively. Random normal deviates given by Wold (1948) were used in groups of 256 deviates. By permutation of these and sign reversal, a distribution of 512 values was obtained from each group for each μ.

Graphical inspection of these distributions indicated that they were well fitted by normal P, α functions and consequently there was little variation in the sensitivity of the combination of probabilities test relative to the efficient test for different significance levels α. To obtain numerical values of the relative sensitivities, P, α functions were taken which had the same areas as the aggregated cumulative frequency distributions, after adjustment of these aggregate areas by covariance of the area for each group on the variance of each group of normal deviates; the expected value of this variance is unity.

Table 19.3 gives the final results. Their accuracy can be assessed from the variation between the areas for the different groups of deviates (with

Table 19.3

Combination of Probabilities Test: Values of the Relative Sensitivity (%) Obtained by a Monte Carlo Investigation

Number of Experiments	Relative Standard Errors	Test Values of μ_1	μ_2	μ_3
2	1:1	92	95	96
4*		89	90	91
8		85	86	87
16*		86	86	86
4	3:4:5:6	83	88	90
8		81	84	86
16		79	81	83
2	1:2	88	92	96
4		82	86	89
8		78	80	83
16		77	78	80
4*	1:2:3:4	79	84	87
8		74	77	80
16*		75	75	78
2	1:4	78	87	94
4		71	79	83
8		67	72	76
16		65	68	71

μ_1, μ_2, μ_3 are the values of μ that give 50% significant results with the efficient (normal) test for levels of significance α of .1, .01, .001 respectively.

Lines marked * are based on 8 groups of 256 random normal deviates, the others on 4 groups.

Table 19.4
Estimated Standard Errors (6 degrees of freedom) for the * Lines

Set	μ_1	μ_2	μ_3
4, 1:1	.53	.21	.25
16, 1:1	1.77	.61	.19
4, 1:2:3:4	1.13	.42	.50
16, 1:2:3:4	1.80	1.13	.67

allowance for the covariance adjustment) in those cases where eight groups were used. The standard errors (6 degrees of freedom) are shown in Table 19.4.

As is to be expected, the relative sensitivity decreases with increasing number of experiments (the tests are identical for a single experiment) and with increasing inequality in the weights. The greater efficiency for μ_3 compared with μ_1 is also expected, because with large μ the probabilities on the null hypothesis will be very small; $\log[P(x)|H_0]$ will then be almost linearly related to x, and consequently $S[-2\log(P(x)|H_0)]$ will be linearly related to $S(x)$. The two tests will, therefore, then be nearly equivalent.

Table 19.5
Arrangement of the Values for Four Experiments with Standard Errors 1:2:3:4 in a P', α Table (P' = power of efficient test) Corresponding to Table 19.1

P' \ α	.1	.01	.001
.035			.79
.15		.79	
.22			.87
.50	.79	.87	.94
.78	.87		
.85		.94	
.965	.94		

If the sensitivity is really the same for all α and a given μ, the results for each set of experiments can be set out in a two-way P, α table similar to Table 19.1. Table 19.5 illustrates this for four experiments with standard errors 1:2:3:4, using for convenience the power P' of the efficient test instead of P.

References

Cox, D. R., and A. Stuart. 1955. Some quick sign tests for trend in location and dispersion. *Biometrika* 42:80–95.

Dixon, W. J. 1953. Power functions of the sign test and power efficiency for normal alternatives. *Ann. Math. Stat.* 24:467–73.

———. 1954. Power under normality of several nonparametric tests. *Ann. Math. Stat.* 25:610–14.

Johnson, N. L., and B. L. Welch. 1939. Applications of the non-central *t*-distribution. *Biometrika* 31:362–89.

Neyman, J. 1935. Statistical problems in agricultural experimentation. *Suppl. J. Roy. Statist. Soc.* 2:107–80.

Neyman, J., and B. Tokarska. 1936. Errors of the second kind in testing "Student's" hypothesis. *J. Am. Statist. Assoc.* 31:318–26.

Walsh, J. E. 1949a. Some significance tests for the median which are valid under very general conditions. *Ann. Math. Stat.* 20:64–81.

———. 1949b. On the range-midrange test and some tests with bounded significance levels. *Ann. Math. Stat.* 20:257–67.

———. 1949c. On the power function of the "best" *t*-test solution of the Behrens-Fisher problem. *Ann. Math. Stat.* 20:616–18.

Wold, H. 1948. Random normal deviates. In *Tracts for computers*, no. 25. New York: Cambridge Univ. Press.